Materials Engineering: Concepts and Applications

Materials Engineering: Concepts and Applications

Editor: Reece Hughes

NY RESEARCH
P R E S S

New York

Published by NY Research Press
118-35 Queens Blvd., Suite 400,
Forest Hills, NY 11375, USA
www.nyresearchpress.com

Materials Engineering: Concepts and Applications
Edited by Reece Hughes

International Standard Book Number: 978-1-63238-553-6 (Hardback)

The publisher's policy is to use permanent paper from mills that operate a sustainable forestry policy. Furthermore, the publisher ensures that the text paper and cover boards used have met acceptable environmental accreditation standards.

Trademark Notice: Registered trademark of products or corporate names are used only for explanation and identification without intent to infringe.

Cataloging-in-Publication Data

Materials engineering : concepts and applications / edited by Reece Hughes.
    p. cm.
Includes bibliographical references and index.
ISBN 978-1-63238-553-6
1. Materials. 2. Materials science. 3. Engineering. I. Hughes, Reece.
TA403 .M38 2017
620.11--dc23

Printed in the United States of America.

Contents

Preface

Materials science is the study of materials, their discovery, design and manufacture with special emphasis being given to solid materials. This book on materials science discusses the advanced principles that determine the structure of particular materials. Allied fields of research in this subject include nanomaterial engineering, biomaterial engineering and optical and magnetic materials engineering. Topics included in this book elucidate the various technologies that are applied at this level. There has been rapid progress in this field and its applications are finding their way across multiple industries. The extensive content of this book provides the readers with a thorough understanding of the subject.

This book is a comprehensive compilation of works of different researchers from varied parts of the world. It includes valuable experiences of the researchers with the sole objective of providing the readers (learners) with a proper knowledge of the concerned field. This book will be beneficial in evoking inspiration and enhancing the knowledge of the interested readers.

In the end, I would like to extend my heartiest thanks to the authors who worked with great determination on their chapters. I also appreciate the publisher's support in the course of the book. I would also like to deeply acknowledge my family who stood by me as a source of inspiration during the project.

Editor

Effect of Moisture Exchange on Interface Formation in the Repair System Studied by X-ray Absorption

Mladena Lukovic * and Guang Ye

Academic Editor: Hong Wong

Section of Materials and Environment, Faculty of Civil Engineering and Geosciences,
Delft University of Technology, Delft 2628 CD, The Netherlands; g.ye@tudelft.nl
* Correspondence: m.lukovic@tudelft.nl

Abstract: In concrete repair systems, material properties of the repair material and the interface are greatly influenced by the moisture exchange between the repair material and the substrate. If the substrate is dry, it can absorb water from the repair material and reduce its effective water-to-cement ratio (w/c). This further affects the hydration rate of cement based material. In addition to the change in hydration rate, void content at the interface between the two materials is also affected. In this research, the influence of moisture exchange on the void content in the repair system as a function of initial saturation level of the substrate is investigated. Repair systems with varying level of substrate saturation are made. Moisture exchange in these repair systems as a function of time is monitored by the X-ray absorption technique. After a specified curing age (3 d), the internal microstructure of the repair systems was captured by micro-computed X-ray tomography (CT-scanning). From reconstructed images, different phases in the repair system (repair material, substrate, voids) can be distinguished. In order to quantify the void content, voids were thresholded and their percentage was calculated. It was found that significantly more voids form when the substrate is dry prior to application of the repair material. Air, initially filling voids and pores of the dry substrate, is being released due to the moisture exchange. As a result, air voids remain entrapped in the repair material close to the interface. These voids are found to form as a continuation of pre-existing surface voids in the substrate. Knowledge about moisture exchange and its effects provides engineers with the basis for recommendations about substrate preconditioning in practice.

Keywords: moisture exchange; repair system; interface; void content

1. Introduction

Moisture transport between a cementitious repair material and a concrete (or mortar) substrate determines the microstructural development of the interface and repair material in concrete repair systems [1,2]. Still, this area of research remains scarcely understood, because the dynamics of water exchange is very complicated and strongly influenced by hydration of the repair material. Only a few studies on the moisture exchange in multilayer systems when fresh, newly-cast material was placed on the matured substrate have been reported [3–5]. In all of these studies, nuclear magnetic resonance (NMR) technique was used. Some preliminary studies on layered "Lego blocks" specimens made of freshly cast cement pastes were performed by using X-ray absorption [6]. Most of these studies only focused on the moisture exchange: they indicated the relative change of moisture content in the repair system. Effects of the moisture exchange on the microstructure of the interface and repair material were not studied. Interface microstructure between a brick and mortar after moisture movement was only investigated by Brocken *et al.* [3].

The aim of this paper is to study effects of the moisture exchange and the substrate preparation on the interface and repair material formation in a repair system. The X-ray absorption technique is first used to quantitatively study moisture movement in a repair system. Ordinary Portland cement (OPC) paste with w/c of 0.3 is used as a repair material. Samples were sealed for 3 d and during this period the moisture exchange was monitored. Subsequently, the internal microstructure of the repair system is studied by CT scanning in order to investigate the consequences of moisture exchange on the formed structure.

2. Materials and Methods

2.1. Materials and Sample Prepartion

The substrate used in the study was a two year old mortar. A standard mortar mixture (OPC CEM I 42.5N (ENCI, Rotterdam, The Netherlands), w/c 1:2, cement-to-sand ratio 1:3) was used. After two years of fog curing (100% relative humidity, temperature 20 °C), two small prism specimens ($18 \times 19 \times 40$ mm^3) were cut with a diamond saw from bigger mortar samples. Prior to casting of the repair material, top surface of the substrate was polished to minimize the influence of surface roughness (Figure 1a). Two samples (marked as S1 and S2) with a sealant on the sides were then placed in a mold and covered with aluminum self-adhesive film (AST) to prevent water evaporation from the sides (Figure 1b). A glass reference was placed between two samples in order to account for variations in the beam intensity, as later explained.

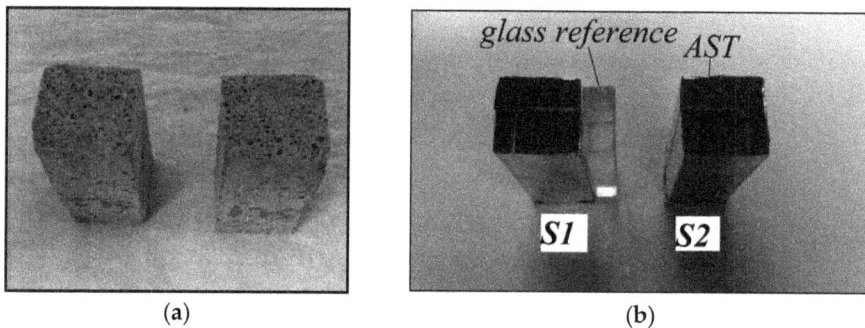

(a)   (b)

Figure 1. Preparation of the mortar substrate sample: (**a**) After cutting and polishing; (**b**) In a mold and covered by aluminum self-adhesive tape (AST).

Before covering with AST, mortar substrate was preconditioned in two ways. Dry substrate (DS) simple was dried in an oven at 105 °C until constant weight was achieved. This was done in order to remove all evaporable water and create a zero initial moisture content at the beginning of the experiment [7]. It has to be noted that these conditions might trigger some microstructural changes and microcracking in the material [8]. The other mortar substrate was kept in the fog room prior to casting of the repair material. This substrate was considered as wet (saturated) substrate (WS). For repair materials, cement paste with OPC CEM 42.5 N and w/c of 0.3 was used.

2.2. X-ray Absorption for Moisture Content Measurements

After the preparation, two samples and a glass reference were placed in a plastic container (Figure 2a). Subsequently, container is placed in a Phoenix Nanotom X-ray system (Phoenix | x-ray, Wunstorf, Germany, Figure 2b) where water exchange was measured. The apparatus is equipped for Computer Tomography (CT scanning), but in this part of the study was limited to X-ray imaging without specimen rotation. A comparison between two X-ray images taken at the beginning and after a certain time step provides a moisture change in the sample at a certain time step. Even though relative humidity (RH) and temperature (T) are not controlled in the X-ray system, RH and temperature measuring devices (see in Figure 2b) were placed in an X-ray system chamber during testing.

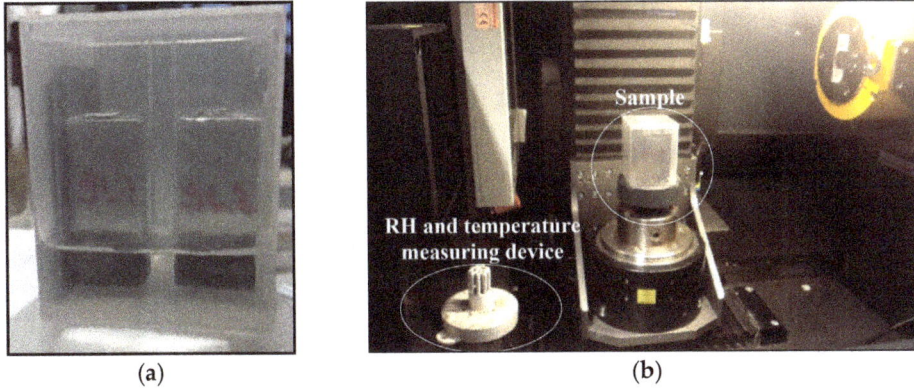

Figure 2. Samples for moisture exchange measurements; (a) two specimens are placed in a plastic container prior to X-ray testing; and (b) the X-ray system with the position of the sample and temperature and RH sensors.

When an object is irradiated with an X-ray, the X-ray is attenuated (scattered and absorbed) due to the interaction with the material. The attenuation behavior of monochromatic X-rays (X-ray photons of a single, consistent energy) can be described by the Beer-Lambert law [9]:

$$I = I_0 e^{-\mu d} \tag{1}$$

where μ is the attenuation coefficient; d is the thickness of the sample; and I_0 is the incident intensity. An attenuated X-ray results in transmitted intensity I. Detector visualizes intensities levels as grey scale values (GSV) [10]. Therefore, the attenuation coefficient of material with a known thickness can be determined by knowing the change in intensity level, $ln(I/I_0)$ (Equation (1)), or by knowing the change in GSV, $ln(GSV/GSV_0)$.

During moisture transport (either wetting or drying), GSV of material is changing. Correlating GSV change with the change in moisture content is done by making use of a simple physical principle. In a dry sample, X-rays are attenuated by the dry material only (Figure 3-left). By scanning through the dry material, a reference image is obtained. If water is added to the porous material, the attenuating material will consist of the dry material plus a thickness of a fictitious water layer (d_w), equivalent to the additional moisture content of the material (Figure 3-right) [10]. Additional moisture content is obtained by logarithmically subtracting a reference image from the image taken during moisture exchange. Therefore, provided that the change in beam intensity, $ln(I/I_0)$, and attenuation coefficient of water are known, the (additional) moisture content inside the material can be determined according to Equation (1). In this case I_0 corresponds to beam intensity after passing through the dry material, while I corresponds to beam intensity after passing through the wet material. For quantification of the moisture content change, the following procedure was used.

- The attenuation coefficient of water can be determined based on the GSV change of empty and water filled container with a known thickness (see Figure 4). The following formula can be used:

$$GSV_{water} = GSV_{air} e^{-\mu_{water} d_{water}} \tag{2}$$

where GSV_{water} is the greyscale value of water; GSV_{air} the grey scale value of air; d_{water} is the thickness of water (which is equal to the thickness of the container); and μ_{water} is the attenuation coefficient of water. The first image is made with a beam passing through an empty container and GSV_{air} (corresponds to I_0) is obtained (Figure 4a). The second image is made with a container filled with water and GSV_{water} (corresponds to I) is obtained (Figure 4b). Thus, the unknown μ_{water} can be calculated based on Equation (2). Note that this equation is equivalent to Equation (1). Further discussion about the attenuation coefficient of water when polychromatic X-ray is used, is given in the following subsection.

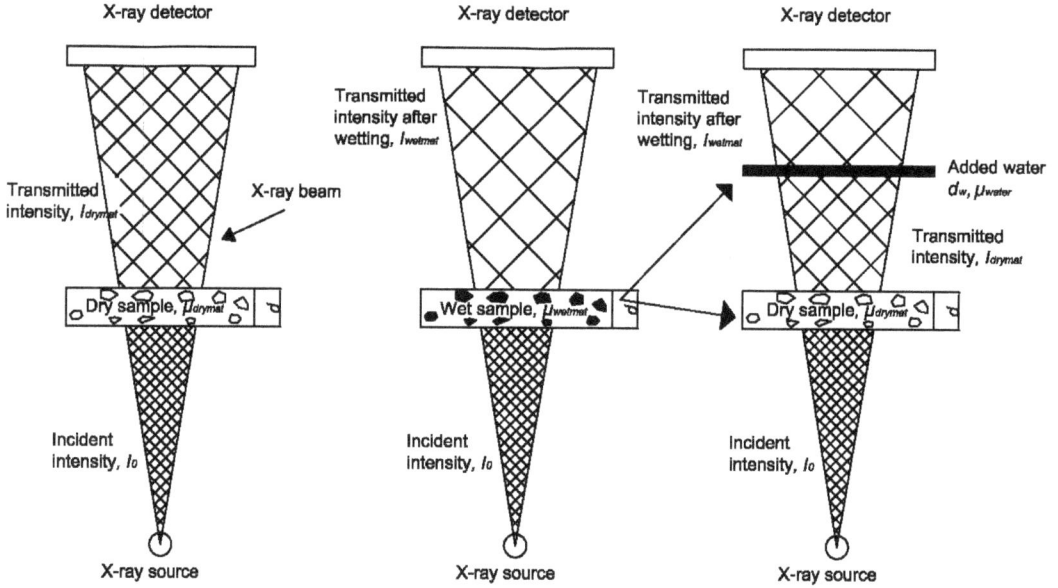

Figure 3. The moisture distribution is obtained by logarithmically subtracting an image of the dry sample I_{dry} from the image of the wet sample I_{wet}, adopted from Reference [10].

- Once the attenuation coefficient of water is determined, GSV of the wet porous material can be correlated to GSV of the dry porous material according to the following equation (also derived from Equation (1)):

$$GSV_{wetmat} = GSV_{drymat}e^{-\mu_{water}d_w} \tag{3}$$

where d_w is the thickness of fictitious water layer equivalent to the additional moisture content in sample (see Figure 3) and can be expressed as:

$$d_w = \Delta c_{water}\frac{d}{\rho_{water}} \tag{4}$$

Here, Δc_{water} is the change of water content (g/cm^3); d is the thickness of the sample; and ρ_w is the density of water. From these equations, Δc_{water} can be determined as:

$$\Delta c_{water} = -\frac{\rho_{water}}{\mu_{water}d}\ln\frac{GSV_{wetmat}}{GSV_{drymat}} \tag{5}$$

- In this study, only the middle part of the specimen (around 16 mm) is analyzed in order to exclude the influence of edges (see Figure 5). Obtained moisture profiles are then averaged over the specimen's width. As a result, the change in moisture content is obtained as function of specimen height.

Due to slight variations in the beam intensity, the obtained GSV varies even without a change in the moisture content. In order to account for this effect, a glass reference was used in all analyses (see Figures 1b and 4a). It is considered that glass does not absorb water and therefore, in this region,

GSV should be constant during the analysis. Therefore, each image was normalized according to the variations of GSV of the glass reference in order to account for changes in the beam intensity.

The parameters for X-ray analysis were set as: X-ray tube voltage 130 kV, X-ray tube current 270 µA. The spatial resolution was 30 µm/pixel. Each image used in the analysis (a representative image) is an average of 25 images. With 0.5 s needed for acquisition of an image, a representative image for a certain time step was obtained in 12.5 s.

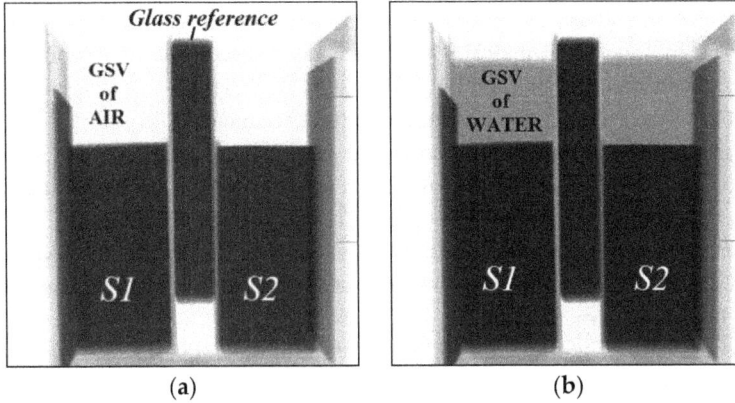

Figure 4. Difference in GSV between an empty and water filled container placed at the top of the samples S1 and S2, used for calculating attenuation coefficient of water: (**a**) GSV of an empty container; and (**b**) GSV of water-filled container.

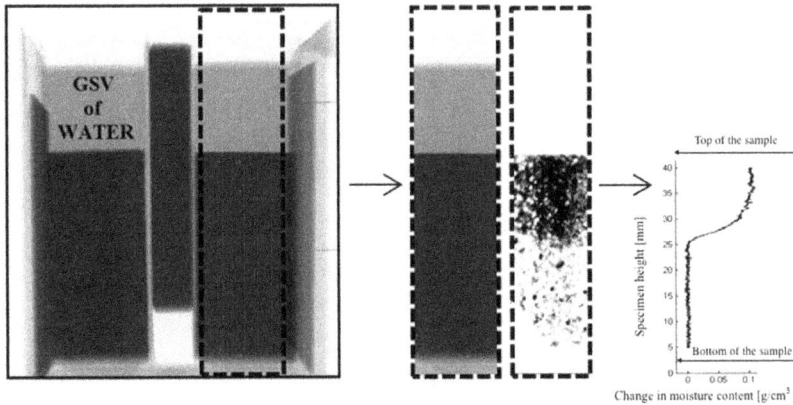

Figure 5. An example of the original X-ray image and analyzed part of the specimen from which the moisture profile is calculated.

2.2.1. Determination of the Attenuation Coefficient

When using polychromatic X-ray source (photons emitted over a spectrum of energy levels), such as the one used in this research, two aspects should be considered in order to determine the attenuation coefficient of water [11].

The first aspect is that the attenuation coefficient of water depends on the water layer thickness. This dependency is a consequence of using polychromatic X-ray source which leads to so called "beam hardening" effect: while photons are passing through the material, lower energy X-ray photons are attenuated easier, the energy spectrum is changing, and progressively, with increasing thickness of material, remaining protons become "harder" to attenuate. As a result, the measured attenuation coefficient will depend on the thickness of the material. Attenuation provided by a certain thickness of a material is described by the term "effective attenuation coefficient". Pease *et al.* [11] observed that, with the increasing thickness of the water (and also other materials such as clay brick, concrete, and wood), the effective attenuation coefficient of material decreases. The effective attenuation coefficient of water in this research, measured for the different water layer thickness (Figure 6), shows the same behavior.

The second aspect is related to the beam-hardening effect and the widely used non interacting composite system (Figure 3), which assumes that the porous parent material does not influence the attenuation coefficient provided by water. In the absorption (or drying) tests, however, photons are attenuated both by the parent material and the water. Therefore, the attenuation coefficient of water cannot be determined independently of the parent material (in this case mortar substrate). In order to account for this, Pease *et al.* [11] introduced the term called "coupled effective attenuation coefficient" of water which is a function of the parent material and its thickness. They observed that, with increasing thickness of different materials (e.g., concrete, cement paste, calcium silicate, *etc*), the coupled effective attenuation coefficient of water decreases. In this research, the same was tested by placing the 18 mm thick mortar substrate in front of the water holder (with inner dimensions of 4 mm, 9 mm, and 33 mm). The obtained coupled effective attenuation coefficient of water was lower than the effective one (Figure 6), similar to findings of Pease *et al.* [11].

As the thickness of the water layer should be similar to the anticipated maximum change of the water content in the parent material, the attenuation coefficient of the 4 mm thick water layer, placed behind the 18 mm thick mortar substrate (0.1728 cm^{-1}) was used further to quantify the change of the water content. Accordingly, the maximum change of the water content in the mortar substrate is assumed to be around 0.22 g/cm^3, which should be reasonably close to the maximum moisture content of the mortar.

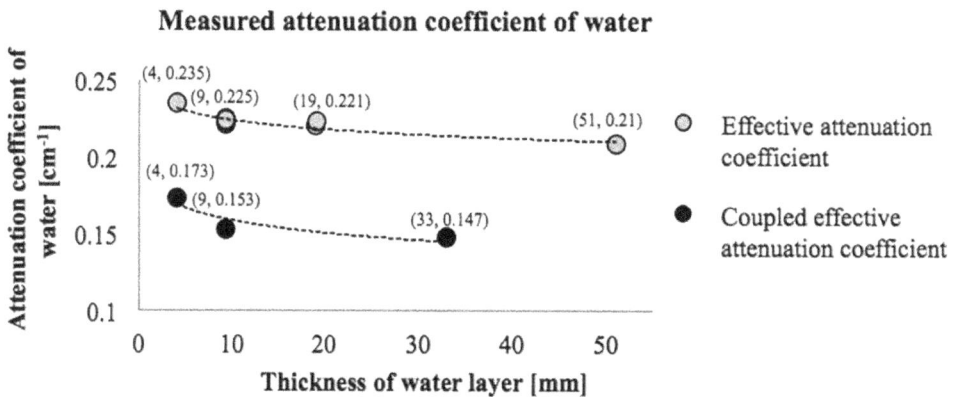

Figure 6. Influence of water layer thickness and parent material (mortar substrate) on measured attenuation coefficient of water.

2.2.2. Limitations of the Experiment

An important limitation of this technique is caused by the conical beam of the X-ray setup. As a result, the thickness of the material through which the beam passes depends on the position of the source with respect to the specific location (*i.e.*, coordinate) in the sample (Figure 7a). Consequently, at the top and the bottom of the sample, the thickness of the attenuating material is larger, and this will affect the effective attenuation coefficient. The thickness of the sample through which the beam is passing varies between 18 mm in the middle of the specimen (*i.e.*, when the beam is perpendicular to the specimen) and 18.16 mm at the top and bottom (Figure 7). In the current research, this was not taken into account, but should be considered for improvement of the technique.

One more consequence of the conical beam is that X-ray photons are not emitted parallel to the repair/substrate interface. The interface location, therefore, is not well defined, and it includes a small zone of both the repair material and the substrate. The scattering of the X-ray beam is, therefore, modified in this zone. The width of this zone was estimated to be around 0.99 mm (Figure 7). This interfacial zone is marked further in the chapter with the dashed line (of the similar thickness) to indicate the unreliability of the results. Similarly, the bottom and the top of the specimen are also affected (4.83 mm and 3.43 mm, respectively, see Figure 7), and these parts were excluded from the graphs. Note that these effects are dependent on the specimen thickness and the distance between the object and the source.

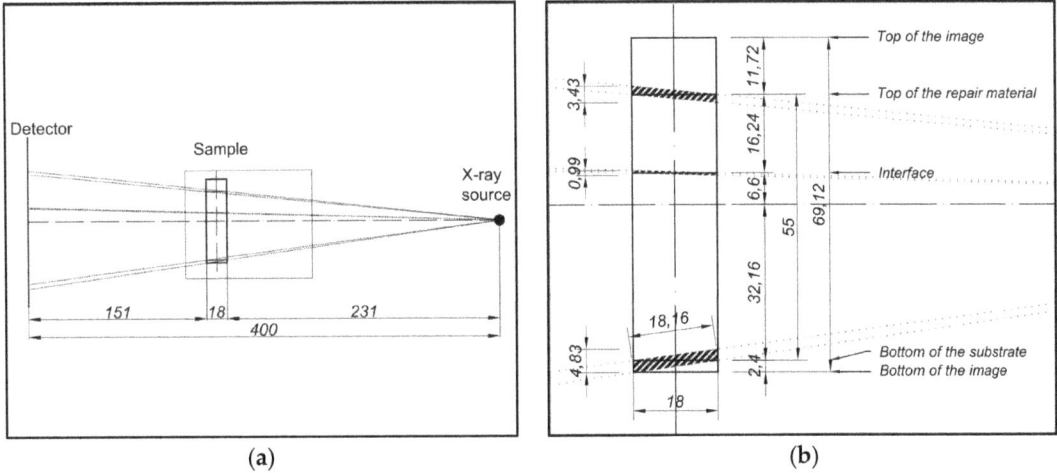

Figure 7. Experimental setup and influence of the conical beam on the measured data, units are in (mm): (**a**) Position of the detector, the specimen and the source; (**b**) Beams passing through the sample, high magnification.

2.3. Degree of Hydration Measurements

Degree of hydration of tested repair materials was measured based on non-evaporable water content and according to the procedure given in [12]. Repair materials were split from the substrates immediately after 3 d of sealed curing. Then, the samples were ground into powder. The powder was dried in an oven at a temperature of 105 °C for 24 h to remove first the evaporable water. Simultaneously, the crucibles were dried in a furnace at a temperature of 1000 °C. Around 2 g of powder was added to the crucibles and the weight was recorded as W_{105}. The crucibles were placed in a furnace at a temperature of 1000 °C and left there for 3 h to remove the non-evaporable water. The weight of the sample was measured and recorded as W_{1000}. Two parallel measurements were done. The degree of hydration, α, was calculated using the following equation [13]:

$$\alpha = \frac{W_{105} - W_{1000}}{0.23 \times W_{1000}} \tag{6}$$

Equation (6) is simplified as it does not take into account cement loss on ignition [14] and assumes that the hydration of 1 g of anhydrous cement produces 0.23 g of non-evaporable water. These simplifications, however, are considered not to have significant influence on results, as long as raw materials and the assumptions were the same throughout all tests. Note that the measured degree of hydration corresponds to the bulk degree of hydration in the repair material. This means that microstructural differences across the repair material thickness as appear in the practice are not captured.

2.4. CT Scanning for the Mesostructure Characterization

After the moisture content measurements (after 3 d of sealed curing), the void content in the repair system was characterized. Tested repair systems were investigated by a CT scanner in order to study effects of moisture exchange on repair system microstructure. During each scan, multiple X-ray images of a specimen are taken at different angles (*i.e.*, 1440 tomographic images were taken over a complete 360° rotation). A single scan took about 90 m to perform. Using a reconstruction algorithm, a 3D image of the internal structure of a specimen is produced. Voxel size was also 30 μm (the same as pixel resolution for measurement of moisture transport).

3. Moisture Movement in the Repair System and Discussion

Cement pastes with a w/c of 0.3 and around 15 mm thickness were cast on the top of the mortar substrates. As already explained, substrates were preconditioned in different ways. In one case, substrate was wet (specimen marked as w/c = 0.3, WS, S3, Figure 8-left specimen) and in the other, substrate was completely dry (specimen marked as w/c=0.3, DS, S3, Figure 8-right specimen). Immediately after casting, repair systems are sealed with aluminum self-adhesive tape and placed in an X-ray system in order to investigate the moisture exchange between repair materials and mortar substrates (Figure 8). Temperature and relative humidity were recorded during the experiment. Average temperature and relative humidity were 27.69 °C and 16.83% with standard deviations of 0.49 °C and 1.18%, respectively.

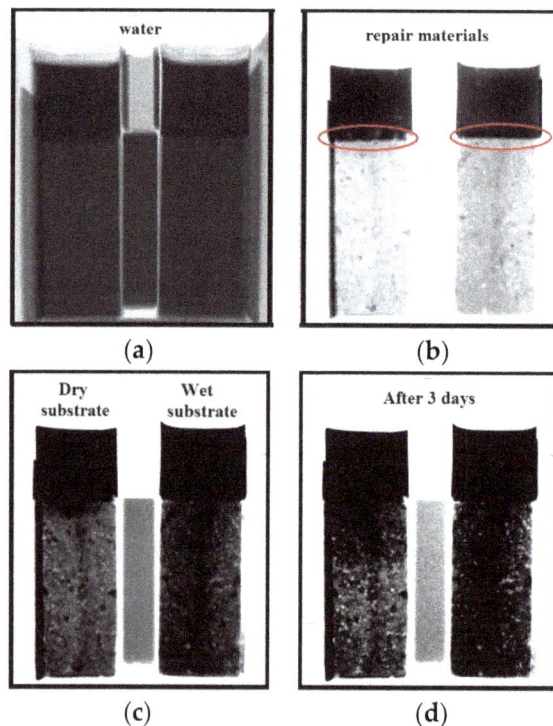

Figure 8. X-ray measurements on repair systems (w/c = 0.3, DS, S3 and w/c = 0.3, WS, S3): (a) Reference image; (b) Chosen GSV range with the emphasis on the repair material; (c) Chosen GSV range with the emphasis on the mortar substrate; and (d) Image after 3 d of sealed curing.

It is assumed that the main component absorbed by the mortar substrate from the repair material is water. Therefore, although cement slurry (and not the pure water) is actually absorbed by the mortar substrate, the coupled effective attenuation coefficient of water is used in the calculations.

The reference image was made 136 min after mixing the repair material. The first measured profile is obtained 14 min after making the reference image and 150 min after mixing the repair material (Figure 9 and magnified moisture profiles in Figure 10). As a consequence, the change in moisture profiles at the top of the substrate (close to the interface) could not be captured. These parts (indicated by a dashed rectangle in Figure 9a) are filled with water during the initial 150 min. As previously discussed, the interface is not parallel to the X-ray beam and, therefore, the scattering process is modified in this zone. Therefore, the results in this zone are not reliable and are indicated by the dashed line. As in first 5 h, moisture absorption is constant and consistent (see for more details [15]), the test was continued.

In order to perform the integration procedure and calculate the cumulative water content for the whole repair system, three integration limits are defined: top of repair material (defined as the repair material height immediately after casting, top arrow in Figure 9), interface (dashed line), and bottom of the mortar substrate (where the moisture profiles begin). In these calculations it is assumed that the coupled effective attenuation coefficient of the water in the repair material is equal to the coupled effective attenuation coefficient of the water in the mortar substrate. In Figure 6 it was shown, however, that the coupled effective attenuation coefficient is sensitive to the porous material thickness and the thickness of water layer. These are probably not the same for the repair material with w/c of 0.3 (*i.e.*, 23% of water and 77% of OPC by mass of the paste) and the mortar substrate with w/c of 0.5 (*i.e.*, 66% of aggregates, 11% of water, and 22% of OPC by mass of mortar). Furthermore, the calculation of the water movement within the repair material is very complex because of the ongoing hydration and the volume changes in the material. The height of the repair material is changing due to the setting in the early stage and later due to shrinkage. In addition, the use of a conical X-ray beam introduces the difficulties in defining the exact location of the interface (as previously explained). Having in mind all these assumptions and simplifications, further calculations for the water content inside the repair material should be taken only as indicative.

The cumulative water content as a function of time is calculated for the repair material and the substrate and curves are given in Figure 11. The dry substrate is absorbing water (cumulative water content is positive) while the repair material is losing water (cumulative water content is negative). If the calculation procedure and sealing are perfect the amount of water that is lost from the repair material should be equal to that absorbed by the substrate. However, amount of water absorbed by the substrate is higher. As both aluminum foil and sealant are used for sealing the specimen, it is considered that sealing is good but that numerous assumptions in the calculation (as previously explained), led to differences in calculated cumulative water contents.

Contrary to the repair material cast on initially dry substrate in w/c = 0.3, DS, S3, the repair material was cast on saturated substrate in w/c=0.3, WS, S3 did not lose water. This is because the substrate was saturated and could not absorb the water from the repair material. Although the repair material is hydrating (some water is becoming chemically bound), the mass of water inside the material is not changing. Therefore, only moisture movement, and not hydration, is monitored with X-ray absorption. Although hydration was not captured, moisture redistribution within the repair material itself (as a consequence of hydration) was. Moisture is moving from the interface zone to the bulk repair material. The interface region (30–50 μm) is typically characterized by a locally higher w/c (less cement) than the bulk repair material. Due to the coarser pore structure, water from this porous region will be taken to the denser bulk paste, resulting in creation of large empty pores within the interface, as explained by Bentz and Hansen for aggregate-paste ITZ [6]. As a result, moisture profiles are becoming concave (indicated by arrows in Figure 9a,b). The same phenomenon is also modelled with hydration model [16].

In a repair system, however, some water for hydration of the interface can be partially taken back from the saturated substrate. Negative shift in cumulative moisture content in Figure 11 (marked with the dashed arrow), indicates moisture movement from the substrate to the repair material. However, water loss in this region (indicated in magnified profiles in Figure 10a) can also be a consequence of water redistribution in the substrate itself, as the moisture profile in the repair material does not clearly confirm this. Although from the moisture profiles water gain in the repair material cannot be clearly seen, it should exist. This is because of the hydration of the repair material and water redistribution in the repair system in order to again reach hygral equilibrium. In the repair material, there are two causes of water loss: water taken by the unsaturated substrate and hydration reaction of the material itself (self-desiccation). Hydration of the repair material causes the consumption of capillary water from the repair material, thereby reducing the relative humidity that, at a certain point in time, was in equilibrium with the substrate. Therefore, a new capillary pressure potential difference is created. In order to restore the equilibrium state, water is driven back from the substrate to the repair material (w/c = 0.3, DS, S3, indicated by arrow in Figure 11). In sample w/c = 0.3, WS, S3 (Figure 10b) this can be clearly seen. The substrate is losing water from the top while the moisture profiles in the repair material indicate water gain close to the interface. In this sample, therefore, moisture profiles both in the substrate and the repair material confirm the moisture exchange.

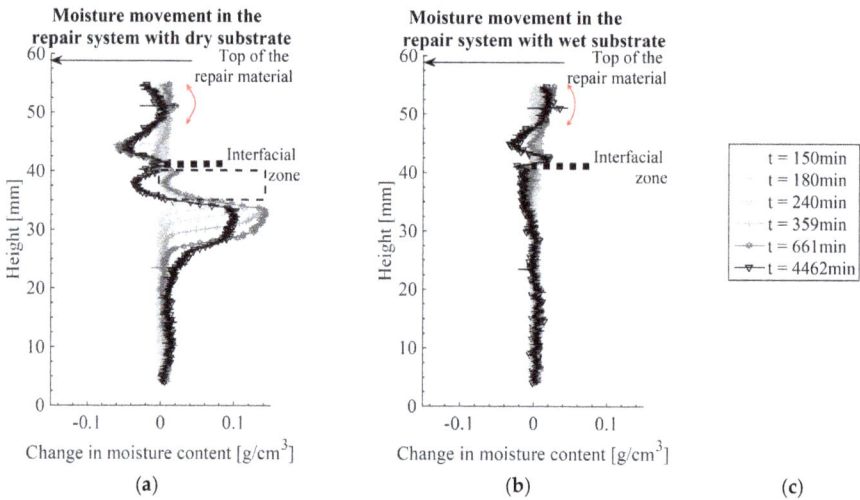

Figure 9. Moisture profiles in the repair systems as a function of time, t (t after mixing the repair material): (a) dry substrate, w/c = 0.3, DS, S3; (b) wet substrate, w/c = 0.3, WS, S3 with the legend; and (c) legend.

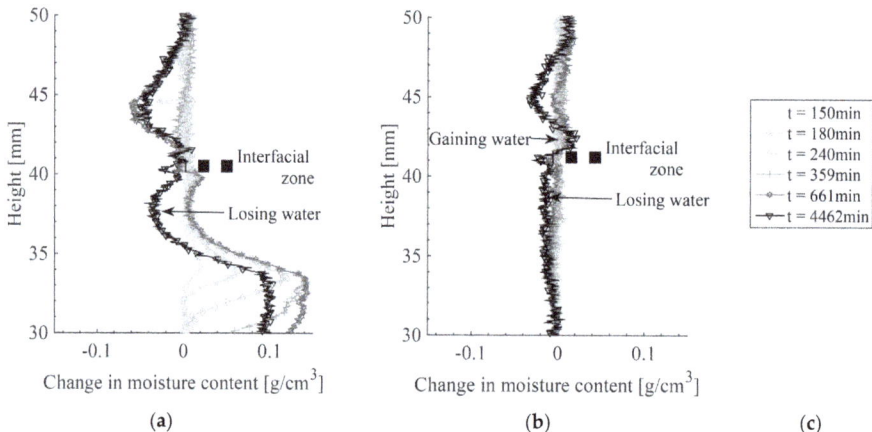

Figure 10. Moisture profiles in the repair systems close to the interface (dashed line): (a) dry substrate, w/c = 0.3, DS, S3; (b) wet substrate, w/c = 0.3, WS, S3; and (c) legend.

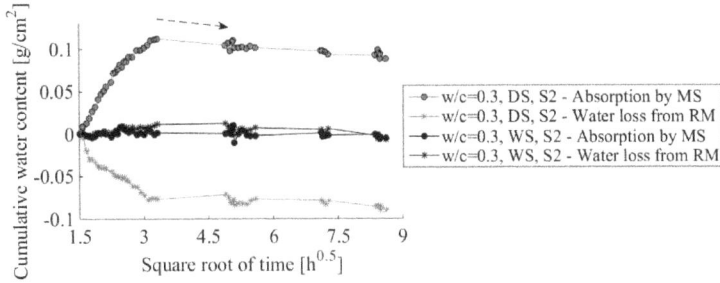

Figure 11. Calculated cumulative water contents in w/c = 0.3, DS, S2 and w/c = 0.3, DS, S3. MS denotes mortar substrate and RM denotes repair material.

4. Effects of Moisture Movement in the Repair System and Discussion

4.1. Effect of Moisture Movement on the Degree of Hydration of the Repair Material

After 3 d, samples w/c = 0.3, DS, S3 and w/c = 0.3, WS, S3 were taken out of the X-ray system. Repair materials were split from the substrates and degree of hydration was measured (75 h after casting). Apart from repair materials that were cast on top of the substrates (DS and WS), bulk repair material was cast in a separate mold (NoS) and placed in the same conditions as repair systems (inside the X-ray system chamber), was also tested. Degrees of hydration for all three samples and standard deviations are given in Figure 12.

Repair material cast on a wet substrate has a lower degree of hydration compared to the bulk repair material (NoS in Figure 12). A possible reason is the sealing procedure. The bulk repair material mold (NoS) was easily closed with the lid immediately after casting. On the other hand, with the repair material cast at the top of the substrate, this sealing took some time (20 m), meaning that during this period water could evaporate. Taking into account standard deviation, differences in measured degrees of hydration in these two samples (with NoS and WS) were not big.

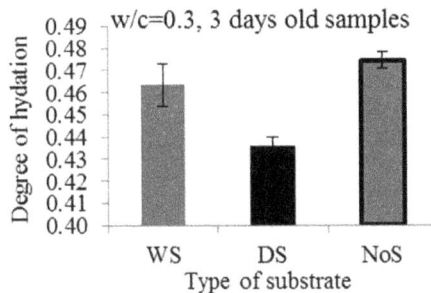

Figure 12. Influence of the mortar substrate saturation on the measured degree of hydration of the repair material, w/c = 0.3, WS, S2 (WS), w/c = 0.3, DS, S3 (DS) and bulk repair material (NoS).

Dry substrate caused a reduction in the degree of hydration of the repair material. This is due to the reduction of the effective w/c caused by moisture loss of the repair material. This reduction would be significantly higher with the higher absorption rates of the substrate. Furthermore, with thinner layer of repair material, the reduction would be probably even higher. This is because the substrate needs less time to absorb water from the repair material. As a result, hydration will be hindered or stopped even earlier.

4.2. Effect of Moisture Movement on Microstructure of the Repair Material and Interface

After measurements of the moisture transfer, the specimens were scanned in a CT scanner to investigate the effects of moisture exchange on resulting microstructure of the repair system (resolution is 30 μm). Two cross sections are given in Figure 13. Section I-I is made directly above the interface in two specimens (w/c = 0.3, DS, S3 and w/c = 0.3, WS, S3). There is a clear difference in void content in the two specimens. In order to investigate this further in 3D, all the slices parallel to the interface are aligned on top of each other. 3D structure is made and images of the scanned system close to the interface (containing around 1.5 mm of the substrate and 1.5 mm of the material) are given in Figures 14 and 15. The repair material can be clearly differentiated from the substrate. By segmenting (thresholding) the dark pixels (representing voids) in the original image, the image showing only voids in the repair system can be obtained (Figures 14b and 15b). A comparison between the original 3D image (whole GSV range) and the thresholded image (chosen GSV range) is given in Figure 14a,b. As the greyscale value of the void and the sealant around the mortar substrate are very close to each other, the sealant can be also observed in thresholded images (Figures 14b and 15b).

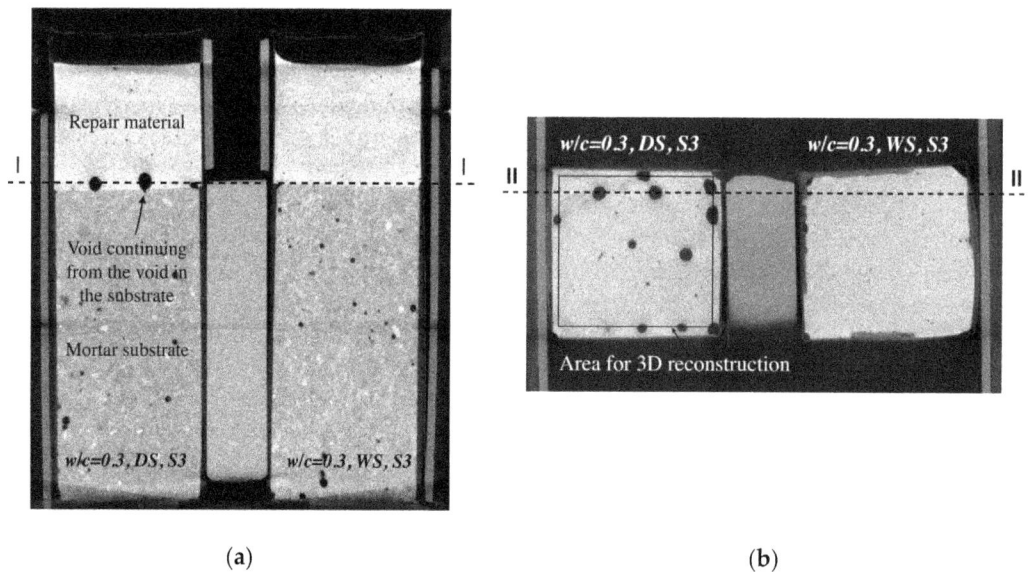

(a)   (b)

Figure 13. CT scan images for specimens w/c = 0.3, DS, S3 and w/c = 0.3, WS, S3 made at the age of 3 d: (a) section II-II (Figure b), front view; and (b) section I-I (Figure a), top view.

The moisture exchange between the repair material and substrate seems to have a significant influence not only on the effective w/c, degree of hydration and uniformity of properties of the repair material and interface, but also on void content close to the interface. The void content in two samples is compared in Figure 16. Significantly more voids are observed when the substrate is dry. Dry substrate absorbs water from the repair material (Figure 9a). When the water is absorbed, air from the substrate is being released. This air stays entrapped at the interface resulting in high void content.

It was also observed that voids from the substrate are "initiation" points for voids in the repair material (Figures 13a and 16). A void from the substrate is continuing in the repair material, making this an intrinsically weaker zone. In this system, the mortar substrate was a well-cured specimen with low porosity. It is anticipated that, with a more porous substrate, this influence would be significantly higher. In addition, it would be interesting to investigate if the same would be observed around the microcracks or cracks that are usually present in a substrate. In substrates used in this research, with a resolution of 30 μm, no cracks were observed.

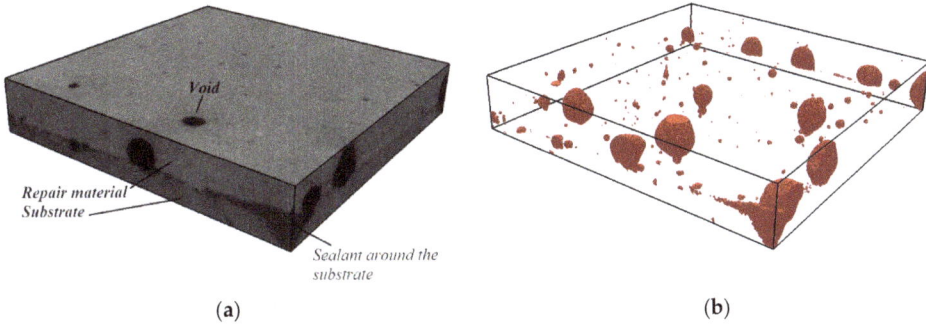

Figure 14. Void content in the repair system close to the interface, w/c = 0.3, DS, S3 (3mm thickness): (a) original structure (whole GSV range); and (b) thresholded voids (chosen GSV range).

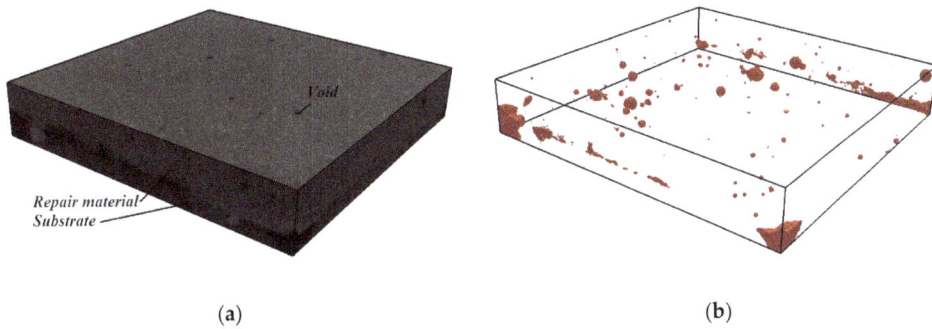

Figure 15. Void content in the repair system close to the interface, w/c = 0.3, WS, S2 (3mm thickness): (a) original structure (whole GSV range); and (b) thresholded voids (chosen GSV range).

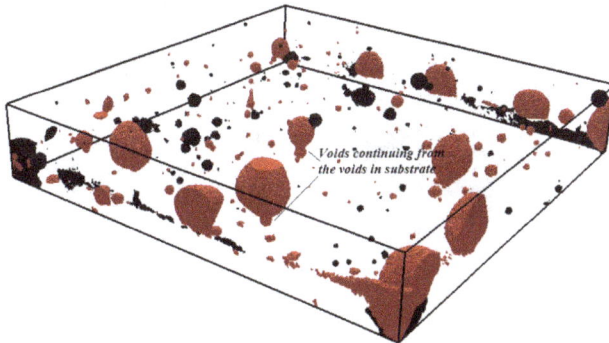

Figure 16. Void content in the repair system close to the interface w/c = 0.3, WS, S2 and w/c = 0.3, DS, S3, influence of the substrate saturation (black if the substrate is wet, red if the substrate is dry).

The void content close to the contact of two materials can be also quantified. In order to calculate the void content, an area of interest is chosen inside the images (Figure 13b). Every image, starting from the substrate approximately 1 mm below the interface, is analyzed. The percentage of voids in thresholded images is calculated by dividing the void area (number of red pixels in Figures 14b and 15b) by the total area (all pixels). Spacing between the images is 30 μm. As a result, void content as a function of distance from the interface is obtained. In w/c = 0.3, DS, S3, 2.5 mm of the repair material closest to the interface is affected by moisture absorption and result in higher volume content compared to the repair material cast on the fully saturated substrate (Figure 17).

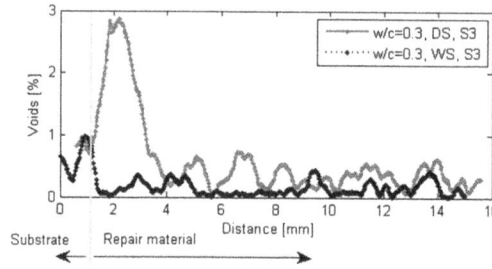

Figure 17. Void content in the repair system close to the interface (influence of the substrate saturation).

High content of air voids at the interface of the cement paste cast on initially dry brick was also observed in the study of Brocken *et al.* [3]. They analyzed the whole-section specimen and concluded that only 20%–30% of surface area of the paste is in good contact with the fired-clay brick. With sand-lime brick they found a better bond: 55%–70%. They suggested that poor bond and high void content might be attributed to "ineffective contact of the cement paste during laying on the brick: presumably due to fast compaction and poor laying". Another reason might be the phenomenon observed in this research: the brick is absorbing water, hereby releasing air which stays entrapped at the interface.

Air release from the substrate due to the moisture movement is not only related to the repair material. It is also observed when water is being absorbed by the substrate (Figures 18 and 19). Original X-ray images were thresholded so that voids can be easily observed (Figure 18). In Figure 19, air void development at the surface of mortar substrate as a function of time is given. With time, void content in the contact zone between water and mortar is also increasing.

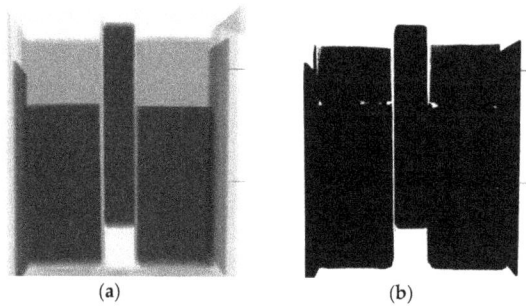

(a)   (b)

Figure 18. Images from X-ray absorption measurement when water is absorbed by the substrate: (a) original; and (b) thresholded image.

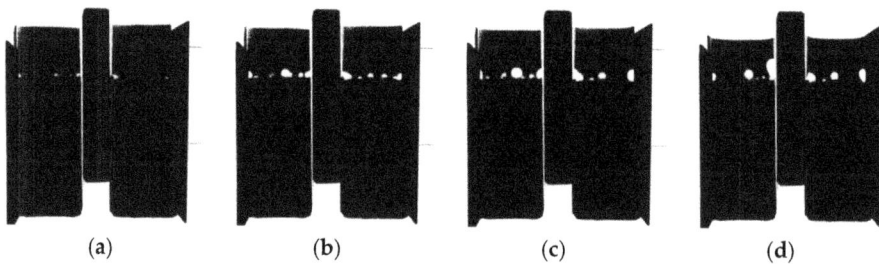

(a)   (b)   (c)   (d)

Figure 19. Void development when water is absorbed by the mortar substrate: (a) Reference image; (b) After 7 min; (c) After 17 min; (d) After 240 min.

This was not previously reported for the repair system performance (or any multilayer system with porous materials). However, similar performance as in repair systems is observed in light weight concretes [17]: "Water absorption of the light weight aggregates (LWA) after the concrete leaves the mixer, is accompanied by an expulsion of air from the particles. This air will form a rim of bubbles at the LWA surface. For some types of LWA, this amount of expelled air during setting is considerable and the aggregate-paste bond might be weakened. This is particularly the case when using LWA with a medium water absorption, that means neither very open capillary and fast absorption nor very impermeable. A normal procedure to repair this damage is remixing the concrete after a period and thus getting the bubbles evenly distributed in the paste."

"Medium water absorption (neither impermeable nor very open capillary)" is probably the condition to be expected for most of the concretes that should be repaired. The only practical solution is saturating concrete substrate prior to repair application. For practical application, it is important to know how long the substrate should be saturated prior to casting of the repair material. It seems important, therefore, to investigate whether the void content at the interface would differ with different saturation regimes of the substrate.

If the substrate is not saturated, water will be absorbed from the repair material. As a result, effective w/c of the repair material will be reduced. Therefore, bond strength due to lower resulting w/c, can be higher compared to the repair system with fully saturated substrate, as previously reported [1]. However, avoiding to pre-saturate the substrate in order to enable higher interlocking and possibly higher bond strength is probably not a good approach. A proper approach would be to directly design a repair material with lower w/c (if needed) and with a pre-saturated substrate with the dry surface. A more uniform and controlled microstructure of the repair material and interface would be obtained. Only then will the repair material reach its designed properties. Otherwise, specifying a certain w/c of repair material is meaningless, because this will change once the repair material is cast.

Implications of these tests for the reflective cracking phenomenon should be investigated. It might be expected that, once the repair material is cast, if the substrate contains microcracks, these cracks will also absorb water and release air. This air might stay entrapped at the interface, making the region around the existing crack weaker area in the repair material as well. Consequently, this area might be susceptible to further cracking.

5. Conclusions

In this paper, X-ray absorption technique was used to investigate the dynamics of moisture exchange in repair systems and its effects on the formed microstructure. Based on the presented study and the investigated parameters, the following conclusions can be drawn:

- Water exchange in a repair system has a critical influence on the microstructure formation of the bulk repair material and the interface. Water loss of the repair material by the substrate absorption reduces the effective w/c and degree of hydration of the repair material.

- The microstructure of the interface is significantly affected by the moisture exchange. Higher absorption of the substrate results in a more porous interface and bulk repair material. Pores and voids in the substrate, which are initially air-filled, are releasing this air to get water. Due to the high viscosity of the repair material and difficulties in compacting, this air remains entrapped at the interface.

- In the repair practice, the substrate should always be pre-saturated with the dry surface. Care should be taken that there is no water layer trapped at the surface of the substrate prior to the application of the material. This will provide a more uniform microstructure development in the repair material and a denser interface. Otherwise, properties of the repair material and the interface will be strongly influenced by the porosity, microcracks, moisture content, and absorption rate of the substrate. This is something that is difficult to control while designing repair systems.

In the future, the X-ray technique presented here can be improved and the same samples should be also tested by different methods (*i.e.*, neutron radiography, NMR) to compare the results. In addition, sensitivity studies of the coupled effective attenuation coefficient for the cement paste and mortar with different w/c and aggregate content should be performed.

Acknowledgments: Financial support by the Dutch Technology Foundation (STW) for the project 10981-"Durable Repair and Radical Protection of Concrete Structures in View of Sustainable Construction" is gratefully acknowledged.

Author Contributions: Mladena Lukovic and Guang Ye designed and performed the experiments, and wrote the paper.

Conflicts of Interest: The authors declare no conflict of interest. The founding sponsors had no role in the design of the study; in the collection, analyses, or interpretation of data; in the writing of the manuscript, and in the decision to publish the results.

References

1. Zhou, J. Performance of Engineered Cementitious Composites for Concrete Repairs. Ph.D. Thesis, Delft University of Technology, Delft, The Netherlands, 11 January 2011.
2. Courard, L.; Lenaers, J.F.; Garbacz, A. Saturation level of the superficial zone of concrete and adhesion of repair systems. *Constr. Build. Mater.* **2011**, *25*, 2488–2494. [CrossRef]
3. Brocken, H.; Spiekman, M.; Pel, L.; Kopinga, K.; Larbi, J. Water extraction out of mortar during brick laying: A NMR study. *Mater. Struct.* **1998**, *31*, 49–57. [CrossRef]
4. Faure, P.; Caré, S.; Po, C.; Rodts, S. An MRI-SPI and NMR relaxation study of drying–hydration coupling effect on microstructure of cement-based materials at early age. *J. Magn. Reson. Imaging.* **2005**, *23*, 311–314. [CrossRef] [PubMed]
5. Kazemi-Kamyab, H.; Denarie, E.; Bruhwiler, E.; Wang, B.; Thiery, M.; Faure, P.F.; Baroghel-Bouny, V. Characterization of Moisture Transfer in UHPFRC-Concrete Composite Systems at Early Age. In Proceedings of the 2nd International Conference on Microstructural-Related durability of Cementations Composites, Amsterdam, The Netherlands, 11–13 April 2012; Sun, W., Breugel, K.V., Miao, C., Ye, G., Chen, H., Eds.;
6. Bentz, D.; Hansen, K. Preliminary observations of water movement in cement pastes during curing using X-ray absorption. *Cem. Concr. Res.* **2000**, *30*, 1157–1168. [CrossRef]
7. Hall, C. Water sorptivity of mortars and concretes: A review. *Mag. Concr. Res.* **1989**, *41*, 51–61. [CrossRef]
8. Ye, G. Experimental Study and Numerical Simulation of the Development of the Microstructure and Permeability of Cementitious Materials. Ph.D. Thesis, Delft University of Technology, Delft, The Netherlands, 18 December 2003.
9. Michel, A.; Pease, B.J.; Geiker, M.R.; Stang, H.; Olesen, J.F. Monitoring reinforcement corrosion and corrosion-induced cracking using non-destructive x-ray attenuation measurements. *Cem. Concr. Res.* **2011**, *41*, 1085–1094. [CrossRef]
10. Roels, S.; Carmeliet, J. Analysis of moisture flow in porous materials using microfocus X-ray radiography. *Int. J. Heat. Mass. Tran.* **2006**, *49*, 4762–4772. [CrossRef]
11. Pease, B.J.; Scheffler, G.A.; Janssen, H. Monitoring moisture movements in building materials using X-ray attenuation: Influence of beam-hardening of polychromatic X-ray photon beams. *Constr. Build. Mater.* **2012**, *36*, 419–429. [CrossRef]
12. Juenger, M.C.G.; Jennings, H.M. Examining the relationship between the microstructure of calcium silicate hydrate and drying shrinkage of cement pastes. *Cem. Concr. Res.* **2002**, *32*, 289–296. [CrossRef]
13. Neville, A.M. *Properties of Concrete*, 4th ed.; Pearson: Harlow, UK, 1997; pp. 35–37.
14. Lam, L.; Wong, Y.; Poon, C. Degree of hydration and gel/space ratio of high-volume fly ash/cement systems. *Cem. Concr. Res.* **2000**, *30*, 747–756. [CrossRef]
15. Lukovic, M.; Schlangen, E.; Ye, G.; Breugel, K.V. Moisture Exchange in Concrete Repair System Captured by X-ray Absorption. In Proceedings of the 4th International Conference on Concrete Repair, Leipzig, Germany, 5–7 October 2015.

16. Koenders, E.A.B. Simulation of Volume Changes in Hardening Cement-Based Materials. Ph.D. Thesis, Delft University of Technology, Delft, The Netherlands, 23 September 1997.

17. *LWAC Material Properties State-of the-Art*; Technical Report for European Union–Brite EuRamIII; European Union: Brussels, Belgium, 1998; pp. 32–34.

2

Preparation and Characterization of Blended Films from Quaternized Hemicelluloses and Carboxymethyl Cellulose

Xian-Ming Qi [1], Shi-Yun Liu [1], Fang-Bing Chu [1], Shuai Pang [1], Yan-Ru Liang [1], Ying Guan [2], Feng Peng [1,*] and Run-Cang Sun [1]

Academic Editor: Jun-ichi Anzai

[1]   Beijing Key Laboratory of Lignocellulosic Chemistry, Beijing Forestry University, Beijing 100083, China; qxmfly@foxmail.com (X.-M.Q.); Liushiyun_Daisy@163.com (S.-Y.L.); chufangbingcfb@163.com (F.-B.C.); pangshuai2012@126.com (S.P.); l8141586@163.com (Y.-R.L.); rcsun3@bjfu.edu.cn (R.-C.S.)

[2]   School of Forestry and Landscape Architecture, Anhui Agricultural University, Hefei 230036, China; xiaomi1231@163.com

*   Correspondence: fengpeng@bjfu.edu.cn

Abstract: Utilization of hemicelluloses from biomass energy is an important approach to explore renewable resources. A convenient, quick, and inexpensive method for the preparation of blended films from quaternized hemicelluloses (QH) and carboxymethyl cellulose (CMC) was introduced into this study. QH and CMC solution were first mixed to form homogeneous suspension, and then were dried under vacuum to fabricate the blended films. The FT-IR and XRD results indicated that the linkage between QH and CMC was due to the hydrogen bonding and electrostatic interaction. From the results of mechanical properties and water vapor permeability (WVP), the tensile strength of the blended films increased with the QH/CMC content ratio increasing in appropriate range, and the WVP of the blended films decreased. The maximum value of tensile strength of blend film achieved was 27.4 MPa. In addition, the transmittances of the blended films increased with the decreasing of QH/CMC content ratio. When the weight ratio (QH: CMC) was 1:1.5, the blend film showed the best light transmittance (45%). All the results suggested that the blended films could be used in areas of application in the coating and packaging fields from the good tensile strength, transmittance, and low WVP.

Keywords: blend film; quaternized hemicelluloses; carboxymethyl cellulose

1. Introduction

With the decrease in fossil fuels, much attention has been paid to investigate the new biodegradable materials. The increased social awareness for sustainable development gather momentum in favor of materials from renewable resources [1–3]. Therefore, polysaccharides have become the main part of natural-based materials which are particularly advantageous due to their biocompatibility, biodegradability, and non-toxicity [4–6]. Production of environmentally friendly packaging materials based on polysaccharides from renewable resources are also expected to substitute for petroleum-based packaging materials [7]. Cellulose and hemicelluloses are the most and the second most abundant polysaccharides in biomass, respectively [8–11]. Some studies have been widely studied for preparing films from them. Biocomposite films based on quaternized hemicelluloses and montmorillonite showed good thermal properties by the addition of clay nanoplatelets [12]. Bacterial cellulose as reinforcement was added in the arabinoxylan films, it could effectively increase the stiffness and strength of the films [13]. In addition, hydrophobic films formed by acetylated bleached hemicelluloses and acetylated cellulose showed the improvement of thermal and mechanical properties [14]. Therefore, the films based on hemicelluloses and cellulose have a prospective application in many fields.

Various derivatives of cellulose have been synthesized and used in practical applications. Carboxymethyl cellulose (CMC) is one of the most widely applied cellulose derivatives and it has good film forming property, which can form transparent films and possess high mechanical strength [15,16]. It contains a hydrophobic polysaccharide backbone and many hydrophilic carboxyl groups, showing amphiphilic characteristic [17]. CMC is an acidic polysaccharide, due to its a number of carboxylic substituents [18]. In addition, hemicelluloses are also the excellent materials for the film preparation [19–21]. Quaternized hemicelluloses (QH) are derivative of hemicelluloses which have highly solubility and cationic property because of the cationic agents and hydroxide radical in QH [22].

Blending is a main method for fabricating films by mixing two components, which is a simple and effective method to improve the properties of films [23,24]. When the two components are compatible, more homogeneous structure and better physicochemical properties of the blended films can be obtained than that of films from individual components [25]. In this article, quaternized hemicelluloses (QH) were obtained from bamboo hemicelluloses by esterified with 2,3-epoxypropyltrimethyl ammonium chloride [26]. Then, QH and CMC as the positive and negative charged polyelectrolyte, respectively, could form the polyelectrolyte complex through intermolecular hydrogen bonding and strong electrostatic attraction. A series of novel blended films that are non-toxic, renewable, and biodegradable were prepared from the two kinds of biopolymer according to a predetermined ratio. The structural analyses of the blended films were determined by FT-IR and XRD. The morphological structure was demonstrated by scanning electron microscopy (SEM) and atomic force microscopy (AFM). The mechanical properties and light transmittance were studied by tensile test and UV/Vis spectrophotometer. The water vapor permeability (WVP) of the blended films was performed according to the standard ASTM E 96/E 96M-05 [27]. The relationship between the structure and their physicochemical properties of the blended films was also discussed in this study.

2. Experimental Section

2.1. Materials

The hemicelluloses was extracted and purified from the *phyllostachys pubescens* (Meishan, Sichuan Province, China). The sugar composition of hemicelluloses was 83.6% xylose, 5.1% arabinose, 4.2% glucose, 0.4% galactose, and 6.8% glucuronic acid (relatively molar percent). The molecular weight obtained by gel permeation chromatography (GPC) showed that the native hemicelluloses had an average molecular weight (Mw) of 13,420 $g \cdot mol^{-1}$ and a polydispersity of 4.1, corresponding to a degree of polymerization of 88. Sodium hydroxide was purchased from Beijing Chemical Works (Beijing, China). 2,3-epoxypropyltrimethyl ammonium chloride (ETA) was obtained from Sigma-Aldrich Co., Ltd. (St. Louis, MO, USA) CMC, the degree of substitution is 0.78, was supplied by Tianjin Jinke Fine Chemical Research Institute (Tianjin, China). The weighed average molecular weight of CMC is 1.9×10^4, and the viscosity is 800–1200 Pa· s. All reagents mentioned above were directly used without further purification.

2.2. Preparation of Quaternized Hemicelluloses

The synthesis of QH was carried out in a three-necked flask fitted with a mechanical stirrer and a reflux condenser. 0.66 g of dried hemicelluloses were dissolved in 5 mL distilled water at 60 °C for 30 min. An aqueous solution of sodium hydroxide (the molar ratio of NaOH to ETA, 0.75) was added, followed by adding of ETA (the molar ratio of ETA to anhydroxylose units in hemicelluloses, 2.0). The mixture was stirred at 60 °C, for a total period of 5 h. Upon completion of the reaction, the mixture was cooled to ambient temperature. Then, the reaction mixture was filtered off and thoroughly washed with ethanol, and dried in a vacuum oven at 45 °C for 24 h.

2.3. Preparation of Blended Films

QH were dissolved in distilled water and stirred for 6 h. The suspension was centrifuged at 4000 rpm for 10 min, and the supernatant was stored in refrigerator at 4 °C for the use of film preparation. The insoluble solids were dried to calculate concentration of the supernatant. CMC was dissolved in distilled water at 60 °C for 30 min under magnetic stirring to prepare the solution with concentration of 2 wt %. The QH and CMC solution concentrations were all 2 wt %, and they were mixed under magnetic stirring energetically at 60 °C. The mixing solution was degassed by rotary evaporator at room temperature for 2 h, after which per 10 mL solution was cast into a 5.5 cm diameter plastic petri dishes and dried under vacuum at 40 °C for 36 h. Then the blended films were obtained and peeled off carefully. The different volume ratios of QH to CMC are shown in Table 1.

Table 1. Composition of the blended films.

Sample Code	QH: CMC (V/V)
Film 1	1.0:1.0
Film 2	1.0:1.5
Film 3	2.0:1.0
Film 4	1.5:1.0

2.4. FT-IR Spectroscopy

FT-IR spectra of QH, CMC, and QH/CMC blended films were recorded using a Nicolet iN 10 Fourier transform infrared spectrometer (Thermo Nicolet Corporation, Madison, WI, USA). QH and CMC powder were prepared by grinding with potassium bromide and laminating. FT-IR spectra were recorded in the spectra range from 4000–650 cm^{-1}. The blended film samples were measured under the ATR mode with spectral range of 4000–650 cm^{-1}.

2.5. X-ray Diffraction

X-ray diffraction patterns of QH, CMC, and QH/CMC blended films were recorded with a XRD-6000 X-ray diffractometer (Shimadzu, Kyoto, Japan), using Nickel-filtered Cu Kα radiation (λ = 0.154 nm) at 40 kV and 30 mA in the 2θ range of 5°–45° at a speed of 5°·min^{-1}.

2.6. Atomic Force Microscopy (AFM)

Surface roughness of the films was measured using a Multimode 8 (Bruker, Billerica, MA, USA), Atomic force microscope (AFM), operated in tapping mode. Silicon cantilever was used throughout the study with the nominal tip radius of 10 nm curvature. All scans were performed in air, at room temperature. Height and phase contrast images were recorded simultaneously with a resonance frequency between 250 and 300 kHz and a scan angle of 0°.

2.7. Scanning Electron Microscopy (SEM)

Scanning electron micrographs of the films, brittle fractured in liquid nitrogen, were taken with S-3400N scanning electron microscope (Hitachi, Tokyo, Japan). Prior to the SEM analysis, the films were sputter coated with a thin layer of gold. The acceleration voltage was set to 15 kV and magnification was 1500×.

2.8. Light Transmittance

The light transmittance of the blended films were performed by a UV-Vis spectrophotometer (UV 2300, Tech Comp, Shanghai, China) at wavelength of 200–800 nm. The films strips were attached to the surface of the cuvette with the size of 12 mm × 60 mm, taking cuvette as a reference.

2.9. Tensile Properties

Tensile strength and breaking elongation of films were measured using a miniature universal testing machine (CMT6503, SANS Technology stock Co., Ltd., Shenzhen, China), with a load cell of 100 N, operating in tension mode. The initial grip distance was 20 mm, and the rate of grip separation was 5 mm/min. The specimens were cut to the size of 30 mm × 10 mm. At least three repeated measurements of each film were tested and the average value was used to determine mechanical properties.

2.10. Water Vapor Permeability

The standard ASTM E 96/E 96M-05 was used to determine water vapor permeability (WVP) [27]. Aluminum cups were filled with 15 g anhydrous calcium chloride as desiccant and the film samples were mounted over the cups at room temperature. The desiccant was dried at 240 °C for 3–5 h before use. Aluminum cups were then placed in a cabinet containing water. WVP was determined by

$$WVP = \frac{qd}{ts\Delta P}$$

where $q/t, d$, s, ΔP are the slope of the weight increase *versus* time (g/s), the film thickness (cm), the film area exposed to moisture (4.91 cm^2), and the difference of the vapor pressure between the two sides of films (23.76 mmHg), respectively.

3. Results and Discussion

3.1. FT-IR Analysis

Figure 1 shows the FT-IR spectra of CMC and QH. As can be seen, the absorption band at 1592 cm^{-1} is assigned to asymmetrical COO– stretching of CMC [28]. The signals at 1414 and 1320 cm^{-1} are related to the symmetrical stretching vibrations of the carboxylate groups and C–H bending, respectively [29]. The broad peak at 1059 cm^{-1} is attributed to the stretching of C–O–C [30]. In the spectrum of QH, the absorption band at 3354 cm^{-1} corresponds to –OH stretching vibration, and the vibration at 1609 cm^{-1} reflects the blending mode of absorbed water [31]. The stretching vibration at 1034 cm^{-1} and 894 cm^{-1} which can be ascribed the C–O–C and the β-glycosidic bond, respectively [32]. In addition, a new absorption peak at 1475 cm^{-1} appeared in QH, which originates from the bending vibration of –CH$_3$ and –CH$_2$ in quaternary ammonium group. It suggests that the hemicelluloses were modified successfully [33]. Compared with the FT-IR spectra of pure CMC and QH, the spectra of blended films are shown in Figure 2. The stretching vibration of the C–O–C was broadened and shifted evidently to the 1100 cm^{-1}, which was caused by the overlap of the ether bond in both CMC and QH. The absorption peaks at 1597 cm^{-1} and 896 cm^{-1} of blended films are attributed to asymmetrical COO– stretching and β-glycosidic bond, respectively. The broad absorption band at 3300–3500 cm^{-1} is also assigned to –OH stretching of hydrogen bonds. The blended films were fabricated by mixing the QH with CMC, abundant hydrogen bonds were therefore formed between the QH and CMC chains. Furthermore, plenty of ammonium cations are on the surface of QH, and there are many carboxyl anions in CMC chains. This resulted in the electrostatic interactions that were formed between the QH and CMC. Therefore, the blending of QH and CMC was mainly due to the hydrogen bonds and electrostatic interactions between them.

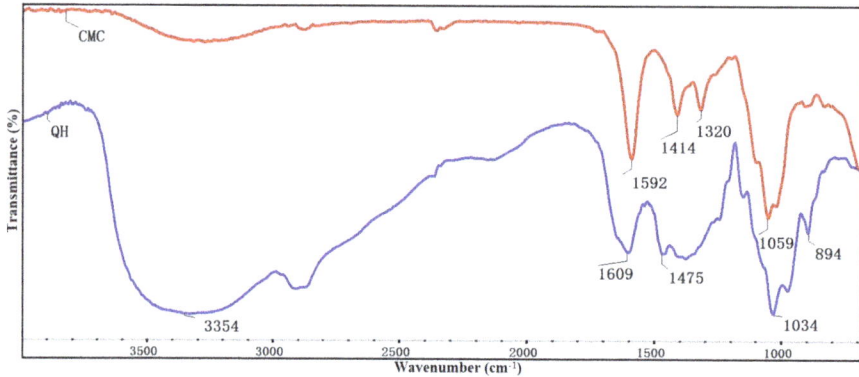

Figure 1. The FT-IR spectra of CMC and QH.

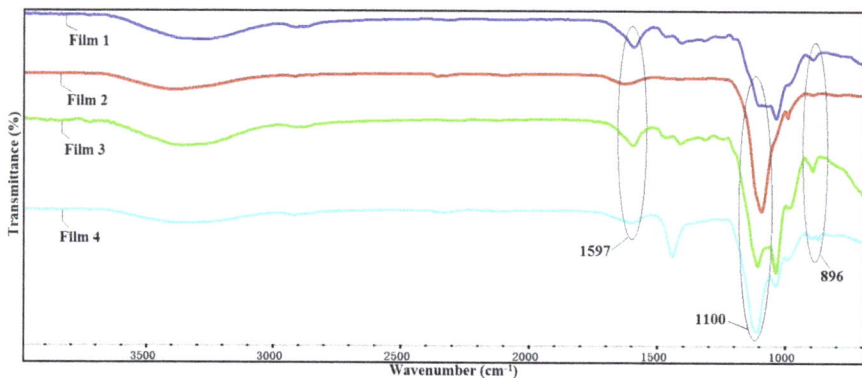

Figure 2. The FT-IR spectra of four blended films.

3.2. X-ray Analysis

The impact of chemical modification on the crystal structure was further evaluated by using X-ray diffraction analysis. The X-ray diffraction patterns of QH and CMC are shown in Figure 3. The diffraction peak of QH is at $2\theta = 19.9°$ [34]. It was found that the diffraction peak intensity of QH was lower than that of CMC, which was due to the amorphism of hemicelluloses. The diffraction of CMC consists of two major crystalline peaks are at $2\theta = 18.9°$ and $32.6°$, respectively, which are the characteristic of cellulose I [35,36]. Moreover, the intensity of CMC diffraction peak at $2\theta = 28.2°$ and $38.7°$ become weak, because the hydroxyl in cellulose are substituted by carboxyl groups. It could be partially explained by molecular rearrangement of CMC. The XRD patterns of the blended films showed that the major diffraction peaks are at $2\theta = 19.0°, 33.0°, 34.2°$, and $37.8°$ in Figure 4. It suggested that the diffraction peaks of composite films were overlapped and shifted. The degree of crystallinities had the following order: CMC > Film 4 > Film 3 > Film 2 > Film 1 > QH, and the crystallinity values of them were present in Table 2. Theoretically, the crystallinity value of the blended films should increase with the increase of CMC amount, because CMC is crystalline and QH is amorphous. However, the XRD results of blended films indicate that the crystallinity value of the blended films do not show the obvious positive correlation with CMC amount. The intermolecular hydrogen bonding and electrostatic interaction are the driving force of blended films between QH and CMC, which may change the crystal structure of the blended films. The intensity of the driving force is different with the increase of QH/CMC content. Moreover, the crystalline region of CMC and the amorphous region of QH also have influence on the final degree of crystallinity of the blended films. Therefore, the crystallinity values of the blended films do not show obvious regular change with the QH/CMC content ratio.

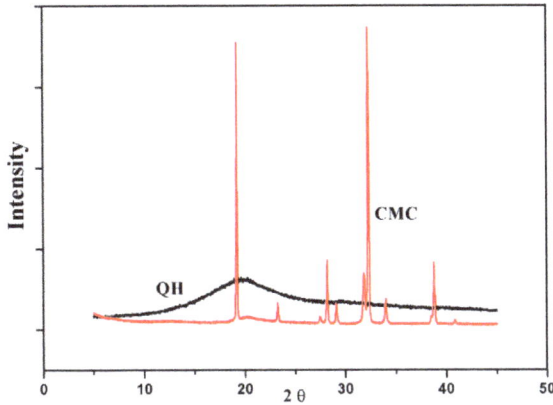

Figure 3. The XRD patterns of QH and CMC.

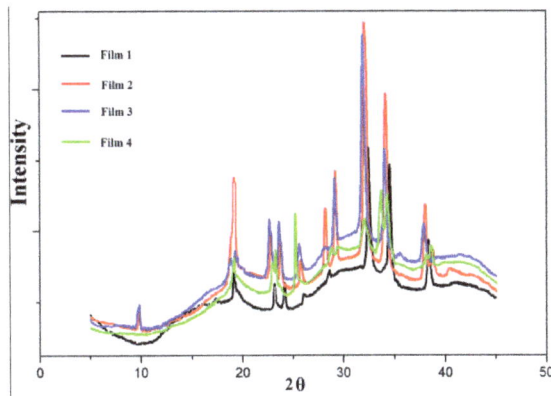

Figure 4. The XRD patterns of Film 1, Film 2, Film 3, and Film 4.

Table 2. The degree of crystallinity of QH, CMC, and the blended films.

Sample Code	QH	CMC	Film 1	Film 2	Film 3	Film 4
Crystallinity(%)	9	84.6	27.7	33.4	34	35.9

3.3. Morphology of Blended films

The SEM images of the surface and cross section of the blend film are present in Figure 5. As can be seen from Figure 5a, the surface of the film is smooth and homogeneous without any aggregation. It suggested that the QH and CMC were diffused evenly in the film. In the microstructure of the film cross section in Figure 5b, a dense and smooth structure of blended film suggested that strong interaction force existed between QH and CMC. QH and CMC have a positive charge and negative charge, respectively. In addition, abundant of hydroxyl groups were existed in the surface of QH and CMC. Therefore, hydrogen bonding and electrostatic attraction between the oppositely charged polymers are the driving force for the formation of tight and smooth films.

The surface nanotopography and toughness of the four blended films were investigated by AFM. The height images, 3D topography, and phase topography of the films are shown in Figure 6. From the height images and 3D topography of films, it was obviously found that the surfaces of blended Film 1 and Film 2 were more flat than those of blended Film 3 and Film 4. The roughness values were calculated in the areas of 2 μm × 2 μm, and the roughnesses of Film 1, Film 2, Film 3, and Film 4 were 11.4, 3.48, 40.7, and 57.3 nm, respectively. The roughnesses of the blended films increased with the increase of QH relative content ratio from 1:1.5 (Film 2), 1:1 (Film 1) to 1.5:1 (Film 4). It was due to that QH have many branch chains on its backbone, which resulted in the roughness of the blended films increased. When the ratio of QH/CMC reached up to 2:1, the roughness of blend Film 3 had a little decrease compared with Film 4, but still higher than that of Film 1 and Film 2.

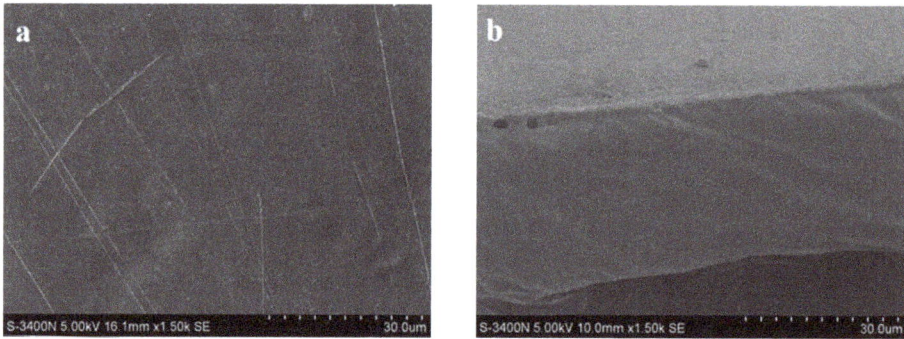

Figure 5. SEM images of the surface (**a**) and cross section (**b**) of Film 4.

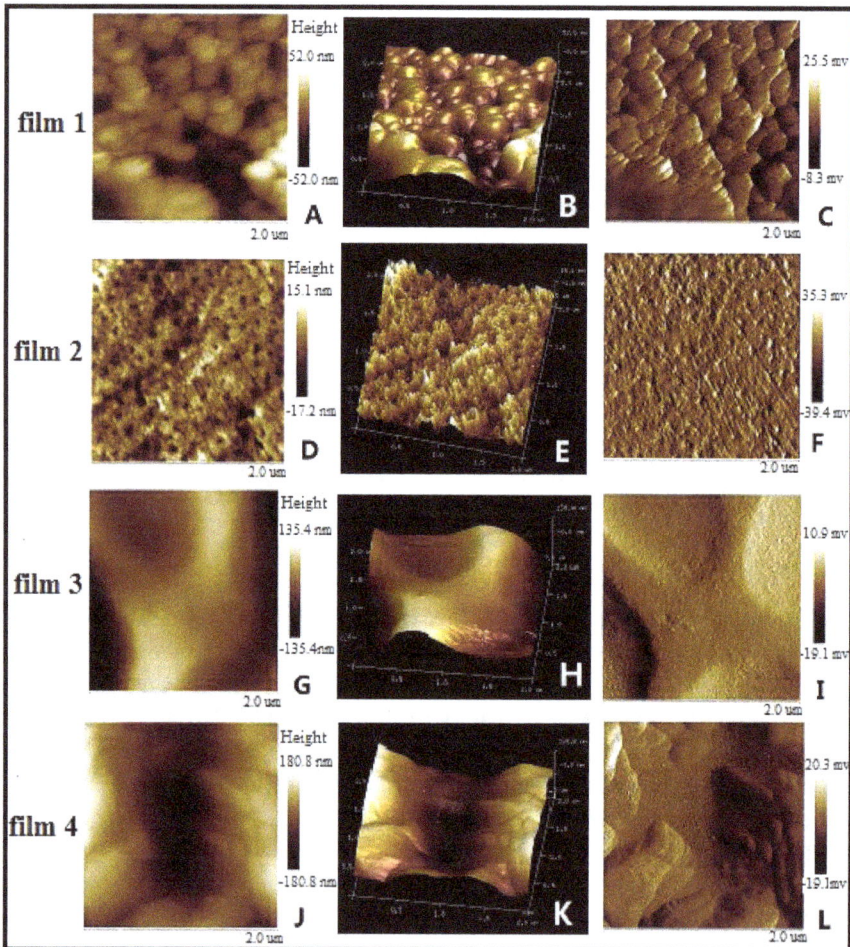

Figure 6. AFM height images (**A,D,G,J**), 3D topography (**B,E,H,K**), and phase topography (**C,F,I,L**) of the blend Film 1, Film 2, Film 3, and Film 4.

3.4. Mechanical Properties

The tensile stress, tensile strain at break, and Young's modulus of films are shown in Table 3, and the tensile stress-strain curves of the blended films are shown in Figure 7. As can been seen, it was found that the tensile strength of blended films increased with the increase of QH/CMC content ratio from 1:1.5 (Film 2), 1:1 (Film 1) to 1.5:1 (Film 4), and the maximum value 27.4 MPa was obtained in Film 4; nevertheless, the tensile strength of Film 3 (QH/CMC, 2:1) reduced obviously when the QH relative content further increased. It was mainly caused by the intermolecular hydrogen bonding and electrostatic interaction between QH and CMC. That is to say, the miscibility of the films increases with the QH/CMC content in the appropriate range. From Film 2 to Film 1 to Film 4, the quantity and intensity of the hydrogen bonding and electrostatic interaction increased with the QH content increasing. Therefore, the tensile strength of the films increased consequently. However, with the relative content of QH further increasing, the miscibility of the films decreased, which resulted in the hydrogen bonding and electrostatic force were not formed enough between QH and CMC because of the relatively small amount of CMC. That is to say, the formed interaction force was too weak to support the structure of the Film 3, resulting in the sharp decrease of tensile strength of Film 3. The results discussed above indicated that the mechanical properties of the blended films were greatly affected by the QH content, and the excess QH would reduce the mechanical property of the blended films.

Table 3. Tensile testing results of the four blended films.

Sample	Tensile Strength (MPa)	Tensile Strain at Break (%)	Young's Modulus (MPa)
Film 1	25.2 ± 2.3	2.8 ± 0.7	1319.9 ± 102.7
Film 2	14.2 ± 0.9	2.6 ± 0.9	842.0 ± 11.1
Film 3	14.8 ± 1.6	1.7 ± 0.5	1177.2 ± 126.5
Film 4	27.0 ± 1.5	3.9 ± 0.8	1118.5 ± 53.5

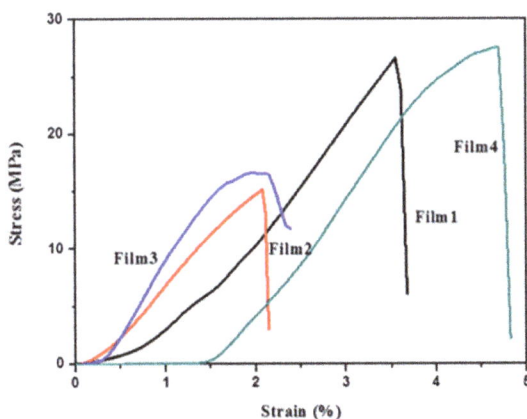

Figure 7. The tensile stress-strain curves of Film 1, Film 2, Film 3, and Film 4.

3.5. UV-Vis Transparency of Films

Generally, transparency of film is an assistant criterion to judge the miscibility of blended films. The light transmittances of the blended films at wavelength of 200–800 nm are shown in Figure 8. It can be seen obviously that the transmittance curve of each film rose gradually with the increase of wavelength. Comparing the four blended films, the transparencies of the blended films increased with the QH/CMC content ratio decreasing in visible light spectrum, and it had the following order: Film 2 > Film 1 > Film 4 > Film 3. In other words, a relatively high content of CMC in the blended films is beneficial for the transparency of the blended films.

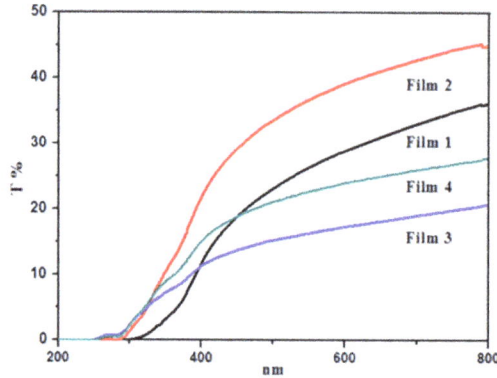

Figure 8. UV-Vis spectra of blend Film 1, Film 2, Film 3, and Film 4.

3.6. Water Vapor Permeability (WVP)

The WVP values of the blended films are important measures for the applications of packaging materials. One of the main functions of food packaging is to avoid or minimize moisture transfer between food and the surrounding atmosphere. Low WVP widens the application of the composite packaging film, especially in a highly humid environment. The WVP curves of the four blended films are shown in Figure 9. It was found that all the four blended films showed a relatively lower WVP, suggesting that the fabricated films have the capability to withstand water vapor. The WVP values of blended films had the following order: Film 2 > Film 3 > Film 1 > Film 4, indicating that the WVP values decreased with the increase of QH/CMC content ratio from 1:1.5 (Film 2), 1:1 (Film 1) to 1.5:1 (Film 4). However, an interesting result was observed. When the QH/CMC content ratio was further increased to 2:1 (Film 3), the WVP value of Film 3 increased instead. The causes of this phenomenon are the different densification of the four blended films. As discussed in the analysis of the blended films' mechanical properties, the quantity and intensity of the hydrogen bonding and electrostatic force increased with the QH/CMC content ratio increasing, which led to the tightness of internal structure of blended films. It also prevented water molecules from diffusing through the films. Therefore, the WVP of films decreased with QH/CMC content ratio increasing. When the QH/CMC content ratio was further increased to 2:1 (Film 3), the hydrogen bonding and electrostatic interaction were not formed enough between QH and CMC because of the relatively small amount of CMC. Then, the structure of blend Film 3 became loose again and the WVP of Film 3 increased. Therefore, the interactions between components have obvious influence on the densification of the blended films [37]. The results indicated that the WVP of the blended films is related to the intensity of interaction force between QH and CMC, and it is in accordance with the tensile strength of the blended films.

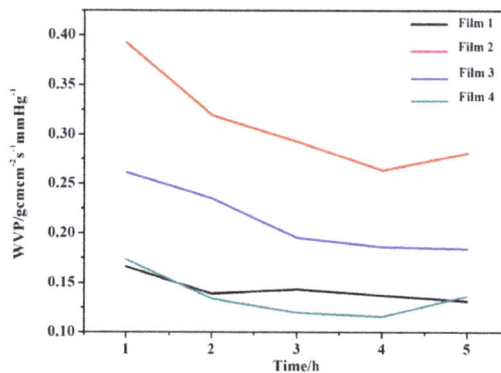

Figure 9. The water vapor permeability of blend Film 1, Film 2, Film 3, and Film 4.

4. Conclusions

A series of blended films of QH and CMC were successfully fabricated in this study, and they were linked by the hydrogen bonding and electrostatic interaction. The blended films with different proportions of QH and CMC were also studied to determine their properties. From the results of mechanical properties and water vapor permeability, it was found that the intensity of hydrogen bonding and electrostatic interaction between QH and CMC increased with the increasin QH/CMC content ratio in the appropriate range, which made the film structurally dense, resulting in the tensile strength increase and WVP film decrease, respectively. However, hydrogen bonding and electrostatic interaction were not formed enough to support the structure of the film when excess QH was added, which led to the film tensile strength decrease and the WVP increase. The light transmittances of the blended films increased with the decrease of the QH/CMC content ratio. The prepared films show good strength, low WVP, and good light transmittances, which are beneficial for the applications of films as packaging materials. Besides, compared with the petroleum-based plastics, the films fabricated from QH and CMC have biocompatible, nontoxic, and biodegradable properties. Therefore, the films fabricated by blending CMC with QH might become attractive in the application of packaging materials to replace the traditional petroleum-based packaging materials.

Acknowledgments: This work was supported by Grants from State Forestry Administration (201404617), 2014 Beijing Student Research Training Program (S201410022044), National Natural Science Foundation of China (21406014), Ministries of Education (NCET-13-0670), and Author of National Excellent Doctoral Dissertations of China (201458).

Author Contributions: Shiyun Liu and Fangbing Chu performed the experiments; Yanru Liang and Shuai Pang analyzed the data; the paper was written under the direction and supervision of Feng Peng and Runcang Sun; Xianming Qi and Ying Guan are responsible for the writing of this work.

Conflicts of Interest: The authors declare no conflict of interest.

References

1. Ragauskas, A.J.; Williams, C.K.; Davison, B.H.; Britovsek, G.; Cairney, J.; Eckert, C.A.; Frederick, W.J.; Hallett, J.P.; Leak, D.J.; Liotta, C.L. The path forward for biofuels and biomaterials. *Science* **2006**, *311*, 484–489. [CrossRef] [PubMed]

2. Qi, X.M.; Guan, Y.; Chen, G.G.; Zhang, B.; Ren, J.L.; Peng, F.; Sun, R.C. A non-covalent strategy for montmorillonite/xylose self-healing hydrogels. *RSC Adv.* **2015**, *5*, 41006–41012. [CrossRef]

3. Edlund, U.; Ryberg, Y.Z.; Albertsson, A.C. Barrier films from renewable forestry waste. *Biomacromolecules* **2010**, *11*, 2532–2538. [CrossRef] [PubMed]

4. Sadeghi, M.; Soleimani, F. Synthesis of novel polysaccharide based superabsorbent hydrogels via graft copolymerization of vinylic monomers onto kappa carrageenan. *Int. J. Chem. Eng. Appl.* **2011**, *2*, 304–306.

5. Pérez, J.; Munoz-Dorado, J.; de la Rubia, T.; Martinez, J. Biodegradation and biological treatments of cellulose, hemicellulose and lignin: An overview. *Int. Microbiol.* **2002**, *5*, 53–63. [CrossRef] [PubMed]

6. Rodrigues, F.H.; Spagnol, C.; Pereira, A.G.; Martins, A.F.; Fajardo, A.R.; Rubira, A.F.; Muniz, E.C. Superabsorbent hydrogel composites with a focus on hydrogels containing nanofibers or nanowhiskers of cellulose and chitin. *J. Appl. Polym. Sci.* **2014**, *131*, 39725. [CrossRef]

7. Kochumalayil, J.J.; Zhou, Q.; Kasai, W.; Berglund, L.A. Regioselective modification of a xyloglucan hemicellulose for high-performance biopolymer barrier films. *Carbohydr. Polym.* **2013**, *93*, 466–472. [CrossRef] [PubMed]

8. Siró, I.; Plackett, D. Microfibrillated cellulose and new nanocomposite materials: A review. *Cellulose* **2010**, *17*, 459–494. [CrossRef]

9. Mamman, A.S.; Lee, J.M.; Kim, Y.C.; Hwang, I.T.; Park, N.J.; Hwang, Y.K.; Chang, J.S.; Hwang, J.S. Furfural: Hemicellulose/xylosederived biochemical. *Biofuels Bioprod. Biorefin.* **2008**, *2*, 438–454. [CrossRef]

10. Saha, B.C. Hemicellulose bioconversion. *J. Ind. Microbiol. Biotechnol.* **2003**, *30*, 279–291. [CrossRef] [PubMed]

11. Stevanic, J.S.; Bergström, E.M.; Gatenholm, P.; Berglund, L.; Salmén, L. Arabinoxylan/nanofibrillated cellulose composite films. *J. Mater. Sci.* **2012**, *47*, 6724–6732. [CrossRef]

12. Guan, Y.; Zhang, B.; Tan, X.; Qi, X.M.; Bian, J.; Peng, F.; Sun, R.C. Organic–inorganic composite films based on modified hemicelluloses with clay nanoplatelets. *ACS Sustain. Chem. Eng.* **2014**, *2*, 1811–1818. [CrossRef]

13. Stevanic, J.S.; Joly, C.; Mikkonen, K.S.; Pirkkalainen, K.; Serimaa, R.; Rémond, C.; Toriz, G.; Gatenholm, P.; Tenkanen, M.; Salmén, L. Bacterial nanocellulose-reinforced arabinoxylan films. *J. Appl. Polym. Sci.* **2011**, *122*, 1030–1039. [CrossRef]

14. Gordobil, O.; Egüés, I.; Urruzola, I.; Labidi, J. Xylan–cellulose films: Improvement of hydrophobicity, thermal and mechanical properties. *Carbohydr. Polym.* **2014**, *112*, 56–62. [CrossRef] [PubMed]

15. Arnon, H.; Zaitsev, Y.; Porat, R.; Poverenov, E. Effects of carboxymethyl cellulose and chitosan bilayer edible coating on postharvest quality of citrus fruit. *Postharvest Biol. Technol.* **2014**, *87*, 21–26. [CrossRef]

16. Ghanbarzadeh, B.; Almasi, H.; Entezami, A.A. Physical properties of edible modified starch/carboxymethyl cellulose films. *Innov. Food Sci. Emerg. Technol.* **2010**, *11*, 697–702. [CrossRef]

17. Zhang, S.W.; Liu, W.T.; Liang, J.; Li, X.Y.; Liang, W.N.; He, S.Q.; Zhu, C.S.; Mao, L.Y. Buildup mechanism of carboxymethyl cellulose and chitosan self-assembled films. *Cellulose* **2013**, *20*, 1135–1143. [CrossRef]

18. Habibi, N. Preparation of biocompatible magnetite-carboxymethyl cellulose nanocomposite: Characterization of nanocomposite by FTIR, XRD, FESEM and TEM. *Spectrochim. Acta A* **2014**, *131*, 55–58. [CrossRef] [PubMed]

19. Eronen, P.; Österberg, M.; Heikkinen, S.; Tenkanen, M.; Laine, J. Interactions of structurally different hemicelluloses with nanofibrillar cellulose. *Carbohydr. Polym.* **2011**, *86*, 1281–1290. [CrossRef]

20. Egüés, I.; Eceiza, A.; Labidi, J. Effect of different hemicelluloses characteristics on film forming properties. *Ind. Crop. Prod.* **2013**, *47*, 331–338. [CrossRef]

21. Hansen, N.M.; Plackett, D. Sustainable films and coatings from hemicelluloses: A review. *Biomacromolecules* **2008**, *9*, 1493–1505. [CrossRef] [PubMed]

22. Ren, J.L.; Sun, R.C.; Liu, C.F. Etherification of hemicelluloses from sugarcane bagasse. *J. Appl. Polym. Sci.* **2007**, *105*, 3301–3308. [CrossRef]

23. Bourtoom, T.; Chinnan, M.S. Preparation and properties of rice starch-chitosan blend biodegradable film. *LWT Food Sci. Technol.* **2008**, *41*, 1633–1641. [CrossRef]

24. Mathew, S.; Abraham, T.E. Characterisation of ferulic acid incorporated starch-chitosan blend films. *Food Hydrocoll.* **2008**, *22*, 826–835. [CrossRef]

25. Zhang, P.Y.; Whistler, R.L. Mechanical properties and water vapor permeability of thin film from corn hull arabinoxylan. *J. Appl. Polym. Sci.* **2004**, *93*, 2896–2902. [CrossRef]

26. Ren, J.L.; Sun, R.C.; Liu, C.F. A view of etherification of hemicelluloses. *J. Cellul. Sci. Technol.* **2007**, *2*, 74–78.

27. Kumaran, M. Interlaboratory comparison of the ASTM standard test methods for water vapor transmission of materials (E 96–95). *J. Test. Eval.* **1998**, *26*, 83–88.

28. Hatanaka, D.; Yamamoto, K.; Kadokawa, J.I. Preparation of chitin nanofiber-reinforced carboxymethyl cellulose films. *Int. J. Biol. Macromol.* **2014**, *69*, 35–38. [CrossRef] [PubMed]

29. Fan, L.H.; Peng, M.; Zhou, X.Y.; Wu, H.; Hu, J.; Xie, W.G.; Liu, S.H. Modification of carboxymethyl cellulose grafted with collagen peptide and its antioxidant activity. *Carbohydr. Polym.* **2014**, *112*, 32–38. [CrossRef] [PubMed]

30. Yadollahi, M.; Namazi, H.; Barkhordari, S. Preparation and properties of carboxymethyl cellulose/layered double hydroxide bionanocomposite films. *Carbohydr. Polym.* **2014**, *108*, 83–90. [CrossRef] [PubMed]

31. Peng, F.; Ren, J.L.; Xu, F.; Bian, J.; Peng, P.; Sun, R.C. Fractional study of alkali-soluble hemicelluloses obtained by graded ethanol precipitation from sugar cane bagasse. *J. Agric. Food Chem.* **2009**, *58*, 1768–1776. [CrossRef] [PubMed]

32. Ren, J.L.; Sun, R.C.; Liu, C.F.; Lin, L.; He, B.H. Synthesis and characterization of novel cationic SCB hemicelluloses with a low degree of substitution. *Carbohydr. Polym.* **2007**, *67*, 347–357. [CrossRef]

33. Katsura, S.; Isogai, A.; Onabe, F.; Usuda, M. NMR analyses of polysaccharide derivatives containing amine groups. *Carbohydr. Polym.* **1992**, *18*, 283–288. [CrossRef]

34. Peng, X.W.; Ren, J.L.; Zhong, L.X.; Sun, R.C. Nanocomposite films based on xylan-rich hemicelluloses and cellulose nanofibers with enhanced mechanical properties. *Biomacromolecules* **2011**, *12*, 3321–3329. [CrossRef] [PubMed]

35. Tingaut, P.; Zimmermann, T.; Lopez-Suevos, F. Synthesis and characterization of bionanocomposites with tunable properties from poly (lactic acid) and acetylated microfibrillated cellulose. *Biomacromolecules* **2009**, *11*, 454–464. [CrossRef] [PubMed]

36. Svagan, A.J.; Azizi Samir, M.A.; Berglund, L.A. Biomimetic polysaccharide nanocomposites of high cellulose content and high toughness. *Biomacromolecules* **2007**, *8*, 2556–2563. [CrossRef] [PubMed]
37. Xu, Y.X.; Kim, K.M.; Hanna, M.A.; Nag, D. Chitosan–starch composite film: Preparation and characterization. *Ind. Crop. Prod.* **2005**, *21*, 185–192. [CrossRef]

3

Inactivated Sendai Virus (HVJ-E) Immobilized Electrospun Nanofiber for Cancer Therapy

Takaharu Okada [1,2,3], Eri Niiyama [1,2], Koichiro Uto [2], Takao Aoyagi [2] and Mitsuhiro Ebara [2,4,*]

Academic Editor: Nicole Zander

[1]   Graduate School of Pure and Applied Sciences, University of Tsukuba, 1-1-1, Tennodai, Tsukuba, Ibaraki 305-8577, Japan; OKADA.Takaharu@nims.go.jp (T.O.); NIYAMA.Eri@nims.go.jp (E.N.)
[2]   Biomaterials Unit, Nano-Life Field, International Center for Materials Nanoarchitectonics (WPI-MANA), National Institute for Materials Science (NIMS), 1-1 Namiki, Tsukuba, Ibaraki 305-0044, Japan; flatronl1753s@gmail.com (K.U.); aoyagi.takao@nihon-u.ac.jp (T.A.)
[3]   Japan Society for the Promotion of Science (JSPS), 8 Ichibancho, Chiyoda-ku, Tokyo 102-0083, Japan
[4]   Graduate School of Tokyo University of Science, 6-3-1 Niijuku, Katsushika-ku, Tokyo 125-8585, Japan
*   Correspondence: EBARA.Mitsuhiro@nims.go.jp

Abstract: Inactivated Hemagglutinating Virus of Japan Envelope (HVJ-E) was immobilized on electrospun nanofibers of poly(ε-caprolactone) by layer-by-layer (LbL) assembly technique. The precursor LbL film was first constructed with poly-L-lysine and alginic acid via electrostatic interaction. Then the HVJ-E particles were immobilized on the cationic PLL outermost surface. The HVJ-E adsorption was confirmed by surface wettability test, scanning laser microscopy, scanning electron microscopy, and confocal laser microscopy. The immobilized HVJ-E particles were released from the nanofibers under physiological condition. *In vitro* cytotoxic assay demonstrated that the released HVJ-E from nanofibers induced cancer cell deaths. This surface immobilization technique is possible to perform on anti-cancer drug incorporated nanofibers that enables the fibers to show chemotherapy and immunotherapy simultaneously for an effective eradication of tumor cells *in vivo*.

Keywords: nanofiber; layer-by-layer; inactivated Sendai virus; HVJ-E

1. Introduction

According to the world healthcare organization's (WHO) report in 2012, 8.2 million deaths worldwide could be attributed to cancer [1]. Of these, nearly 45% of new cases were diagnosed in Asia. In Japan, cancer has been the top cause of death since 1981. Strategies to decrease the number of deaths caused by cancer, therefore, remain a high priority in medical research. The cancer treatment options depend on the type and stage of cancer, possible side effects, and the patient's preferences and overall health. There are many cancer treatment methods that have been developed such as surgery, radiotherapy, and chemotherapy. Depending on the cancer type and location, severances, and potential to metastasize, medical doctors will choose one or a combination of the treatments. For example, systematic chemotherapy along with surgery or radiotherapy are the most commonly used therapeutic strategies for cancer [2]. Hyperthermia has also been combined with chemotherapy or radiotherapy, because the application of hyperthermia has been known to cause manifold cytotoxic effects to tumor cells [3,4]. We have recently developed a unique nanofiber mesh that can run both chemotherapy and hyperthermia treatment at the same time [5]. The fibers are composed of a temperature responsive N-isopropylacrylamide (NIPAAm) [6,7] based copolymer, anti-cancer drugs, and magnetic nanoparticles (MNPs). When an alternative magnetic field is applied to the fibers, the fibers shrink because of self-generated heat from the incorporated MNPs this induces the deswelling of polymer networks in the nanofiber [5,8]. The contraction of the fibers expels the incorporated anti-cancer drugs directly into the cancer area. Applying an alternative magnetic field induces both direct heating of cancer cells and the subsequent release of anti-cancer drugs. The synergistic effect of chemotherapy and hyperthermia treatment showed virtual eradication of the tumor cells *in vitro* compared with individual treatment.

In recent years, cancer immunotherapy has emerged as the fourth treatment modality, in addition to surgery, chemotherapy, and radiotherapy [9,10]. The main types of immunotherapy now being used include monoclonal antibodies, immune checkpoint inhibitors, cancer vaccines, or other non-specific immunotherapies. Kaneda *et al.* have recently shown a new type of immune treatment using inactivated Hemagglutinating Virus of Japan Envelope (HVJ-E). HVJ-E is prepared by irradiating UV light to destroy the internal RNA of the HVJ (or Sendai) virus [11]. Although the virus is no longer cytotoxic and is unable to proliferate, however, the cell fusion functionality remains active. Recently, this fusion ability has been demonstrated in the literature for the delivery of artificially encapsulated plasmids or proteins to cells [12–14]. It has been also reported that HVJ-E can induce tumor specific apoptosis and anti-cancer immunity *in vitro* and *in vivo*. Kurooka *et al.* reported an injected HVJ-E particle induces interleukin-6 (IL-6) production that enhances a strong anti-tumor immunity *in vivo* [15,16]. The HVJ-E is also able to induce cell selective tumor apoptosis *in vitro* which was reported by Kawaguchi *et al.* [17]. They seeded HVJ-E on prostatic tumor cells, PC-3 and LN-Cap, and then the PC-3 cells were selectively killed by HVJ-E because the expression level of HVJ-E receptors on PC-3 cells is higher than that on LN-Cap cells. Thus, HVJ-E has been attracting considerable attention as a new type of therapeutic material for cancer therapy. From these perspectives, we develop a novel anti-cancer nanofiber mesh using HVJ-E (Figure 1). The nanofibers are developed by electrospinning poly(ε-caprolactone) (PCL). The surface of PCL nanofiber is coated with polyelectrolytes via layer-by-layer (LbL) assembly to construct the precursor film for HVJ-E immobilization. Release of the HVJ-E particles in physiological conditions and inductions of tumor cell deaths are studied through *in vitro* tests.

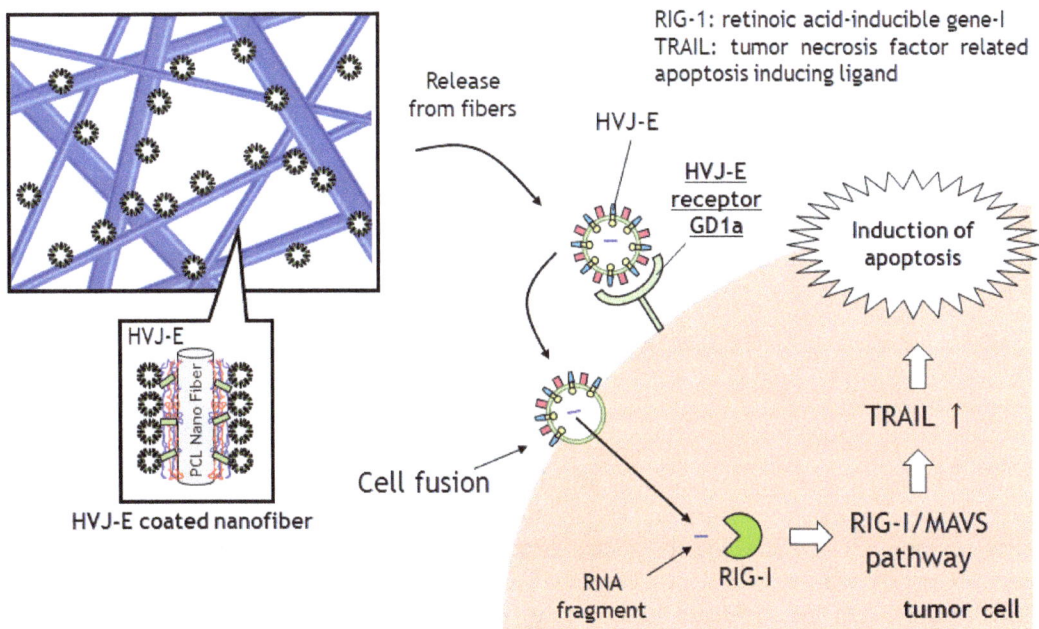

Figure 1. Schematic illustrations of HVJ-E coated nanofiber meshes and the mechanism of cancer apoptosis induced by the released HVJ-E particles from nanofiber.

2. Results and Discussion

2.1. Preparations of HVJ-E Immobilized PCL Nanofibers via LbL Method

We choose to make nanofiber meshes with poly(ε-caprolactone) (PCL) in this study because PCL is biocompatible, biodegradable, and widely used in the biomaterial/biomedical fields such as tissue engineering and drug delivery [18–20]. Firstly, ε-caprolactone monomer was polymerized by a method described previously [21]. The PCL nanofibers were prepared by the electrospinning method. Briefly, PCL was dissolved in 1,1,1,3,3,3-hexafluoro-2-propanol (HFIP) (20 wt/v %) and electrospun for 2 h onto an aluminum substrate using 20 kV at a flow rate of 1 mL/h [22]. To immobilize HVJ-E particles on PCL nanofiber meshes, layer-by-layer (LbL) assembly technique was employed (Figure 2a). The LbL method is widely used because it can be applied on many substrates such as glass, gold, plastic, and fibers [23–26]. The LbL films have been extensively used in the biological/biomedical fields because the components of the films can be taken from biocompatible precursors such as hyaluronic acid, chitosan, alginic acid (ALG), and poly-L-lysine (PLL). Some researchers have chosen to use LbL as a method to retain protein activity in an immobilized state to deliver the protein to a desired location. Guilot *et al.*, for example, reported that the bone morphogenetic protein 2 (BMP-2) immobilized PLL/hyaluronic acid based LbL films on a bone implant enhanced the bone regeneration in an embedded site [27]. The result showed that BMP-2 was delivered to an embedded site without any inactivation of protein activity. We have recently reported the HVJ-E immobilization on LbL films and confirmed the viral protein activity of the HVJ-E [28]. In this study, the PCL nanofibers were immersed into polyethylene amine (PEI) solution to prepare a precursor layer on the fibers. Then the PEI adsorbed fibers were immersed into ALG solutions, followed by washing with HEPES/NaCl buffer. The ALG adsorbed fibers were dipped into PLL solutions to form the polyion complex. In order to make a stable structure in the cell culture media, the LbL layer was crosslinked. Both 1-Ethyl-3-(3-dimethylaminopropyl) carbodiimide (EDC) and *N*-hydroxysuccinimide (NHS) were used to crosslink the carbonic acid groups of ALG and amino groups of PLL in LbL multilayers [29]. The FTIR spectrum was measured to confirm the existence of crosslinked LbL multilayer on PCL nanofibers (Figure S1). The broad weak peak in the spectrum at 3200–3400 cm^{-1} and at 1515–1600 cm^{-1} were assigned to the N-H stretching and amide II in crosslinked PLL-ALG structure in LbL multilayer, respectively. This observation suggested the successful immobilization of LbL multilayer on PCL nanofibers. After the crosslinking reaction, PLL outermost crosslinked nanofibers were immersed into HEPES/NaCl solutions containing 3000 Hemagglutinating unit/mL (HAU/mL) of HVJ-E to obtain HVJ-E adsorbed LbL coated PCL nanofibers. HVJ-E has an anionic zeta potential in physiological pH and could adsorb on the final PLL cationic positively charged layer [30]. As shown in Figure 2b, the nanofiber mesh became hydrophilic after the HVJ-E immobilizations due to the surface proteins located on HVJ-E envelope [31].

2.2. Characterizations of HVJ-E Coated PCL Nanofibers

2.2.1. SLM and SEM Observations.

Firstly, prepared PCL fibers, LbL coated PCL nanofibers and HVJ-E coated PCL nanofibers were observed by scanning laser microscopy (SLM) and scanning electron microscopy (SEM). SLM images of PCL, LbL coated PCL nanofibers, and HVJ-E immobilized LbL coated PCL nanofibers are shown in Figure 3a–c, respectively. Some membrane-like structures are observed between the adjacent nanofibers after LbL coating, which could be caused by the assembly of multilayer films in the limited space in between the adjacent fibers. In order to observe the immobilized HVJ-E particles on the nanofibers, the magnified images of nanofibers were also observed by SEM and shown in Figure 3d–f. The average diameter of PCL nanofibers was estimated to be 153 $\pm$ 42 nm (an average of 28 fibers). Although it was difficult to identify the layered structure and HVJ-E particles, the surface morphology significantly changed after the immobilization of HVJ-E particles [30]. These results clearly indicate that HVJ-E particles were successfully coated onto the surface of PCL nanofibers.

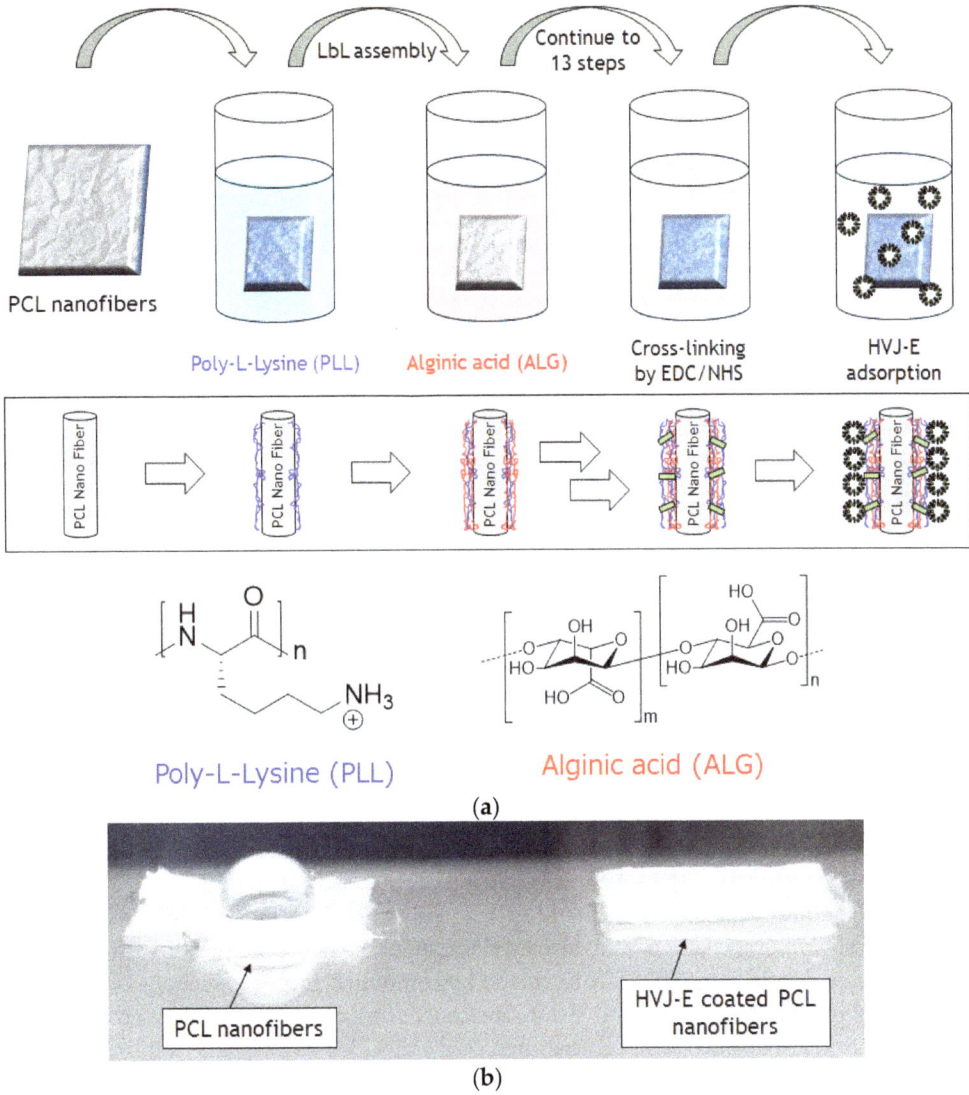

Figure 2. (a) Layer-by-Layer coating on PCL nanofibers with poly-L-Lysine (PLL) and alginic acid (ALG) via alternative deposition of polymers. After LbL coating, the fibers were crosslinked by EDC/NHS. HVJ-E was then adsorbed on the PLL outermost surface; (b) Photograph of water droplet on the PCL nanofiber mesh with (right) and without (left) HVJ-E immobilization.

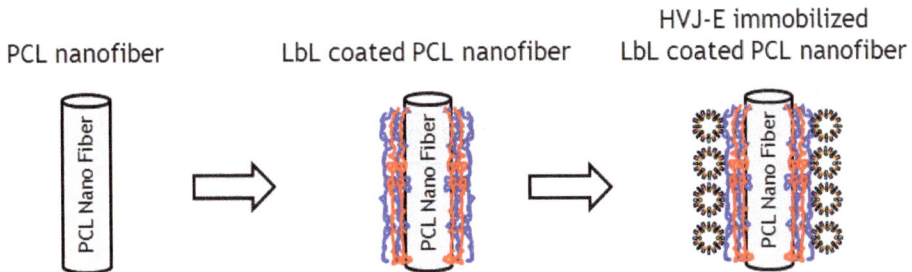

Figure 3. *Cont.*

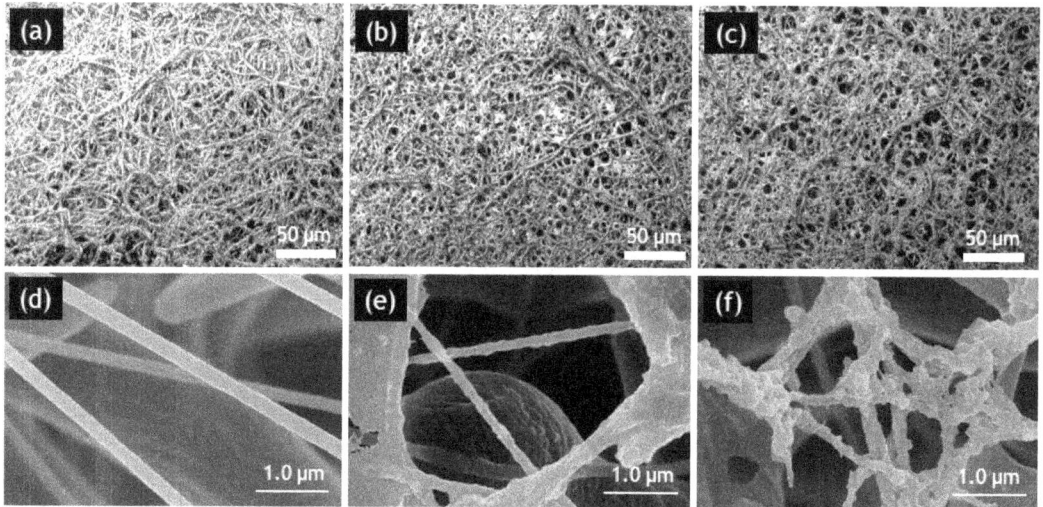

Figure 3. SLM (**a–c**) and SEM (**d–f**) images of PCL(**a,d**), LbL coated (**b,e**), and HVJ-E coated PCL nanofibers (**c,f**).

2.2.2. CLM Observations

One of the great advantages of nanofiber structure compared to two dimensional substrates is the large surface area and porosity of the nanofiber [32]. Therefore, a large amount of molecules can be deposited onto the fiber surface. But it is normally difficult to achieve homogeneous coating within a thick nonwoven nanofiber mesh because the concentration is gradually changed in the thickness direction. Therefore, confocal laser microscopy (CLM) was used to observe the distribution of HVJ-E particles within the nanofiber. Figure 4 shows the confocal microscope images of HVJ-E immobilized nanofibers. PLL and HVJ-E were stained with FITC (green) and pKH-26 (red), respectively. It can be seen that relatively uniform coatings were formed throughout the nanofiber mesh. The bottom images are vertical images for the nanofiber reconstructed by combining multiple sections of CLM images. These images show that both green and red fluorescence were observed not only on the surface of PCL-nanofibers but also inside the fiber structures.

2.3. HVJ-E Releasing from the Nanofibers

The release of HVJ-E particles from the nanofibers was examined by measuring the absorbance at 280 nm by UV-vis (Figure 5). The desorption of HVJ-E particles was achieved by using ionic strength and temperature changes to interfere with the electrostatic interactions between HVJ-E and PLL. Therefore, the HVJ-E release experiment was conducted in Dulbecco's phosphate buffer saline (PBS) at 37 °C. The HVJ-E particles were released over time, and it reached equilibrium at around 60 hemagglutinin unit (HAU) of HVJ-E after 8 h. Figure S2 in the supplementary material shows the effect of the amount of HVJ-E on the cytotoxicity of PC-3 cells. According to this data, 60 HAU of HVJ-E is enough to induce the cell cytotoxicity. The inserted graph plots the released HVJ-E during the first four hours against the square root of a time. Interestingly, it fits to the Higuchi model which describes the release rate of drugs from a matrix system [33].

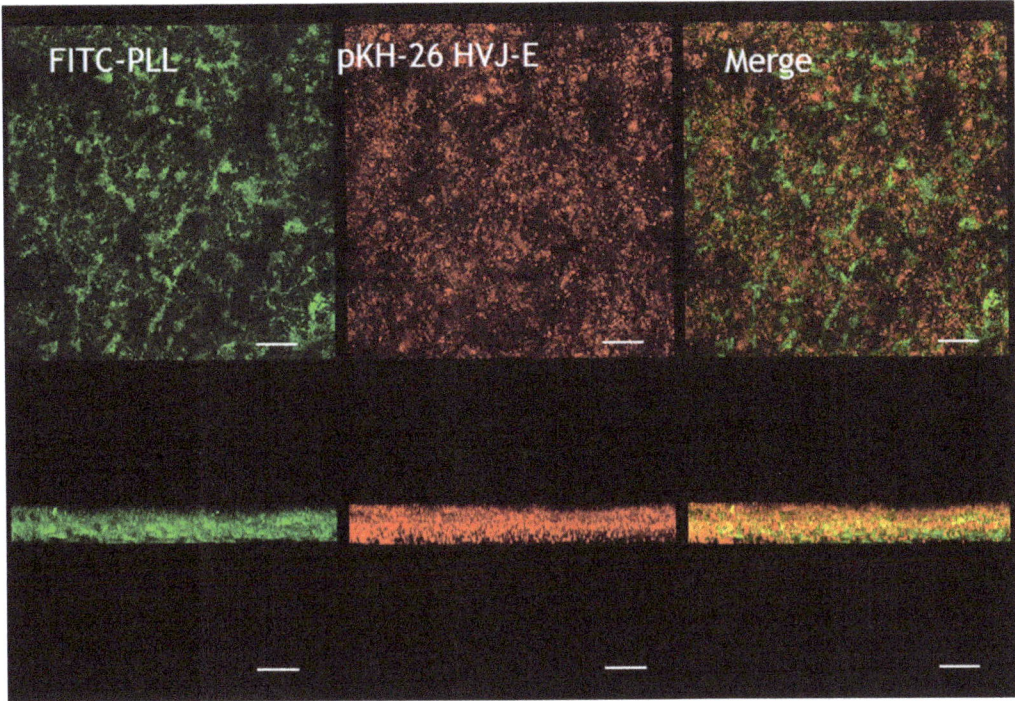

Figure 4. Fluorescent images of HVJ-E immobilized nanofibers observed by CLM. PLL and HVJ-E were stained with FITC (green) and pKH-26 (red), respectively. Top-view (top) and cross-sectional (bottom) images of the nanofiber mesh (scale bars = 50 μm).

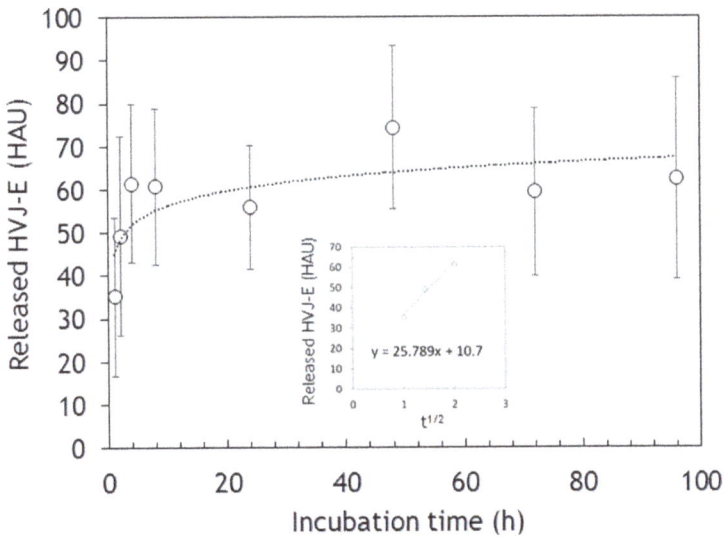

Figure 5. The HVJ-E release kinetics from nanofiber meshes measured by UV-vis. The inserted graph plots the released HVJ-E against the square root of a time.

2.4. Cytotoxic Assay

In this part, we attempted to confirm whether the released HVJ-E from nanofibers induce tumor cell death. We conducted *in vitro* tests to observe the viability and proliferation ability of tumor cells. Metastatic prostate cancer PC-3 cells were cultured in 24-well plates. PC-3 cells were chosen because previous study suggested that the PC-3 cells have a HVJ-E receptor GD1a that enables them to capture HVJ-E particles [17]. A 1 cm × 1 cm of nanofiber mesh was added to the cell culture wells for co-cultivation with nanofibers and cells were cultured for 72 h (Figure 6a). Compared with the control (medium only), there were no significant decreases in the proliferation of the cells in the presence of PCL and LbL nanofibers. This result indicates that nanofiber itself shows no cytotoxicity for PC-3 cells. By contrast, the cell proliferation was significantly suppressed in the presence of HVJ-E immobilized nanofiber mesh. This result corresponds well to the *in vitro* HVJ-E release study and cytotoxic assay as shown in Figure 5 and Figure S1, respectively. Figure 6a shows the phase contrast image of PC-3 cells without HVJ-E immobilized nanofiber mesh. The cells kept their circular shape. On the other hand, dead cells were observed when cells were co-cultured with HVJ-E immobilized nanofiber mesh, indicating that released HVJ-E particles caused tumor cell death (Figure 6b).

Because Matsushima-Miyagi *et al.* reported that the inside RNA fragments of HVJ-E were recognized by retinoic acid-inducible gene I (RIG-I) [34]. Although the molecular mechanisms of apoptosis induction are still under investigation, the paper hypothesized that the mechanism involved is the RNA fragments of HVJ-E induced a signaling process that attacks the cells via the RIG-I/mitochondrial antiviral signaling protein (MAVS). Signaling downstream led to the cells upregulating tumor necrosis factor related apoptosis ligand (TRAIL) expressions. Then the TRAIL could conjugate with tumor necrosis factor receptor (TNF-R) and cause cell apoptosis. TNF-R4 and TNF-R5, the apoptosis inducing receptors, are strongly expressed on tumor cells. Therefore, HVJ-E could selectively induce apoptosis in cancer cells. These mechanisms could occur via released HVJ-E from PCL nanofibers in this experiment.

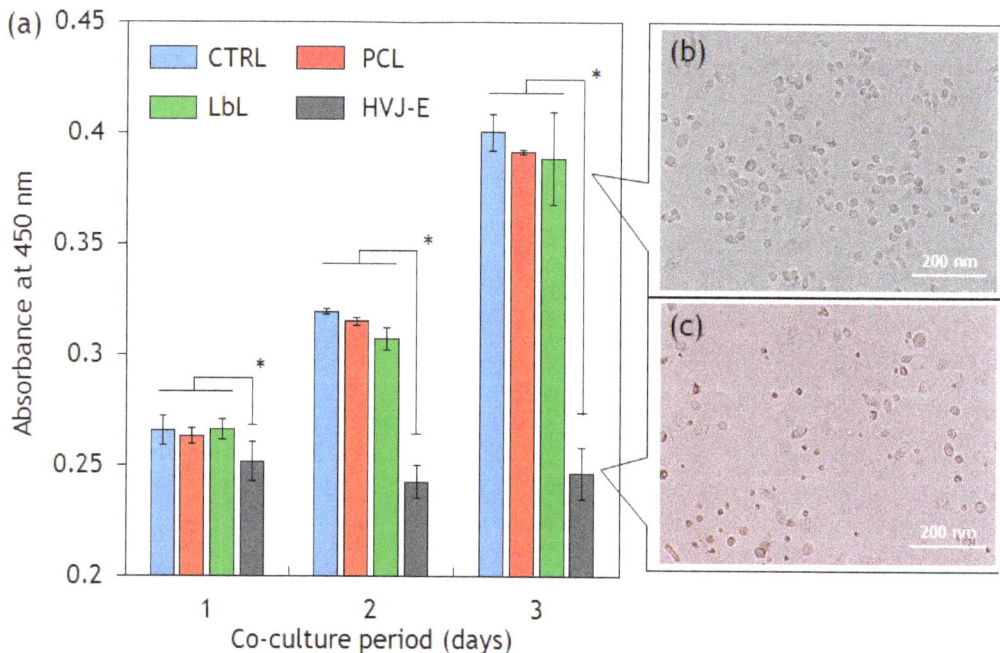

Figure 6. (a) Cytotoxic assays. The viability of PC-3 cells was tested using cell counting kit-8 assay. The cells were co-cultured with PCL, LbL coated (PLL outermost), HVJ-E immobilized (HVJ-E) nanofiber mesh (student's t test $p < 0.01$). The phase contrast images of PC-3 cells (b) without HVJ-E immobilized nanofiber mesh; and (c) with HVJ-E immobilized nanofiber mesh.

3. Experimental Section

3.1. Materials

ε-Captolactone monomer was purchased from Tokyo Kasei (Tokyo, Japan). Poly-L-lysine hydrobromide (M_w 30,000–70,000), sodium alginate (low viscosity), and pKH-26 staining kit were purchased from Sigma Aldrich (St. Louis, MO, USA). Freeze dried HVJ-E (GE-016) was purchased from Ishihara Sangyo Kaisha, Ltd. (Osaka, Japan). RPMI-1640 and Dulbecco's phosphate buffer saline (without calcium) were purchased from Nacalai tesque (Kyoto, Japan). HEPES and cell counting kit-8 were purchased by Dojindo (Kumamoto, Japan). Prostate cancer PC-3 cell line was provided by RIKEN Cell Bank (Tsukuba, Japan).

3.2. Fabrication of Nanofibers

PCL was synthesized by ring-opening polymerization with tin octanoate [21]. The obtained PCL was dissolved in 1,1,1,3,3,3-Hexafluoro-2-propanol (HFIP) solution at the concentration of 20 wt/v%. The solution was then electrospun into nanofibers through 23 G syringe-type needle at a constant flow rate of 1.0 mL/h by 20 kV onto an aluminum foil plate. After electrospinning, the fibers were vacuum dried to remove HFIP completely.

3.3. Preparation of Crosslinked Layer-by-Layer Assembled Films on PCL Nanofibers

Poly-L-lysine (PLL) and alginic acid (ALG) were dissolved in pH 7.4 of HEPES/NaCl buffer (10 mM HEPES and 150 mM NaCl). The concentration of each solution was adjusted to the concentration of 1.0 mg/mL and 5.0 mg/mL respectively. Firstly, PCL nanofibers were dipped into 0.1 mg/mL of poly ethylene imine (PEI) solution for 10 min to fabricate a precursor film on the fiber. The fibers were then dipped into HEPES/NaCl buffer for 2 min. The PEI coated nanofibers were dipped into the ALG solution for 10 min, followed by washing with HEPES/NaCl buffer for 2 min. The ALG outermost fiber was dipped into PLL solution for 10 min to form the LbL multilayer films. This adsorption/washing process was continued alternatively to assemble the LbL multilayer on PCL nanofibers. The EDC and NHS were dissolved in HEPES/NaCl buffer and the concentration was adjusted to 800 mM and 200 mM. These solutions were mixed together before the crosslinking reaction. The PCL nanofiber with PLL on the outermost fibers was dipped into an EDC-NHS mixed solution for 4 h at 4 °C. After the reaction, the fibers were dipped into HEPES/NaCl buffer overnight to remove unreacted EDC. Attenuated total reflection Fourier transform infrared (ATRFTIR) spectrum was measured on a FTIR spectrophotometer (Perkin-Elmer Spectrum One, Waltham, MA, USA) equipped with universal ATR sampling accessory for surface analysis. The samples had been dried *in vacuo* at room temperature before ATR-FTIR measurement. All of the measurements were performed under identical conditions (number of scans: 24 times, resolution: 8 cm^{-1}).

3.4. HVJ-E Adsorption on the Crosslinked Layer-by-Layer Film on PCL Nanofibers

The HVJ-E fibers were prepared first by crosslinking PLL outermost LbL coated fibers, followed by sterilization with UV light for 15 min. The crosslinked LbL nanofibers were then dipped into 3000 HAU/mL of HVJ-E solution (HEPES/NaCl buffer) for 2 h. The fibers were then washed two times with HEPES/NaCl buffer.

3.5. Characterization of PCL Nanofibers with/without HVJ-E Coating

Scanning Electron Microscope (SEM) image was taken by SU-8000 (Hitachi, Japan) for the examination of nanofiber shapes after platinum coating. Confocal microscope analysis was performed by SP-5 (Leica, Germany). FITC labelled PLL and ALG were assembled on PCL nanofibers as the above described method. Then the sample was observed by Ar laser and He/Ne laser.

3.6. HVJ-E Release from HVJ-E Immobilized LbL Coated Nanofibers

The 1 cm × 1 cm of HVJ-E immobilized LbL coated nanofibers were immersed in 5 mL of PBS. The solution was incubated at 37 °C in the stirring machine (55 times/min). 1 mL of the solution was collected at 1, 2, 4, 8, 24, 48, and 72 h and 1 mL of PBS was added each time, respectively. The absorbance at 280 nm of each sample was measured to examine the concentration of HVJ-E solution and the value was converted to concentration by the prepared standard curve.

3.7. Cytotoxic Ability of HVJ-E Coated PCL Nanofibers

The 10,000 cells of PC-3 cells were cultured in a 24-well plate for 24 h. 1cm × 1 cm of PCL fibers, LbL coated nanofibers and HVJ-E fibers were put in each well and again cultured for 24, 48, and 72 h independently. After each time period, the fibers were removed and 50 μL of cell counting kit-8 reagent was then added to each well and incubated for 2 h at 37 °C. The absorbance was measured at 450 nm by an UV plate reader.

4. Conclusions

In this study, we fabricated HVJ-E immobilized electrospun nanofiber meshes by using the LbL technique. As the HVJ-E particles are fragile, first we prepared precursor LbL film on the surface of PCL nanofibers with PLL and ALG via electrostatic interaction. Then the HVJ-E particles were immobilized on the cationic PLL outermost surface. These immobilized HVJ-E particles were released from the nanofibers under physiological condition. *In vitro* cytotoxic assays demonstrated that the released HVJ-E from nanofibers induced cancer cell deaths. We believe that the HVJ-E nanofiber offers significant promise for the cancer immunotherapy because the proposed method perhaps regarded as a simple, versatile, and an inexpensive bottom-up nanofabrication technique. In addition, this surface immobilization technique is possible to perform on anti-cancer drug incorporated nanofibers that enables the fibers to show chemotherapy and immunotherapy at the same time to effectively eradicate tumor cells *in vivo*.

Acknowledgments: The authors would like to express their gratitude for a Grant-in-Aid for Young Scientists (A) (25702029) from the Ministry of Education, Culture, Sports, Science, and Technology (MEXT), Japan. The authors would also like to acknowledge the financially support by JSPS KAKENHI (Grant Number 264086). We also acknowledge RIKEN-cell bank for providing the prostate cancer cell line.

Author Contributions: Fiber fabrication and LbL coating were performed by Takaharu Okada, Eri Niiyama, and Koichiro Uto. Fiber characterizations were performed by Takaharu Okada and Eri Niiyama. HVJ-E release test and cell culture were performed by Takaharu Okada. Paper was written by Takaharu Okada, Takao Aoyagi, and Mitsuhiro Ebara.

Conflicts of Interest: The authors declare no conflict of interest.

References

1.   Latest World Cancer Statistics Global Cancer Burden Rises to 14.1 Million New Cases in 2012: Marked Increase in Breast Cancers Must Be Addressed. Available online: http://www.iarc.fr/en/media-centre/pr/2013/pdfs/pr223_E.pdf (accessed on 30 October 2015).
2.   Reight, A.; Come, S.E.; Henderson, C.; Gelman, R.S.; Silver, B.; Hayes, D.F.; Shulman, L.N.; Harris, J.R. The sequencing of chemotherapy and radiation therapy after conservative surgery for early-stage breast cancer. *N. Engl. J. Med.* **1996**, *334*, 1356–1361. [CrossRef] [PubMed]
3.   Hurwitz, M.D.; Hansen, J.L.; Prokopios-Davos, S.; Manola, J.; Wang, Q.; Bornstein, B.A.; Hynynen, K.; Kaplan, I.D. Hyperthermia combined with radiation for the treatment of locally advanced prostate cancer. *Cancer* **2011**, *117*, 510–516. [CrossRef] [PubMed]
4.   Hu, S.; Liao, B.; Chiang, C.; Chen, P.; Chen, I.; Chen, S. Core-shell nanocapsules stabilized by single-component polymer and nanoparticles for magneto-chemotherapy/hyperthermia with multiple drugs. **2012**, *24*, 3627–3632. [CrossRef] [PubMed]

5. Kim, Y.-J.; Ebara, M.; Aoyagi, T. A smart hyperthermia nanofiber with switchable drug release for inducing cancer apoptosis. *Adv. Funct. Mater.* **2013**, *23*, 5753–5761. [CrossRef]
6. Okano, T.; Yamada, N.; Sakai, H.; Sakurai, Y. A novel recovery system for cultured cells using plasma-treated polystyrene dishes grafted with poly(n-isopropylacrylamide). *J. Biomed. Mater. Res.* **1993**, *27*, 1243–1251. [CrossRef] [PubMed]
7. Ebara, M.; Hoffman, J.M.; Hoffman, A.S.; Stayton, P.S.; Lai, J.J. A photoinduced nanoparticle separation in microchannels via ph-sensitive surface traps. *Langmuir* **2013**, *29*, 5388–5393. [CrossRef] [PubMed]
8. Hergt, R.; Dutz, S.; Müller, R.; Zeisberger, M. Magnetic particle hyperthermia: Nanoparticle magnetism and materials development for cancer therapy. *J. Phys. Condens. Matter* **2006**, *18*, S2919–S2934. [CrossRef]
9. Blattman, J.N.; Greenberg, P.D. Cancer immunotherapy: A treatment for the masses. *Science* **2004**, *305*, 200–205. [CrossRef] [PubMed]
10. Mellman, I.; Coukos, G.; Dranoff, G. Cancer immunotherapy comes of age. *Nature* **2011**, *480*, 480–489. [CrossRef] [PubMed]
11. Kaneda, Y.; Nakajima, T.; Nishikawa, T.; Yamamoto, S.; Ikegami, H.; Suzuki, N.; Nakamura, H.; Morishita, R.; Kotani, H. Hemagglutinating virus of Japan (HVJ) envelope vector as a versatile gene delivery system. *Mol. Ther.* **2002**, *6*, 219–226. [CrossRef] [PubMed]
12. Kobayashi, Y.; Mercado, N.; Barnes, P.J.; Ito, K. Defects of protein phosphatase 2A causes corticosteroid insensitivity in severe asthma. *PLOS ONE* **2011**, *6*. [CrossRef] [PubMed]
13. Matsuda, S.; Okada, N.; Kodama, T.; Honda, T.; Iida, T. A cytotoxic type III secretion effector of vibrio parahaemolyticus targets vacuolar H+-ATPase subunit c and ruptures host cell lysosomes. *PLOS Pathog.* **2012**, *8*. [CrossRef] [PubMed]
14. Salker, M.S.; Christian, M.; Steel, J.H.; Nautiyal, J.; Lavery, S.; Trew, G.; Webster, Z.; Al-Sabbagh, M.; Puchchakayala, G.; Föller, M.; *et al.* Deregulation of the serum- and glucocorticoid-inducible kinase SGK1 in the endometrium causes reproductive failure. *Nat. Med.* **2011**, *17*, 1509–1513. [CrossRef] [PubMed]
15. Kurooka, M.; Kaneda, Y. Inactivated Sendai virus particles eradicate tumors by inducing immune responses through blocking regulatory T cells. *Cancer Res.* **2007**, *67*, 227–236. [CrossRef] [PubMed]
16. Suzuki, H.; Kurooka, M.; Hiroaki, Y.; Fujiyoshi, Y.; Kaneda, Y. Sendai virus F glycoprotein induces IL-6 production in dendritic cells in a fusion-independent manner. *FEBS Lett.* **2008**, *582*, 1325–1329. [CrossRef] [PubMed]
17. Kawaguchi, Y.; Miyamoto, Y.; Inoue, T.; Kaneda, Y. Efficient eradication of hormone-resistant human prostate cancers by inactivated Sendai virus particle. *Int. J. Cancer* **2009**, *124*, 2478–2487. [CrossRef] [PubMed]
18. Coombes, A.G.A.; Rizzi, S.C.; Williamson, M.; Barralet, J.E.; Downes, S.; Wallace, W.A. Precipitation casting of polycaprolactone for applications in tissue engineering and drug delivery. *Biomaterials* **2004**, *25*, 315–325. [CrossRef]
19. Rai, B.; Teoh, S.H.; Hutmacher, D.W.; Cao, T.; Ho, K.H. Novel PCL-based honeycomb scaffolds as drug delivery systems for rhBMP-2. *Biomaterials* **2005**, *26*, 3739–3748. [CrossRef] [PubMed]
20. Uto, K.; Ebara, M.; Aoyagi, T. Temperature-responsive Poly(ε-Caprolactone) cell culture platform with dynamically tunable nano-roughness and elasticity for control of myoblast morphology. *Int. J. Mol. Sci.* **2014**, *15*, 1511–1524. [CrossRef] [PubMed]
21. Ebara, M.; Uto, K.; Idota, N.; Hoffman, J.M.; Aoyagi, T. Shape-memory surface with dynamically tunable nanogeometry activated by body heat. *Adv. Mater.* **2012**, 273–278. [CrossRef] [PubMed]
22. Garrett, R.; Niiyama, E.; Kotsuchibashi, Y.; Uto, K.; Ebara, M. Biodegradable nanofiber for delivery of immunomodulating agent in the treatment of basal cell carcinoma. *Fibers* **2015**, *3*, 478–490. [CrossRef]
23. Decher, G. Fuzzy nanoassemblies: Toward layered polymeric multicomposites. *Science* **1997**, *277*, 1232–1237. [CrossRef]
24. Mansouri, S.; Merhi, Y.; Winnik, F.M.; Tabrizian, M. Investigation of Layer-by-Layer assembly of polyelectrolytes on fully functional human red blood cells in suspension for attenuated immune response. *Biomacromolecules* **2011**, *12*, 585–592. [CrossRef] [PubMed]
25. Caruso, F.; Trau, D.; Möhwald, H.; Renneberg, R. Enzyme encapsulation in Layer-by-Layer engineered polymer multilayer capsules. *Langmuir* **2000**, *16*, 1485–1488. [CrossRef]
26. Lvov, Y.; Ariga, K.; Ichinose, I.; Kunitake, T. Assembly of multicomponent protein films by means of electrostatic Layer-by-Layer adsorption. *J. Am. Chem. Soc.* **1995**, *117*, 6117–6123. [CrossRef]

27. Guillot, R.; Gilde, F.; Becquart, P.; Sailhan, F.; Lapeyrere, A.; Logeart-Avramoglou, D.; Picart, C. The stability of BMP loaded polyelectrolyte multilayer coatings on titanium. *Biomaterials* **2013**, *34*, 5737–5746. [CrossRef] [PubMed]

28. Okada, T.; Niiyama, E.; Uto, K.; Aoyagi, T.; Ebara, M. A biomimetic approach to hormone resistant prostate cancer cell isolation using inactivated Sendai virus (HVJ-E). *Biomater. Sci.* **2016**, *4*, 96–103. [CrossRef] [PubMed]

29. Kunjukunju, S.; Roy, A.; Ramanathan, M.; Lee, B.; Candiello, J.E.; Kumta, P.N. Acta biomaterialia a Layer-by-Layer approach to natural polymer-derived bioactive coatings on magnesium alloys Q. *Acta Biomater.* **2013**, *9*, 8690–8703. [CrossRef] [PubMed]

30. Okada, T.; Uto, K.; Sasai, M.; Lee, C.M.; Ebara, M.; Aoyagi, T. Nano-decoration of the hemagglutinating virus of Japan Envelope (HVJ-E) using a Layer-by-Layer assembly technique. *Langmuir* **2013**, 7384–7392. [CrossRef] [PubMed]

31. Communie, G.; Crépin, T.; Maurin, D.; Jensen, M.R.; Blackledge, M.; Ruigrok, R.W.H. Structure of the tetramerization domain of measles virus phosphoprotein. *J. Virol.* **2013**, *87*, 7166–7169. [CrossRef] [PubMed]

32. Namekawa, K.; Schreiber, M.T.; Aoyagi, T.; Ebara, M. Fabrication of zeolite–polymer composite nanofibers for removal of uremic toxins from kidney failure patients. *Biomater. Sci.* **2014**, *2*, 674–679. [CrossRef]

33. Siepmann, J.; Peppas, N.A. Higuchi equation: Derivation, applications, use and misuse. *Int. J. Pharm.* **2011**, *418*, 6–12. [CrossRef] [PubMed]

34. Matsushima-Miyagi, T.; Hatano, K.; Nomura, M.; Li-Wen, L.; Nishikawa, T.; Saga, K.; Shimbo, T.; Kaneda, Y. Trail and noxa are selectively upregulated in prostate cancer cells downstream of the RIG-I/MAVS signaling pathway by nonreplicating Sendai virus particles. *Clin. Cancer Res.* **2012**, *18*, 6271–6283. [CrossRef] [PubMed]

4

Investigation of High-Energy Ion-Irradiated MA957 Using Synchrotron Radiation under *In-Situ* Tension

Kun Mo [1,*], Di Yun [1,2], Yinbin Miao [1], Xiang Liu [3], Michael Pellin [1], Jonathan Almer [4], Jun-Sang Park [4], James F. Stubbins [3,5], Shaofei Zhu [6] and Abdellatif M. Yacout [1]

Academic Editor: Jie Lian

[1] Nuclear Engineering Division, Argonne National Laboratory, Lemont, IL 60439, USA; diyun1979@mail.xjtu.edu.cn (D.Y.); ymiao@anl.gov (Y.M.); pellin@anl.gov (M.P.); yacout@anl.gov (A.M.Y.)
[2] Department of Nuclear Engineering, Xi'an Jiaotong University, Xi'an 710049, Shaanxi, China
[3] Department of Nuclear, Plasma, and Radiological Engineering, University of Illinois at Urbana-Champaign, Urbana, IL 61801, USA; xliu128@illinois.edu (X.L.); jstubbin@illinois.edu (J.F.S.)
[4] Advanced Photon Source, Argonne National Laboratory, Lemont, IL 60439, USA; almer@aps.anl.gov (J.A.); parkjs@aps.anl.gov (J.-S.P.)
[5] International Institute for Carbon-Neutral Energy Research (I2CNER), Kyushu University, Fukuoka 819-0395, Japan
[6] Physics Division, Argonne National Laboratory, Lemont, IL 60439, USA; zhu@anl.gov
* Correspondence: kunmo@anl.gov

Abstract: In this study, an MA957 oxide dispersion-strengthened (ODS) alloy was irradiated with high-energy ions in the Argonne Tandem Linac Accelerator System. Fe ions at an energy of 84 MeV bombarded MA957 tensile specimens, creating a damage region ~7.5 μm in depth; the peak damage (~40 dpa) was estimated to be at ~7 μm from the surface. Following the irradiation, *in-situ* high-energy X-ray diffraction measurements were performed at the Advanced Photon Source in order to study the dynamic deformation behavior of the specimens after ion irradiation damage. *In-situ* X-ray measurements taken during tensile testing of the ion-irradiated MA957 revealed a difference in loading behavior between the irradiated and un-irradiated regions of the specimen. At equivalent applied stresses, lower lattice strains were found in the radiation-damaged region than those in the un-irradiated region. This might be associated with a higher level of Type II stresses as a result of radiation hardening. The study has demonstrated the feasibility of combining high-energy ion radiation and high-energy synchrotron X-ray diffraction to study materials' radiation damage in a dynamic manner.

Keywords: synchrotron radiation; oxide dispersion-strengthened (ODS); ion irradiation; *in situ* tensile test

1. Introduction

MA957 is a ferritic oxide dispersion-strengthened (ODS) alloy originally developed by INCO at the end of the 1970's [1,2]. As with many ODS materials, MA957 has exceptional high-temperature strength, creep resistance, and oxidation resistance [3]. The mechanical superiority of MA957 over traditional ferritic materials stems from its unique nano-feature: a high density of nano-scale particles or nano-clusters embedded in the metallic matrix. Under irradiation, these nano-scale particles serve as traps of He and point defects, and thus significantly improve the materials' swelling resistance [4]. MA957 is considered one of most promising structural materials for advanced reactor systems (e.g., fast reactors and fusion systems) due to its outstanding radiation resistance [5]. To better understand the properties of MA957 and further improve its mechanical performance and radiation resistance, decades have been spent performing numerous studies. The mechanical properties including tensile properties, fracture toughness, and creep resistance have been studied at various temperatures and under various test conditions [5–8]. Multiple advanced materials characterization techniques, including atom probe tomography (APT), small angle neutron scattering (SANS), and high-resolution transmission electron microscopy (HRTEM), have been employed to systematically investigate the stoichiometry, crystal structure, size, density, and stability of nano-particles/nano clusters within MA957 [9–14]. Microstructural development of MA957 after neutron radiation, an important reference for possible nuclear applications of the alloy, has been extensively studied at different temperature and doses [1,15–20].

Development of radiation resistant materials requires long-term and systematic in-reactor tests to evaluate irradiation damage as a function of temperature and dose. However, the high cost and long irradiation required for in-reactor testing hinders the progress of nuclear materials development. Recent advances in studying heavy ion irradiations have shown great potential for emulating reactor irradiation damage using ion beams [21,22]. One of the main disadvantages of using heavy ion irradiation is the limited depth of penetration with the ion beam [23]. To considerably extend the depth of ion penetration and thereby attain a larger radiation-damaged region, we employed a 84 MeV Fe ion beam, of an energy much higher than those available from lab-based accelerators. The irradiated specimen was then tensile tested with an *in-situ* synchrotron diffraction measurement. The dynamic responses of un-irradiated and irradiated regions within one single specimen were obtained. This set of experiments demonstrated the feasibility of combining high-energy ion radiation and high-energy synchrotron X-ray diffraction to study materials' radiation damage in a dynamic manner.

2. Materials and Experimental Procedure

The nominal composition of MA957 is 14Cr-1Ti-0.3Mo-0.25Y_2O_3 (wt %). The material contains ~5 nm nanoclusters [24] in grains approximately 500 nm in size. Figure 1 shows the typical microstructure of MA957. The material was machined into SS-J1 type miniature tensile specimens with a gauge section of 1.2 mm × 0.25 mm × 5 mm (width × thickness × length). The tensile specimens were polished first with 600 grit SiC abrasive paper, then by 3 μm alumina suspensions, and finished with 0.02 μm colloidal silica. The high-energy (84 MeV) Fe ion irradiation of the specimens was performed in the Argonne Tandem Linac Accelerator System (ATLAS) at Argonne National Laboratory (ANL). The Fe ion beam with a charge state of +11 was tuned to a Gaussian shape with a full width half maximum (FWHM) of 5.2 mm. The exposure area was controlled to be ~10 mm in diameter by setting up a collimator in front of the sample stage. Two specimens were loaded on the sample stage and irradiated simultaneously (Figure 2a). The large beam size allowed high-energy ion exposure of the entire gauge and part of the grip section of the miniature tensile specimens (Figure 2b). Each of the tensile specimens was placed 1.7 mm from the ion beam center in order to achieve relatively uniform radiation damage in the gauge part of the sample. The gauge region (shown as the pale blue rectangular region in Figure 2b) of each sample absorbed ~9.4% of the total beam current. The average dose rate was ~8 × 10^{11} ions/(cm^2·s), and the achieved dose in the gauge region was ~4.4 × 10^{16} ions/cm^2. A thermocouple was attached to the back side of each tensile specimen in order to provide continuous temperature measurement during the experiment. The stabilized sample temperature during the experiment was measured to be ~250 °C.

Based on the average dose achieved in the gauge portion of the tensile specimens, the implanted Fe ion concentrations (in ppm) and radiation damage (in displacements per atom, dpa) were calculated by the SRIM (stopping and range of ions in matter) computer code [25,26]. A value of 40 eV displacement threshold energy and the Kinchin-Pease option were used in the SRIM calculation following the recommendations in reference [27]. The ion irradiation produced a damaged region of ~7.5 μm in depth, while the peak damage (~40 dpa) was estimated to be ~7 μm from the surface (Figure 3). The average damage levels from the surface to the depth of ~7.5 μm (over the damaged region) and 10 μm (over a single X-ray scan step) were ~7.4 and ~5.7 dpa, respectively.

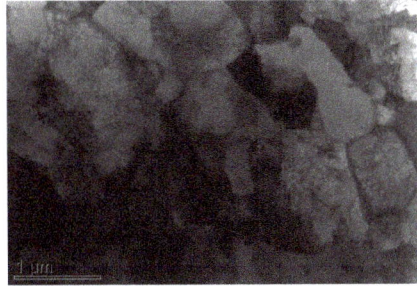

Figure 1. Transmission electron microscopy (TEM) image of the MA957 sample before ion-irradiations.

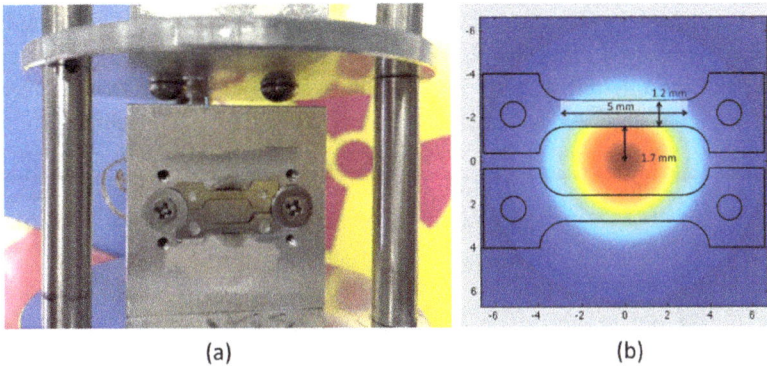

(a)   (b)

Figure 2. (a) Experimental set-up for irradiating tensile specimens at Argonne Tandem Linac Accelerator System (ATLAS); (b) Schematic of the Gaussian beam exposure profile of tensile specimens (axes units: mm).

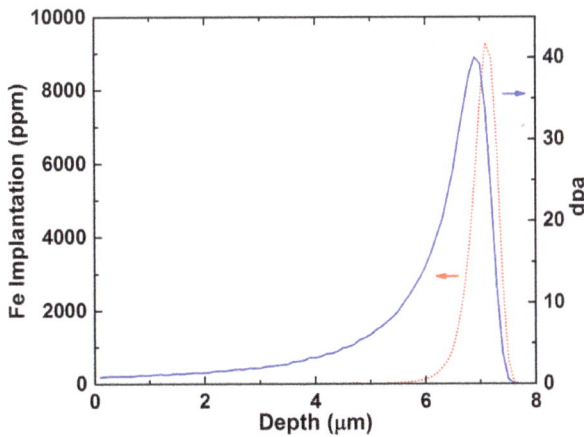

Figure 3. Implanted Fe ion concentration (in ppm) and radiation damage (in dpa) of the irradiated MA957 tensile sample.

The *in-situ* high-energy X-ray diffraction experiment was conducted at the 1-ID beamline of the Advanced Photon Source (APS), ANL. One of the irradiated specimens and one un-irradiated MA957 tensile specimen (as a control sample) were deformed in uniaxial tension using an MTS closed-loop servo-hydraulic test frame (model 858) at room temperature. To perform the X-ray diffraction measurements during tensile testing, the specimen was installed on the grips without any applied load from the MTS test frame. After specimen placement, a 0.1 N load (stress of ~88 MPa) was applied to fix the specimen in position. During tensile testing, the gauge part of the specimen was exposed to the high-energy monochromatic X-ray beam (86 keV), and the diffraction patterns were collected by an amorphous Si detector from General Electric (GE) with a pixel size of 200 μm. Two different types of *in-situ* measurements were performed on the two tensile specimens: (1) a continuous (non-stop) tensile test with X-ray diffraction to measure the bulk response of the un-irradiated MA957 and (2) an intermittent tensile test with X-ray diffraction scanning across the cross-section in order to capture the response in the radiation-damaged region near the sample surface of the ion-irradiated MA957. Figure 4 shows the experimental setup and the sample orientation during the *in-situ* tensile tests. A schematic of the X-ray diffraction measurement of the un-irradiated and ion-irradiated samples is shown in Figure 5. The bulk measurement of the un-irradiated specimen was done using a typical continuous *in-situ* tensile test with X-ray diffraction characterization. The strain rate was approximately 2×10^{-4} s^{-1} based on the displacement rate of the load frame crosshead and the length of the gauge section. The size of X-ray beam was 100×100 μm^2. The distance between the sample and the detector was ~1.4 m. Similar experimental setup and procedures can be found in references [28–33]. A unique X-ray diffraction scanning routine was designed specifically to characterize the ion-irradiated specimen. After loading the specimen on the MTS machine, the surface of irradiated specimen was oriented to be parallel to X-ray beam, *i.e.*, 90° off from the orientation of the un-irradiated specimen during its continuous tensile test. To accurately align the sample, its orientation was adjusted by using the rotation stage under the MTS machine at each stress-strain state during tensile testing. This orientation adjustment relied on the measurement of attenuated X-rays when scanning the cross-section of the sample. The beam size was reduced to 10×10 μm^2 using defining slits. The beam energy was set to ~65 keV to attain a better resolution, while the distance between the sample and the detector was maintained to be ~1.4 m. During the early stage of the tensile test (before sample yielding), load control mode was used to strain the sample to a preset load, then the sample was held at that load to allow X-ray diffraction scanning. The X-ray diffraction scanned from the sample surface (Figure 5b) to ~75 μm depth with a step size of 10 μm. The first step measurement was used to represent the irradiated region. Once sample yielding was observed, the load control mode was changed to strain control mode to avoid overloading the tensile specimen. This intermittent tensile test has been shown to yield similar material performance to the continuous tensile test at low temperatures, and has been applied to study materials' behavior with *in-situ* wide-angle X-ray and neutron scattering [32,34–40].

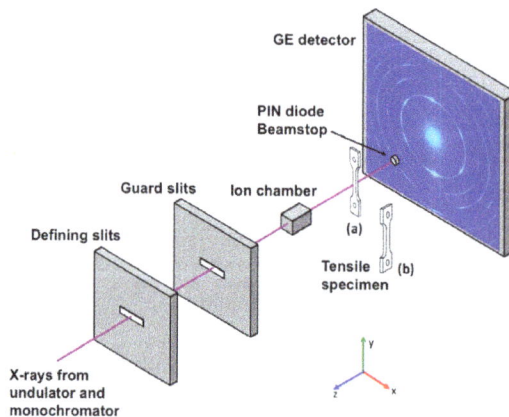

Figure 4. Schematic of the synchrotron experimental setup; the diffraction pattern in the schematic is from the bulk measurement of the un-irradiated MA957 tensile specimen. Sample position (a) is for bulk measurement of the un-irradiated MA957 tensile specimen; and sample position (b) is for X-ray diffraction scan of the ion-irradiated MA957 tensile specimen.

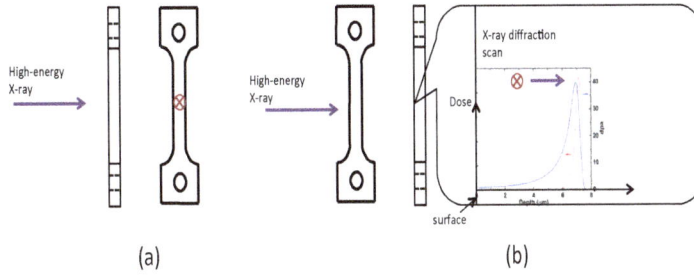

Figure 5. Schematic of the X-ray diffraction measurement of: (**a**) un-irradiated MA957, *i.e.*, the bulk measurement; and (**b**) ion-irradiated MA957 by in-depth cross-section scanning. The red circles with cross show the X-ray inlet direction.

3. Results

Figure 6 shows the stress-strain diagram of the MA957 alloy. The engineering stress, σ_e, was calculated by $\sigma_e = F/A$, where F is the load on the sample and A is the area of the cross-section. F was measured by the MTS loading frame, while A was measured before the tensile test. The entire stress-strain curve was only achieved by tensile testing the un-irradiated bulk MA957 sample. The 0.2% yield strength (YS) was measured to be ~810 MPa. The ultimate tensile strength (UTS) was measured to be ~918 MPa, developed at a strain of ~13.4%. Sample necking began after the sample reached the UTS and continued until sample failure. The total elongation of the 9Cr ODS sample was ~24.1%. For the tensile test of the irradiated specimen, only six individual measurements along the stress-strain curve were conducted because the sample failed when loading to the seventh position in plastic deformation. However, all the measured points fell along the stress-strain curve developed by the continuous test, and this directly confirmed the material's similar mechanical response for intermittent and continuous tensile tests.

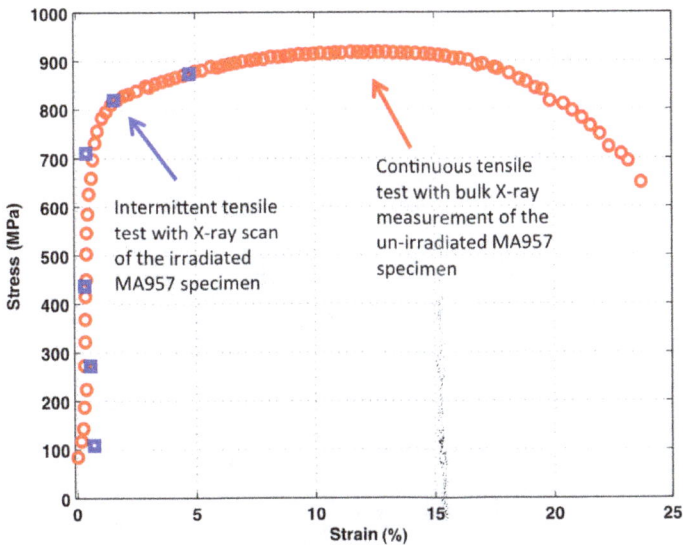

Figure 6. Engineering stress-strain diagram of MA957: the blue squares show stress and strain values during the intermittent tensile test for the irradiated MA957 specimen; the red circles show stress and strain values during the continuous tensile test for the un-irradiated MA957 specimen.

Figure 7 shows the diffraction pattern of the un-irradiated MA957. A strong {110} texture was observed. The lattice constant of MA957 was measured to be 2.873 Å. Since a small beam size was used, a limited X-ray diffraction volume was obtained when measuring the ion-irradiated specimen. Moreover, due to the developed texture in the MA957 sample, only the {110} reflection of the α-Fe matrix provided sufficient statistics in the tensile direction (*i.e.*, 90° in Figure 7) when measuring the irradiated sample. To study and compare the dynamic response of both materials during tensile testing, the strain in the {110} family of planes in the α-Fe matrix was calculated based on the equation [32,41–45]:

$$\varepsilon_{\text{lattice}} = \frac{d_\sigma - d_0}{d_0} \tag{1}$$

where d_σ is the d-spacing of a {110} reflection measured at a specific stress σ and at the 90° (*i.e.*, tensile direction) of the {110} Debye ring. The d_0 is the reference d-spacing measured before loading the specimen. Figure 8 shows the results of the lattice strain evolution for both the un-irradiated and irradiated MA957 samples. As the testing was done continuously, the lattice strain of un-irradiated MA957 evolves smoothly. With applied load, the lattice strain linearly increases in the elastic regime until yielding (red circles in Figure 8). Similar to many materials with multiple phases [18–20,26], in the transition process from elastic to plastic deformation, the lattice strain of the metallic matrix decreases during early yielding but increases afterwards until necking. Compared to the homogeneous microstructure within the un-irradiated sample, the ion-irradiated MA957 has two regions that need to be separately characterized. By using X-ray diffraction scanning with small beam size, both the radiation-damaged region near the sample surface and the un-irradiated region at the inner part of the sample were measured; their lattice strains were calculated and are shown in Figure 8. The lattice strain development in the un-irradiated regions is consistent with the bulk measurement of the un-irradiated specimens, although the deviation from the elastic linearity was not caught due to the small number of measurements during sample yielding. During the early stage of elastic deformation, the loading behavior of the radiation-damaged and the un-irradiated regions in the irradiated sample are also consistent with the bulk measurement of the un-irradiated specimens. The divergence of the lattice strain development in the radiation-damaged region starts at a stress of 700 MPa, about 100 MPa below the yield strength. Upon plastic deformation, the difference between the lattice strain developed in the irradiated and un-irradiated regions increases significantly; its value jumped from ~0 at 430 MPa to 4×10^{-4} at 815 MPa. This indicates that the internal stress of the metallic matrix in the radiation-damaged region is much smaller than that in the un-irradiated region of the same specimen.

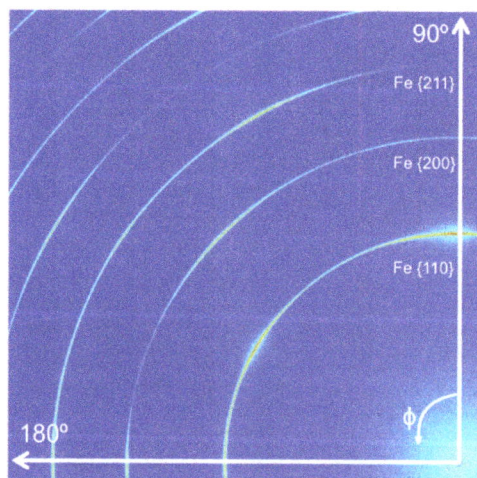

Figure 7. Diffraction pattern of un-irradiated MA957; the tensile direction is at 90°.

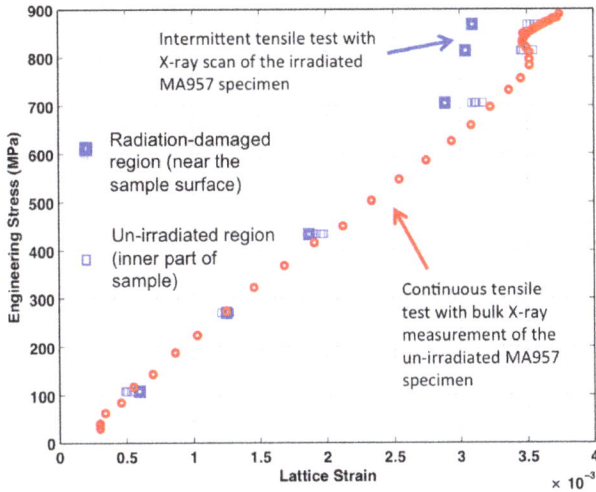

Figure 8. {110} lattice strain evolutions for both un-irradiated and irradiated MA957 samples: the blue squares show lattice strain values during the intermittent tensile test for the irradiated MA957 specimen; the red circles show lattice strain values during the continuous tensile test for the un-irradiated MA957 specimen. The uncertainty in lattice strains is about $\pm 4 \times 10^{-4}$.

4. Discussion

The surface condition is critical for ion-irradiation study, especially when using a lab-based accelerator wherein the induced irradiation damage is concentrated in the near-surface region. The high-energy ions used in the present study produced a ~7.5 μm deep damage region. Thus there is much less sensitivity to the surface condition than when irradiating with lower energy ions. However, electropolishing should be applied to the specimens in further irradiation experiments to provide a smooth and hardening-free sample surface.

As shown in Figure 6, bulk stress-strain responses of the irradiated and un-irradiated regions are consistent, indicating that the impact of the high-energy ion irradiation on the overall tensile sample is negligible. Even though the Fe ion energy is much higher than regular lab-based accelerators, the radiation-damaged region is limited to <10 μm from the surface. Similar lattice strain developments in the un-irradiated sample and the un-irradiated region of the irradiated sample also suggest that the high-energy radiation damage is localized in the near-surface region. This provides us an opportunity to investigate the dynamic responses of un-irradiated and irradiated materials by applying a high-energy X-ray scan to both un-irradiated and irradiated regions within a single specimen. Note that the un-irradiated region in the irradiated sample experienced similar temperature exposure during the high-energy ion irradiation; the temperature gradient along the range of X-ray diffraction scan (surface to the 75 μm depth) was insignificant based on the heat transfer calculation. Thus, the comparison between these two regions within a sample is even better than the comparison to the un-irradiated/control sample.

The radiation effect on the tensile specimen was not evident during the early stage of elastic deformation, indicating that the elastic constant of the MA957 was not changed after high-energy ion irradiation. The difference in lattice strain between the radiation-damaged region and the un-irradiated region became significant when the specimen was loaded near the YS, where the lattice strains began to deviate from the elastic linearity. This phenomenon of deviation from elastic linearity during yielding (Figure 8) is often attributed to the development of Type II stresses [46]. Type II stresses universally exist in polycrystalline metals, and are more significant in multi-phase materials because of the larger discrepancy between phases [47,48]. Compared to the un-irradiated region, the ion-irradiated MA957 region developed higher levels of Type II stresses that resulted in a more significant deviation from the elastic linearity. The radiation induced hardening within the damaged region, caused by a high density of radiation induced defects (*i.e.*, point defects and dislocation loops), may have contributed to the increase in type II stresses when external load was applied. Another possible reason for the radiation induced hardening is the instability of nano-scale particles within the material. For example, the size of nano-scale particles can be significantly decreased after ion irradiation (the radiation damage level is less than 60 dpa) [49].

A major difficulty in this synchrotron experiment is setting the sample alignment in order to enable the high-energy X-ray to accurately penetrate the small radiation-damaged region. This alignment must be conducted at every stress-strain state during the intermittent tensile test; optimizing the sample orientation relative to the X-ray took 30–60 min. During the alignment, the tensile specimen was held by controlling load or displacement, and X-ray absorption measurements were performed periodically at different rotation-angles to search for the minimum absorption near the sample surface. This procedure is difficult to apply to the high-temperature tensile tests because the sample may creep during the alignment. To avoid a time-intensive alignment, a smaller specimen, for example, a wire sample with a diameter of 10 μm or less, would be ideal for the high-energy X-ray measurement. Another option is a thin film specimen with a thickness of 30 μm or less. In this case, the alignment time would be greatly reduced.

As shown in Figure 3, the radiation damage region is not uniform within the volume of measurement (*i.e.*, to a 10 μm depth from the sample surface). The lattice strains of the radiation-damaged region are the result of mixed and complicated responses to both injected interstitial and radiation damage that has huge variation over the region being analyzed. To interpret damage-level (or dpa level) dependent information amidst this complexity is extremely difficult. To better develop the capability to characterize radiation damage in materials by combining high-energy ion radiation and high-energy X-ray diffraction, a micron/submicron sized X-ray beam will be needed to probe into a specific region of interest in an irradiated sample. For example, an X-ray beam can be focused to <2 μm vertically at the beamline sector 1 at APS [50], and beamline sector 34 at APS can provide an X-ray beam with a beam size <500 nm [51]. These focused X-ray beams without much loss in flux will significantly benefit the future studies of materials' ion-irradiation damage.

5. Conclusions

In this paper we have demonstrated the feasibility of combining high-energy ion radiation and high-energy synchrotron X-ray diffraction to study materials' radiation damage in a dynamic manner. The high-energy ion radiation produced a relatively large radiation-damaged region that demonstrates the possibility for study of the mechanical dynamic response of the region with applied stresses. Through *in-situ* X-ray measurement during the tensile test of ion-irradiated MA957, different loading behaviors of the irradiated and un-irradiated regions were observed within the specimen. With the same amount of applied stresses, lower lattice strains were found in the radiation-damaged region compared with those in the un-irradiated region. This difference might be associated with a higher level of Type II stresses as a result of the radiation hardening in the MA957 sample.

Acknowledgments: This work was supported by the U.S. Department of Energy under Contract No. DE-AC-02-06CH11357 between UChicago Argonne, LLC and the Department of Energy. The authors gratefully acknowledge the support of the International Institute for Carbon Neutral Energy Research (WPI-I2CNER), sponsored by the World Premier International Research Center Initiative (WPI), Ministry of Education, Culture, Sports, Science and Technology (MEXT), Japan. The authors would like to thank Carolyn Tomchik for editing the manuscript. This research used resources of the Advanced Photon Source, a U.S. Department of Energy (DOE) Office of Science User Facility operated for the DOE Office of Science by Argonne National Laboratory under Contract No. DE-AC02-06CH11357.

Author Contributions: Kun Mo and Michael Pellin conceived and designed the experiments; Di Yun, Yinbin Miao, Xiang Liu, Jonathan Almer, and Jun-Sang Park performed the APS experiment. Shaofei Zhu arranged and developed the plan for the ATLAS experiment. James F. Stubbins and Abdellatif M. Yacout contributed materials and analysis tools for the experiments.

Conflicts of Interest: The authors declare no conflict of interest.

References

1. Ribis, J. Structural and chemical matrix evolution following neutron irradiation in a MA957 oxide dispersion strengthened material. *J. Nucl. Mater.* **2013**, *434*, 178–188. [CrossRef]
2. Fischer, J.J. Dispersion Strengthened Ferritic Alloy for Use in Liquid-Metal Fast Breeder Reactors (LMFBRS). U.S. Patent 4,075,010 A, 21 February 1978.
3. El-Genk, M.S.; Tournier, J.M. A review of refractory metal alloys and mechanically alloyed-oxide dispersion strengthened steels for space nuclear power systems. *J. Nucl. Mater.* **2005**, *340*, 93–112. [CrossRef]
4. Odette, G.R.; Alinger, M.J.; Wirth, B.D. Recent developments in irradiation-resistant steels. *Annu. Rev. Mater. Res.* **2008**, *38*, 471–503. [CrossRef]
5. Hadraba, H.; Kazimierzak, B.; Stratil, L.; Dlouhy, I. Microstructure and impact properties of ferritic ODS ODM401 (14%Cr-ODS of MA957 type). *J. Nucl. Mater.* **2011**, *417*, 241–244. [CrossRef]
6. Wilshire, B.; Lieu, T.D. Deformation and damage processes during creep of incoloy MA957. *Mater. Sci. Eng. A* **2004**, *386*, 81–90. [CrossRef]
7. Alinger, M.J.; Odette, G.R.; Lucas, G.E. Tensile and fracture toughness properties of MA957: Implications to the development of nanocomposited ferritic alloys. *J. Nucl. Mater.* **2002**, *307*, 484–489. [CrossRef]
8. Yang, W.J.; Odette, G.R.; Yamamoto, T.; Miao, P.; Alinger, M.J.; Hribernik, M.; Lee, J.H. A critical stress-critical area statistical model of the $K_{Jc}(T)$ curve for MA957 in the cleavage transition. *J. Nucl. Mater.* **2007**, *367*, 616–620. [CrossRef]
9. Alinger, M.J.; Odette, G.R.; Hoelzer, D.T. On the role of alloy composition and processing parameters in nanocluster formation and dispersion strengthening in nanostuctured ferritic alloys. *Acta Mater.* **2009**, *57*, 392–406. [CrossRef]
10. Marquis, E.A. Core/shell structures of oxygen-rich nanofeatures in oxide-dispersion strengthened Fe-Cr alloys. *Appl. Phys. Lett.* **2008**, *93*, 181904. [CrossRef]
11. Miller, M.K.; Hoelzer, D.T.; Kenik, E.A.; Russell, K.F. Stability of ferritic MA/ODS alloys at high temperatures. *Intermetallics* **2005**, *13*, 387–392. [CrossRef]
12. Wu, Y.; Haney, E.M.; Cunningham, N.J.; Odette, G.R. Transmission electron microscopy characterization of the nanofeatures in nanostructured ferritic alloy MA957. *Acta Mater.* **2012**, *60*, 3456–3468. [CrossRef]
13. Sakasegawa, H.; Legendre, F.; Boulanger, L.; Brocq, M.; Chaffron, L.; Cozzika, T.; Malaplate, J.; Henry, J.; de Carlan, Y. Stability of non-stoichiometric clusters in the MA957 ODS ferrtic alloy. *J. Nucl. Mater.* **2011**, *417*, 229–232. [CrossRef]
14. Sakasegawa, H.; Chaffron, L.; Legendre, F.; Boulanger, L.; Cozzika, T.; Brocq, M.; de Carlan, Y. Correlation between chemical composition and size of very small oxide particles in the MA957 ODS ferritic alloy. *J. Nucl. Mater.* **2009**, *384*, 115–118. [CrossRef]
15. Yamamoto, T.; Odette, G.R.; Miao, P.; Hoelzer, D.T.; Bentley, J.; Hashimoto, N.; Tanigawa, H.; Kurtz, R.J. The transport and fate of helium in nanostructured ferritic alloys at fusion relevant He/dpa ratios and dpa rates. *J. Nucl. Mater.* **2007**, *367*, 399–410. [CrossRef]
16. Toloczko, M.B.; Gelles, D.S.; Garner, F.A.; Kurtz, R.J.; Abe, K. Irradiation creep and swelling from 400 to 600 °C of the oxide dispersion strengthened ferritic alloy MA957. *J. Nucl. Mater.* **2004**, *329*, 352–355. [CrossRef]
17. Toloczko, M.B.; Garner, F.A.; Maloy, S.A. Irradiation creep and density changes observed in MA957 pressurized tubes irradiated to doses of 40–110 dpa at 400–750 °C in FFTF. *J. Nucl. Mater.* **2012**, *428*, 170–175. [CrossRef]
18. Ribis, J.; Lozano-Perez, S. Nano-cluster stability following neutron irradiation in MA957 oxide dispersion strengthened material. *J. Nucl. Mater.* **2014**, *444*, 314–322. [CrossRef]
19. Miller, M.K.; Hoelzer, D.T. Effect of neutron irradiation on nanoclusters in MA957 ferritic alloys. *J. Nucl. Mater.* **2011**, *418*, 307–310. [CrossRef]
20. Bailey, N.A.; Stergar, E.; Toloczko, M.; Hosemann, P. Atom probe tomography analysis of high dose MA957 at selected irradiation temperatures. *J. Nucl. Mater.* **2015**, *459*, 225–234. [CrossRef]
21. Was, G.S.; Jiao, Z.; Getto, E.; Sun, K.; Monterrosa, A.M.; Maloy, S.A.; Anderoglu, O.; Sencer, B.H.; Hackett, M. Emulation of reactor irradiation damage using ion beams. *Scr. Mater.* **2014**, *88*, 33–36. [CrossRef]

22. Toloczko, M.B.; Garner, F.A.; Voyevodin, V.N.; Bryk, V.V.; Borodin, O.V.; Mel'nychenko, V.V.; Kalchenko, A.S. Ion-induced swelling of ODS ferritic alloy MA957 tubing to 500 dpa. *J. Nucl. Mater.* **2014**, *453*, 323–333. [CrossRef]

23. Was, G.S. *Fundamentals of Radiation Materials Science*; Springer: New York, NY, USA, 2007.

24. Lin, J.L.; Mo, K.; Yun, D.; Miao, Y.; Liu, X.; Zhao, H.; Hoelzer, D.T.; Park, J.S.; Almer, J.; Zhang, G.; *et al. In situ* synchrotron tensile investigations on 14 YWT, MA957, and 9-Cr ODS alloys. *J. Nucl. Mater.* **2015**. in press. [CrossRef]

25. Biersack, J.P.; Haggmark, L.G. A Monte Carlo computer program for the transport of energetic ions in amorphous targets. *Nucl. Instrum. Methods* **1980**, *174*, 257–269. [CrossRef]

26. Ziegler, J.F.; Biersack, J.P. *The Stopping and Range of Ions in Matter*; Springer: New York, NY, USA, 1985; pp. 93–129.

27. Stoller, R.E.; Toloczko, M.B.; Was, G.S.; Certain, A.G.; Dwaraknath, S.; Garner, F.A. On the use of SRIM for computing radiation damage exposure. *Nucl. Instrum. Methods Phys. Res. Sect. B* **2013**, *310*, 75–80. [CrossRef]

28. Pan, X.; Wu, X.; Mo, K.; Chen, X.; Almer, J.; Ilavsky, J.; Haeffner, D.R.; Stubbins, J.F. Lattice strain and damage evolution of 9%–12% Cr ferritic/martensitic steel during *in situ* tensile test by X-ray diffraction and small angle scattering. *J. Nucl. Mater.* **2010**, *407*, 10–15. [CrossRef]

29. Young, M.L.; Almer, J.D.; Daymond, M.R.; Haeffner, D.R.; Dunand, D.C. Load partitioning between ferrite and cementite during elasto-plastic deformation of an ultrahigh-carbon steel. *Acta Mater.* **2007**, *55*, 1999–2011. [CrossRef]

30. Hedström, P.; Lindgren, L.E.; Almer, J.; Lienert, U.; Bernier, J.; Terner, M.; Odén, M. Load partitioning and strain-induced martensite formation during tensile loading of a metastable austenitic stainless steel. *Metall. Mater. Trans. A* **2009**, *40*, 1039–1048. [CrossRef]

31. Cheng, S.; Wang, Y.D.; Choo, H.; Wang, X.L.; Almer, J.D.; Liaw, P.K.; Lee, Y.K. An assessment of the contributing factors to the superior properties of a nanostructured steel using *in situ* high-energy X-ray diffraction. *Acta Mater.* **2010**, *58*, 2419–2429. [CrossRef]

32. Mo, K.; Zhou, Z.; Miao, Y.; Yun, D.; Tung, H.M.; Zhang, G.; Chen, W.; Almer, J.; Stubbins, J.F. Synchrotron study on load partitioning between ferrite/martensite and nanoparticles of a 9Cr ODS steel. *J. Nucl. Mater.* **2014**, *455*, 376–381. [CrossRef]

33. Mo, K.; Tung, H.M.; Li, M.; Almer, J.; Chen, X.; Chen, W.; Hansen, J.B.; Stubbins, J. Synchrotron radiation study on Alloy 617 and Alloy 230 for VHTR application. *J. Press. Vess. Technol. Trans. ASME* **2013**, *135*, 759–766. [CrossRef]

34. Daymond, M.R.; Bourke, M.A.M.; von Dreele, R.B.; Clausen, B.; Lorentzen, T. Use of rietveld refinement for elastic macrostrain determination and for evaluation of plastic strain history from diffraction spectra. *J. Appl. Phys.* **1997**, *82*, 1554–1562. [CrossRef]

35. Clausen, B.; Lorentzen, T.; Bourke, M.A.M.; Daymond, M.R. Lattice strain evolution during uniaxial tensile loading of stainless steel. *Mater. Sci. Eng. A* **1999**, *259*, 17–24. [CrossRef]

36. Clausen, B.; Bourke, M.A.M.; Brown, D.W.; Ustundag, E. Load sharing in tungsten fiber reinforced kanthal composites. *Mater. Sci. Eng. A* **2006**, *421*, 9–14. [CrossRef]

37. Daymond, M.R. The determination of a continuum mechanics equivalent elastic strain from the analysis of multiple diffraction peaks. *J. Appl. Phys.* **2004**, *96*, 4263–4272. [CrossRef]

38. Wang, L.Y.; Li, M.; Almer, J. Investigation of deformation and microstructural evolution in grade 91 ferritic-martensitic steel by *in situ* high-energy X-rays. *Acta Mater.* **2014**, *62*, 239–249. [CrossRef]

39. Wang, L.; Li, M.; Almer, J. *In situ* characterization of grade 92 steel during tensile deformation using concurrent high energy X-ray diffraction and small angle X-ray scattering. *J. Nucl. Mater.* **2013**, *440*, 81–90. [CrossRef]

40. Li, M.; Wang, L.Y.; Almer, J.D. Dislocation evolution during tensile deformation in ferritic-martensitic steels revealed by high-energy X-rays. *Acta Mater.* **2014**, *76*, 381–393. [CrossRef]

41. He, B.B. Introduction to two-dimensional X-ray diffraction. *Powder Diffr.* **2003**, *18*, 71–85. [CrossRef]

42. Miao, Y.; Mo, K.; Zhou, Z.; Liu, X.; Lan, K.C.; Zhang, G.; Miller, M.K.; Powers, K.A.; Mei, Z.G.; Park, J.S.; *et al.* On the microstructure and strengthening mechanism in oxide dispersion-strengthened 316 steel: A coordinated electron microscopy, atom probe tomography and *in situ* synchrotron tensile investigation. *Mater. Sci. Eng. A* **2015**, *639*, 585–596. [CrossRef]

43. Miao, Y.; Mo, K.; Zhou, Z.; Liu, X.; Lan, K.C.; Zhang, G.; Miller, M.K.; Powers, K.A.; Almer, J.; Stubbins, J.F. *In situ* synchrotron tensile investigations on the phase responses within an oxide dispersion-strengthened (ODS) 304 steel. *Mater. Sci. Eng. A* **2015**, *625*, 146–152. [CrossRef]

44. Zhang, G.; Mo, K.; Miao, Y.; Liu, X.; Almer, J.; Zhou, Z.; Stubbins, J.F. Load partitioning between ferrite/martensite and dispersed nanoparticles of a 9Cr ferritic/martensitic (F/M) ODS steel at high temperatures. *Mater. Sci. Eng. A* **2015**, *637*, 75–81. [CrossRef]

45. Zhang, G.; Zhou, Z.; Mo, K.; Miao, Y.; Liu, X.; Almer, J.; Stubbins, J.F. The evolution of internal stress and dislocation during tensile deformation in a 9Cr ferritic/martensitic (F/M) ODS steel investigated by high-energy X-rays. *J. Nucl. Mater.* **2015**, *467*, 50–57. [CrossRef]

46. Stoica, G.M.; Stoica, A.D.; Miller, M.K.; Ma, D. Temperature-dependent elastic anisotropy and mesoscale deformation in a nanostructured ferritic alloy. *Nat. Commun.* **2014**, *5*, 5178. [CrossRef] [PubMed]

47. Hutchings, M.T.; Withers, P.J.; Holden, T.M.; Lorentzen, T. *Introduction to the Characterization of Residual Stress by Neutron Diffraction*; Taylor & Francis Group LLC: Abingdon, UK, 2005.

48. Mo, K. Microstructural Evolution and Mechanical Behavior in Nickel Based Alloys for very High Temperature Reactor. Ph.D. Thesis, University of Illinois at Urbana-Champaign, Urbana, IL, USA, 2011.

49. Chen, T.; Aydogan, E.; Gigax, J.G.; Chen, D.; Wang, J.; Wang, X.; Ukai, S.; Garner, F.A.; Shao, L. Microstructural changes and void swelling of a 12Cr ODS ferritic-martensitic alloy after high-dpa self-ion irradiation. *J. Nucl. Mater.* **2015**, *467*, 42–49. [CrossRef]

50. Shastri, S.D.; Almer, J.; Ribbing, C.; Cederstrom, B. High-energy X-ray optics with silicon saw-tooth refractive lenses. *J. Synchrotron Radiat.* **2007**, *14*, 204–211. [CrossRef] [PubMed]

51. Hofmann, F.; Nguyen-Manh, D.; Gilbert, M.R.; Beck, C.E.; Eliason, J.K.; Maznev, A.A.; Liu, W.; Armstrong, D.E.J.; Nelson, K.A.; Dudarev, S.L. Lattice swelling and modulus change in a helium-implanted tungsten alloy: X-ray micro-diffraction, surface acoustic wave measurements, and multiscale modelling. *Acta Mater.* **2015**, *89*, 352–363. [CrossRef]

5

Activation of Aspen Wood with Carbon Dioxide and Phosphoric Acid for Removal of Total Organic Carbon from Oil Sands Produced Water: Increasing the Yield with Bio-Oil Recycling

Andrei Veksha [1,2,†], Tazul I. Bhuiyan [1,†] and Josephine M. Hill [1,*]

Academic Editor: Alina M. Balu

[1] Department of Chemical and Petroleum Engineering, University of Calgary, 2500 University Drive NW, Calgary, AB T2N 1N4, Canada; tibhuiya@ucalgary.ca

[2] Residues and Resource Reclamation Centre, Nanyang Environment and Water Research Institute, Nanyang Technological University, 1 Cleantech Loop, Clean Tech One, Singapore 637141, Singapore; aveksha@ntu.edu.sg

[*] Correspondence: jhill@ucalgary.ca

[†] These authors contributed equally to this work.

Abstract: Several samples of activated carbon were prepared by physical (CO_2) and chemical (H_3PO_4) activation of aspen wood and tested for the adsorption of organic compounds from water generated during the recovery of bitumen using steam assisted gravity drainage. Total organic carbon removal by the carbon samples increased proportionally with total pore volume as determined from N_2 adsorption isotherms at $-196\ °C$. The activated carbon produced by CO_2 activation had similar removal levels for total organic carbon from the water (up to 70%) to those samples activated with H_3PO_4, but lower yields, due to losses during pyrolysis and activation. A method to increase the yield when using CO_2 activation was proposed and consisted of recycling bio-oil produced from previous runs to the aspen wood feed, followed by either KOH addition (0.48%) or air pretreatment (220 °C for 3 h) before pyrolysis and activation. By recycling the bio-oil, the yield of CO_2 activated carbon (after air pretreatment of the mixture) was increased by a factor of 1.3. Due to the higher carbon yield, the corresponding total organic carbon removal, per mass of wood feed, increased by a factor of 1.2 thus improving the overall process efficiency.

Keywords: activated carbon; adsorption; bio-oil; pore volume; water treatment; yield

---

1. Introduction

Activated carbon from wood is widely used as an adsorbent for water treatment—specifically to adsorb organic compounds. As reported previously, the adsorption capacity of carbon depends on the properties of the adsorbing molecules (e.g., molecular size, structure, solubility and pK_a) and adsorption conditions (e.g., pH, ionic strength and temperature) [1]. Moreover, the adsorption capacity is influenced by the properties of the carbon itself, including porosity (pore volume and pore size distribution), surface functionalities (e.g., oxygen and nitrogen groups) and mineral matter content [1–4].

Both physical and chemical activation techniques have been applied to produce activated carbon for water treatment. Preparation by physical activation consists of pyrolysis of the raw material (*i.e.*, heating in an oxygen deprived atmosphere), and results in a solid carbon (char), which can be further treated (*i.e.*, activated) at high temperature with oxidizing agents, such as steam, CO_2 or air. Alternatively, activated carbon can be produced by chemical activation, in which the feedstock is impregnated with KOH, H_3PO_4, $ZnCl_2$, $FeCl_3$, *etc.* The removal of organic compounds from water by carbon samples with different properties (*i.e.*, produced with different activation methods) was compared previously [5–8]. Okada *et al.* [5] prepared carbon by activation of waste newspaper with K_2CO_3 and steam, and showed that K_2CO_3 activated carbon had higher methylene blue uptake due to larger surface area, pore sizes and higher content of surface oxygen groups. Girgis *et al.* [6] compared the adsorption of methylene blue by activated carbon from peanut hulls prepared by H_3PO_4, $ZnCl_2$, KOH and steam activation, and found the largest adsorption capacities for the H_3PO_4 activated carbon, while KOH activated carbon had the lowest uptake. In a recent study [7], naphthenic acid removal from water was higher on wood activated by H_3PO_4 than by CO_2, probably due to the larger mesopore volumes and surface areas obtained with H_3PO_4. Mestre *et al.* [8], investigated the adsorption of pharmaceutical compounds by activated carbon from cork prepared with steam, KOH and K_2CO_3. The uptakes of ibuprophen, paracetamol, acetylsalicylic acid, caffeine, and clofibric acid were higher by chemically activated carbon, while the uptake of iopamidol was higher by physically activated carbon. These results were attributed to the differences in micro and mesopore structures of the activated carbon.

Cost analysis of carbon production by physical (steam and CO_2) and chemical (H_3PO_4) activation of almond shells, pecan shells and vetiver roots has been performed [9–12], the large capital and reactant investments for H_3PO_4 activation were offset by high activated carbon yields. In contrast, the capital and operational expenses for physical activation are lower but the yields are also lower. The production costs of activated carbon for these materials were estimated as $1.46–2.72\,\$\cdot kg^{-1}$ [10–12], $2.56\,\$\cdot kg^{-1}$ [10] and $1.17–2.89\,\$\cdot kg^{-1}$ [9,11,12] for steam, CO_2 and H_3PO_4 activation, respectively. Low yields of physically activated carbon are attributed to carbon loss during both the pyrolysis and activation stages. Depending on the raw material and operating conditions, only 43%–63% of carbon in biomass is converted to char during pyrolysis while the remaining carbon is contained in bio-oil and pyrolysis gases [13]. To some extent, char yield can be improved by optimizing reaction conditions, such as particle size, moisture content in feedstock, pressure and residence time of volatile pyrolysis products in the reactor as reviewed by Antal and Gronli [14]. Activated carbon yield can also be increased by reducing carbon loss during the activation stage. For this purpose, air gasification followed by carbonization in an inert atmosphere to decompose chemisorbed oxygen has been proposed [15–17]. Although this approach increased the surface area, multiple heat treatment steps are required to produce the activated carbon.

As stated above, the efficacy of activated carbon as an adsorbent for organic compounds is partially determined by its properties and these properties can be manipulated by the activation method [5–8]. Thus, one of the objectives of this study was to investigate which activation method is suitable for the preparation of carbon specifically for the removal of total organic carbon (TOC) from oil sands produced water (specifically from steam assisted gravity drainage, SAGD). SAGD is a common method for bitumen extraction from oil sands. In this technique, two parallel wells are drilled horizontally into the oil sands deposit. Steam generated at a surface plant is pumped into the deposit through the top well to soften the bitumen. The bitumen and condensed steam are then pumped to the surface from the bottom well for processing and separation. The separated water, so called oil sands produced water, contains high concentrations of water soluble organic compounds. In particular, the SAGD water used in this study had a total organic carbon content of ~1000 $mg\cdot L^{-1}$. Because ~90% of this water is recycled, the presence of organic compounds can cause both corrosion and fouling on equipment that result in periodic shutdowns for cleaning and increase operating costs [18]. The activated carbon samples were prepared from locally available aspen wood residues, thus lowering the environmental impact from transportation. CO_2 (physical) and H_3PO_4 (chemical) activation were used and the yield, TOC removal, and, ultimately, the process efficiency of both methods determined. In addition to as prepared activated carbon, acid washed and heat treated samples with reduced ash content and surface functionalities were prepared to determine the effect of these parameters [1–4], if any, on the TOC uptake and clarify the role of porosity.

The second objective of the current study was the investigation of a new method to increase the yield of activated carbon prepared by physical activation. This method is based on the recycling of bio-oil from previous runs to the raw feedstock (*i.e.*, biomass) for pyrolysis in order to increase char production. Bio-oil from biomass is a carbon-rich product with limited application due to acidity, corrosiveness, high oxygen and moisture content, and coking upon reheating [19]. Previous studies have shown that depending on the raw material and operating conditions, the addition of bio-oil to biomass increases the char production by 10%–43% [20–23]. Although the char yield was increased, it is not obvious how the deposited and subsequently pyrolyzed bio-oil will impact the final activated carbon product. Thus, this study included the investigation of the CO_2 activation of char produced from a mixture of biomass and bio-oil. The influence of the char preparation conditions on yield, porosity and removal of organic compounds from SAGD water are discussed.

2. Results and Discussion

2.1. TOC Removal by Activated Carbon from Wood

Activated carbon samples prepared with CO_2 and H_3PO_4 activation of aspen wood were used to investigate the influence of activation method on the TOC removal from SAGD water. The development of porous properties in the activated carbon samples during CO_2 (at 800 °C) and H_3PO_4 (at 500 °C) activation were controlled by activation time and H_3PO_4:wood ratio, respectively. After H_3PO_4 activation, the acid residue was removed by washing with deionized water. In addition to as prepared activated carbon samples, acid washed and heat treated (at 900 °C for 1 h in nitrogen) samples with reduced ash content and surface functionalities were prepared to minimize their effect, if any, on the TOC uptake and clarify the role of porosity. The samples prepared by CO_2 activation are denoted as AC—for activated carbon—followed by the activation time in hours. Samples activated with H_3PO_4 were denoted by the letter P followed by the H_3PO_4:wood ratio of 1:2, 1:1 or 2:1. The abbreviation "HT" denotes samples that were acid washed and heat treated (HT). The TOC removal by the activated carbon samples was compared with a non-activated char (Char-800-HT) prepared from the same feedstock (at 800 °C) and commercial activated carbon (as supplied ColorSorb and acid washed and heat treated ColorSorb-HT) specially designed for the removal of organic molecules and decolorization.

The UV-vis spectrum of the SAGD water in Figure S1 demonstrates light absorbance between 200 and 500 nm, with increasing absorbance at lower wavelength. There was no distinct peak at 254 nm on the spectrum, representative of aromatic compounds. The possible reason is interference from non-aromatic molecules (acetone, 2-butanone, naphthenic acids) contained in large quantities in the water [24].

The porous properties of the samples are listed in Table 1 and N_2 adsorption and desorption isotherms of the selected samples are shown in Figure S2. An increase in CO_2 activation time resulted in increased surface areas and micropore volumes determined by N_2 adsorption (pore sizes up to 20 Å) from 540 $m^2 \cdot g^{-1}$ and 0.18 $cm^3 \cdot g^{-1}$ for 0.3 h treatment (AC0.3-HT) to 910 $m^2 \cdot g^{-1}$ and 0.30 $cm^3 \cdot g^{-1}$ for 3.6 h treatment (AC3.6-HT). The micropore volumes determined by CO_2 (pore sizes less than ~7 Å) also increased with activation time from 0.23 to 0.32 $cm^3 \cdot g^{-1}$. The meso/macropore volume increased during the first hour of activation up to 0.22 $cm^3 \cdot g^{-1}$ and did not change significantly thereafter. As expected, the surface areas and pore volumes of CO_2 activated carbon samples were higher compared to non-activated char (Char-800-HT). With H_3PO_4 activation, higher H_3PO_4:wood ratios resulted in higher surface areas and meso/macropore volumes, while the micropore volume determined by N_2 adsorption was the largest in the sample produced with a H_3PO_4:wood ratio of 1:1. The micropore volume determined by CO_2 adsorption did not change with an increase in H_3PO_4 concentration. Compared to CO_2 activation, H_3PO_4 activation generally resulted in larger surface areas and micropore volumes (determined by N_2 adsorption). Pore size distributions of four of the activated carbon samples are shown in Figure 1. H_3PO_4 activation produced the widest pore diameters. More specifically, the P2:1-HT sample contained pores with diameters up to 100 Å, while the pores in samples AC3.6-HT and ColorSorb-HT were smaller than 40 Å and 60 Å, respectively. In the non-activated Char-800-HT sample no pores larger than 15 Å were observed.

Table 1. Properties of activated carbon prepared by physical and chemical activation of wood, with and without heat treatment.

Sample	Yield (%)	Meso/Macropore Volume-N_2 $(cm^3 \cdot g^{-1})$	Micropore Volume-N_2 $(cm^3 \cdot g^{-1})$	Micropore Volume-CO_2 $(cm^3 \cdot g^{-1})$	BET-N_2 Surface Area $(m^2 \cdot g^{-1})$
		Char			
Char-800-HT	21	0.02	0.17	0.23	440
		CO_2 Activated Carbon			
AC0.3-HT	19	0.09	0.18	0.23	540
AC1.0-HT	15 ± 1 [a]	0.22	0.19	0.28	690
AC1.0		0.21	0.16	0.23	600
AC1.8-HT	11	0.21	0.22	0.27	750
AC3.6-HT	6	0.19	0.30	0.32	910
		H_3PO_4 Activated Carbon			
P1:2-HT	46	0.04	0.35	0.27	870
P1:1-HT	45	0.11	0.46	0.26	1110
P1:1		0.09	0.60	0.24	1240
P2:1-HT	41	0.71	0.40	0.26	1350
		Steam Activated Commercial Carbon			
ColorSorb-HT	-	0.30	0.36	0.37	1140
ColorSorb		0.31	0.30	0.32	988

[a] Average of six measurements $\pm$ standard deviation.

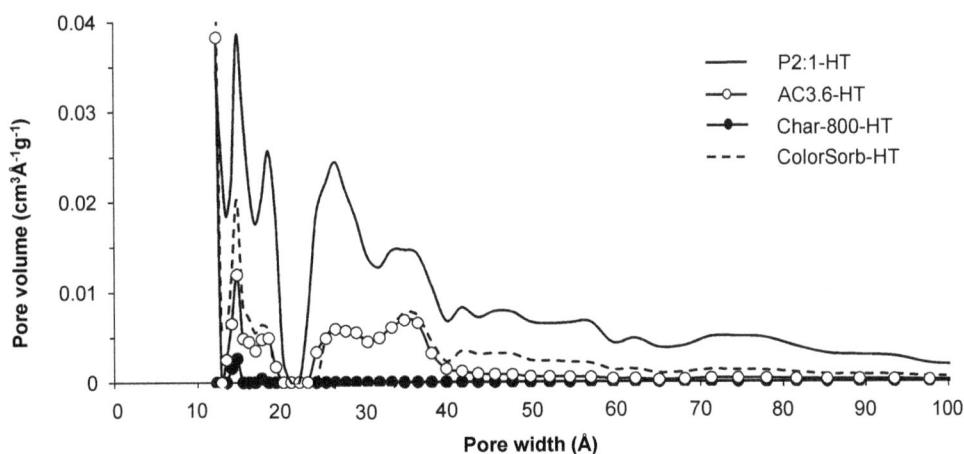

Figure 1. Pore size distributions of selected activated carbon determined with non-local density functional theory.

The percentages of TOC removed from SAGD water by the prepared activated carbon samples are shown in Figure 2. The TOC uptakes by the heat treated samples (AC1.0-HT, P1:1-HT, ColorSorb-HT) were similar to those by the corresponding as prepared samples (AC1.0, P1:1, ColorSorb), suggesting that ash components and surface functionalities removed from the carbon by acid washing and heat treatment, respectively, had little if any influence on the TOC removal. To investigate whether the sample acidity/basicity could influence the TOC uptake, the pH of the water was measured after adsorption. Compared to the feed SAGD water with pH 9.5–9.8, the pH after adsorption on all activated carbon samples did not change more than 0.7 units. Figure S3 shows the influence of pH on the precipitation of TOC. There are no changes in TOC concentration till pH 7 followed by a rapid decrease in TOC content in water at pH below 7. These data suggest that to influence water chemistry and, consequently, adsorption behavior by the samples, a decrease in pH by at least ~2.5–2.8 units from the initial value is required, which is larger than that observed for all samples.

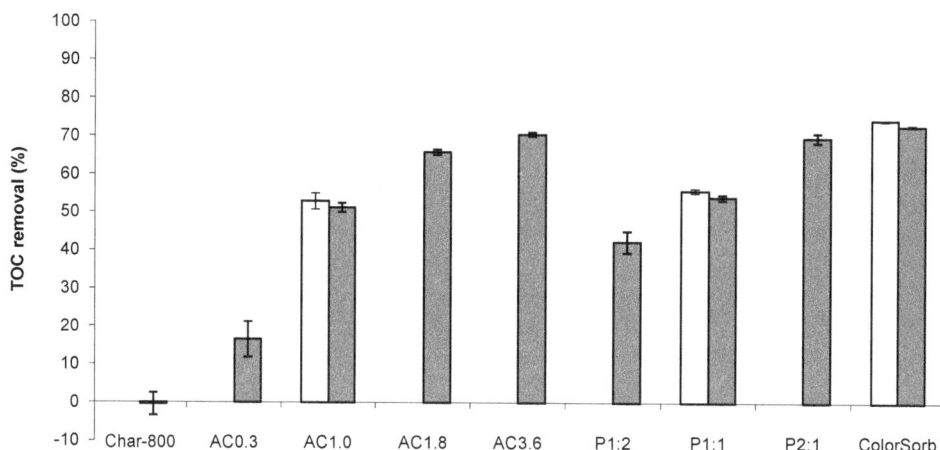

Figure 2. Total organic carbon (TOC) removal by activated carbon from steam assisted gravity drainage (SAGD) water (white and grey bars for as prepared and heat treated (HT) carbon, respectively).

Longer activation times with CO_2 and higher H_3PO_4:wood ratios increased the TOC removal, which could be attributed to the development of porosity during activation as discussed below. In contrast, non-activated Char-800-HT did not remove organic compounds from SAGD water, suggesting that activation was a necessary step for the development of the pores responsible for adsorption. These results are consistent with a previous study [25], in which non-activated char was found to be ineffective for the removal of methyl orange from aqueous solutions. Regardless of the activation process, the maximum TOC removal by the prepared activated carbon samples was ~70%, which was similar to the removal by the commercial steam activated carbon from wood (*i.e.*, ColorSorb).

It is hard to define the range of pore sizes responsible for the removal of TOC by activated carbon, as SAGD water contains a large number of organic compounds with various sizes and chemical structures [24,26]. However, it is likely that micropores probed by CO_2 (smaller than 7 Å) did not contribute to the adsorption of TOC, as these pores were present in all carbon samples, including the non-activated carbon (Char-800-HT) that had no TOC removal (Figure 2). In contrast, pores probed by N_2 were suitable for TOC removal as suggested by Figure 3 in which the TOC uptake is plotted versus total pore volume. The TOC removal increased with total pore volume for the samples activated with either CO_2 or H_3PO_4. The pore volume was increased by activating for longer times or increasing the H_3PO_4:wood ratio (Table 1). According to Figure 3, the data sets for CO_2 and H_3PO_4 activated carbon samples have different slopes, which could be due to either different mechanisms of TOC uptake by the samples or due to removal of different compounds from the water. A more detailed study is required to clarify this phenomenon. Nevertheless, from a practical viewpoint, the linear relationship between TOC removal and total pore volume suggests that the latter can be utilized as a screening parameter during selection of activated carbon prepared by the same activation method for SAGD water treatment.

The TOC removal was comparable for CO_2 and H_3PO_4 activation (Figure 2) but the yields weredifferent—less than 20% for the former compared to greater than 41% for the latter (Table 1). These yields translate to 4.8–17.5 g of wood consumed per 1 g of CO_2 activated carbon and only 2.2–2.4 g of wood consumed per 1 g of H_3PO_4 activated carbon. The low yield of CO_2 activated carbon is attributed to the significant mass loss during pyrolysis. Besides conversion to char, wood components, such as hemicellulose, cellulose and lignin, undergo depolymerization, fragmentation and conversion to volatile pyrolysis species upon heating as reviewed by Di Blasi [27]. Although some of the volatile species can be partially converted to char by the secondary reactions, they mainly form condensable bio-oil and non-condensable gases [13,22,28]. Further losses occur during activation from the partial gasification of the char by the oxidizing agent (in this study CO_2) to develop porosity. In the case of H_3PO_4 activation, higher yields of carbon from lignocellulosic materials, including wood, are attributed to crosslinking reactions between decomposing fragments via phosphate and polyphosphate bridges that prevent release of relatively small molecules from the solid phase [29]. After the removal of (poly)phosphate bridges by water washing, the structure of the produced carbon remains in an expanded state with an accessible pore structure [29].

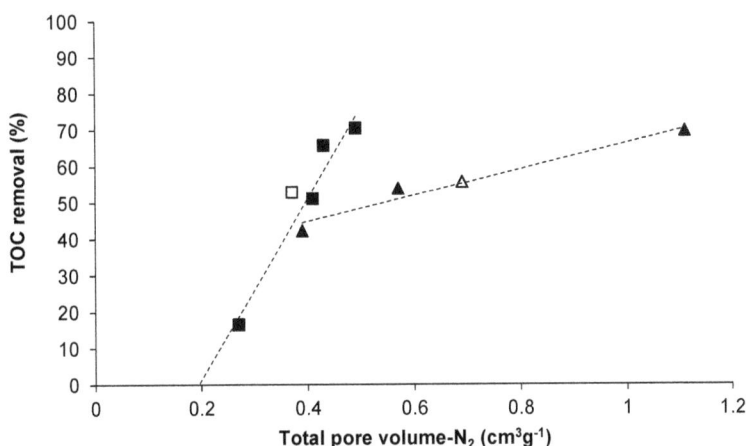

Figure 3. Relationship between TOC removal and total pore volumes of the activated carbon: open and closed squares for as prepared and heat treated CO_2 activated carbon, respectively; open and closed triangles for as prepared and heat treated H_3PO_4 activated carbon, respectively.

2.2. Increasing the Yield of CO2 Activated Carbon

A method based on the recycling of bio-oil from previous runs [23] will now be used to increase the yield of activated carbon. In the results presented thus far, the feed was only wood. Subsequent results are those obtained with a feed of wood and bio-oil. Figure 4 illustrates the two approaches used to prepare porous carbon samples by CO_2 activation. In the first approach, the mixture of wood and bio-oil undergoes air pretreatment at 220 °C for 3 h in order to stabilize the mixture and then the temperature is increased up to 600 °C for the pyrolysis (Figure 4a). A 3 h pretreatment with air at 220 °C is sufficient to achieve the maximum char mass gain [23]. In the second approach, 0.48% KOH is added to the mixture of wood and bio-oil to catalyze char formation upon heating to 600 °C (Figure 4b). The content of KOH was limited to 0.48% to keep the ash content in the char at ~5%, which is considered a maximum for a good quality product [14]. Specifically, ash contents in the char from wood, mixture of wood and bio-oil with air pretreatment and KOH addition were 2.0%, 1.4% and 5.5%, respectively. Mineral matter/ash can block the pores and preferentially adsorb water, hindering adsorption of the desired adsorbate [1]. Previously, it was demonstrated that air pretreatment and KOH addition prior to pyrolysis result in the higher char yields from wood and bio-oil mixture compared to the pyrolysis without these treatments [23].

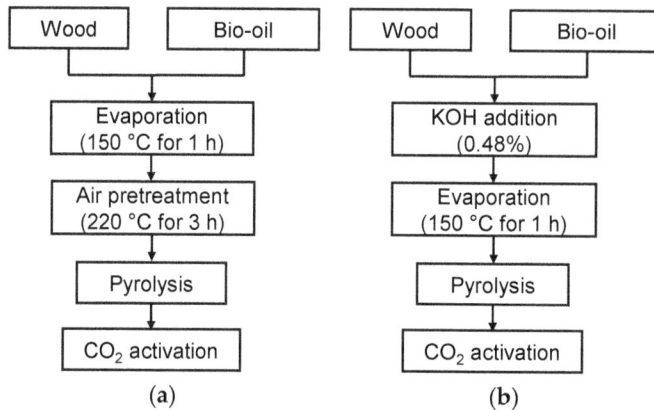

Figure 4. Activated carbon preparation methods with air pretreatment (a) and KOH addition (b) prior to pyrolysis of wood with bio-oil.

In both approaches water and organic compounds with boiling points below 150 °C were removed by evaporation at 150 °C (Figure 4). Otherwise, these compounds accumulate in bio-oil upon continuous recycling to new batches of wood. Similar to the AC1.0 sample, the char samples produced from the mixture of wood and bio-oil were activated with CO_2 for 1 h at 800 °C. The char samples prepared by KOH addition and air pretreatment of wood and bio-oil mixture are denoted as Char-600-KOH and Char-600-Air, respectively. The activated carbon prepared with KOH addition to the wood and bio-oil mixture includes "KOH" in the name followed by HT, if applicable, (e.g., AC1.0-KOH-HT is activated carbon prepared from wood with bio-oil and KOH by CO_2 activation for 1 h followed by acid washing and heat treatment), while the carbon prepared by air pretreatment of wood and bio-oil includes "Air" in the name.

Table 2 lists the properties of the parent char samples and the CO_2 activated carbon. The char yields produced by bio-oil recycling increased to 30.4% and 30.9% (Char-600-Air and Char-600-KOH, respectively) compared to 23.7% for the Char-600 sample. The porous properties of Char-600-Air were similar to Char-600, while the porosity of Char-600-KOH was lower, likely due to pore blockage by potassium compounds as discussed in our previous study [23]. After activation, the yield of the AC1.0-Air carbon was 19.4% (Table 2), which is 1.3 times larger than the yield of AC1.0 from wood (15%, Table 1). In contrast, despite the increased yield of Char-600-KOH (30.9%), the yield of AC1.0-KOH was only 15.4% after activation, suggesting that the addition of potassium hydroxide for the pyrolysis of wood with bio-oil was not beneficial for the production of activated carbon under these conditions. The lower yield of this activated carbon sample compared to AC1.0-Air can be attributed to the enhanced carbon loss due to the reaction with CO_2 in the presence of potassium compounds. Potassium compounds are known catalysts for char gasification by CO_2 [30].

The acid washing and heat treatment of activated carbon prepared from the mixture of wood and bio-oil significantly increased the porosity (Table 2), in contrast to the minor improvement achieved for sample AC1.0 (compare to AC1.0-HT in Table 1). In the AC1.0-KOH-HT carbon, the observed increase in porosity could be partially explained by the presence of potassium compounds, which can enhance the development of pores during CO_2 activation [31]. Total pore volume of as prepared AC1.0-KOH calculated on an ash free basis was larger than those of AC1.0 and AC1.0-Air samples (0.43 cm$^3 \cdot$ g^{-1}, 0.38 cm$^3 \cdot$ g^{-1} and 0.35 cm$^3 \cdot$ g^{-1}, respectively). However, the increase in porosity was also observed in AC1.0-Air-HT, prepared without potassium hydroxide addition (Table 2). Therefore, it is likely that the observed phenomenon was primarily attributed to the use of bio-oil for the preparation of the activated carbon rather than to KOH.

Table 2. Yield and porous properties of carbon produced from wood and bio-oil mixture.

Sample	Yield (%) [a]	Meso/Macropore Volume-N_2 ($cm^3 \cdot g^{-1}$)	Micropore Volume-N_2 ($cm^3 \cdot g^{-1}$)	Micropore Volume-CO_2 ($cm^3 \cdot g^{-1}$)	BET-N_2 Surface Area ($m^2 \cdot g^{-1}$)
Char from Wood					
Char-600	23.7 ± 0.4 [b]	0.02 [b]	0.17 [b]	0.18 [b]	440 [b]
Char from Wood and Bio-Oil Mixture					
Char-600-Air	30.4 ± 0.7	0.03	0.17	0.20	450
Char-600-KOH	30.9 ± 0.1	0.02	0.01	0.15	30
Activated Carbon from Wood and Bio-Oil Mixture					
AC1.0-Air	19.4 ± 0.1	0.15	0.19	0.25	600
AC1.0-Air-HT		0.19	0.32	0.33	930
AC1.0-KOH	15.4 ± 0.1	0.20	0.18	0.24	640
AC1.0-KOH-HT		0.25	0.31	0.36	1020

[a] Yield is calculated per mass of dried wood required for production of carbon based on the average of three measurements ± standard deviation; [b] data for Char-600 are from reference [23].

To investigate whether the bio-oil addition influenced the number of surface oxygen groups, the activated carbon prepared with air pretreatment was characterized with temperature-programmed decomposition (Figure S4). According to the literature [32–34], CO_2 evolution can be attributed to decomposition of carboxylic groups and anhydrides (200–450 °C), peroxides (500–550 °C), lactols, lactones and anhydrides (above 600 °C). CO evolution can be ascribed to anhydrides (350–400 °C), phenols, hydroquinones (600–700 °C), carbonyls, quinones and ether groups (700–800 °C). At temperatures higher than 900 °C, CO evolves from pyrone and chromene groups. There were differences between CO and CO_2 profiles of AC1.0-Air carbon from those of AC1.0 (Figure S4) due to different groups on the surfaces of the samples. However, the total amounts of CO_2 and CO evolved from AC1.0-Air (19 µmol·g^{-1} and 114 µmol·g^{-1}, respectively) were similar to those from AC1.0 (18 µmol·g^{-1} and 110 µmol·g^{-1} of CO_2 and CO, respectively), suggesting no change in the number of surface oxygen groups due to bio-oil recycling.

The porous carbon samples prepared from wood and bio-oil were then tested for TOC removal and the results are shown in Figure 5a. Relative to the removal achieved by sample AC1.0 (results reproduced from Figure 2 for ease of comparison), the percentages of TOC removal were 6% lower and 9% higher for the as prepared AC1.0-Air and AC1.0-KOH samples, respectively. The differences are consistent with the differences in the porosities of the samples (Tables 1 and 2). Acid washing and heat treatment of these samples improved the TOC removal levels up to ~70% (Figure 5a), similar to that achieved with samples AC3.6-HT, P2:1-HT and ColorSorb-HT (Figure 2). Again the increase is consistent with the increase in porosities after the acid washing and heat treatment (Table 2). The pH of the water was measured after contact with the AC samples. In all cases, the pH was between 9.2 and 9.7, with measurement errors of 0.2–0.3 units.

The results for the as prepared carbon have been normalized for their different yields and TOC capacities by expressing the TOC removal in terms of mass of organic carbon removed per mass of wood used to prepare the activated carbon (mg·g-wood^{-1}) as shown in Figure 5b. This normalization indicates that it is beneficial to recycle the bio-oil in terms of TOC removal per mass of wood feed. As mentioned in the Introduction, recycling also eliminates the problem of handling and disposal of the bio-oil. The results were similar for the two activated carbon preparation methods (Figure 4) with slightly more TOC removal obtained with the activated carbon prepared with air pretreatment (AC1.0-Air) than KOH addition (AC1.0-KOH), and with 1.2 times higher removal for AC1.0-Air compared to AC1.0, which was prepared from wood only. Thus, a higher conversion efficiency of the feed (wood) to an adsorbent for water treatment was obtained by bio-oil recycling.

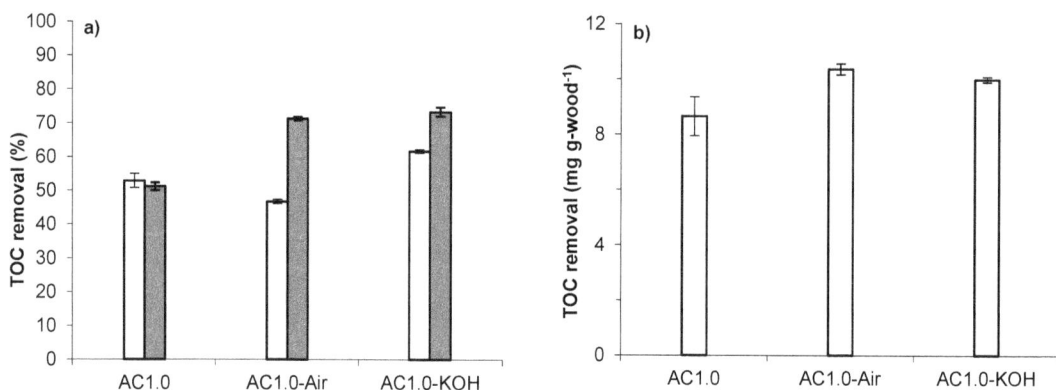

Figure 5. TOC removal by activated carbon prepared from the mixture of wood and bio-oil (white and grey bars for as prepared and HT carbon, respectively): (**a**) in per cent of removed organic carbon from water; and (**b**) in mg of organic carbon per mass of wood used for carbon production.

3. Materials and Methods

3.1. Preparation of Activated Carbon

Aspen (*Populus tremuloides*) wood chips provided by Alberta-Pacific Forest Industries Inc. (Boyle, AB, Canada) were used as the feedstock. According to the proximate analysis, the wood chips contained 18.5% fixed carbon, 80.9% volatile matter and 0.6% ash on a dry basis. The contents of carbon, hydrogen and nitrogen were 48.6%, 6.0% and 0.6%, respectively, and were reported previously [23]. Activated carbon was prepared by several different methods. The first method consisted of wood pyrolysis to produce a char that was then activated in the presence of CO_2. The second method involved pyrolyzing the wood with recycled bio-oil with either KOH addition or with air-pretreatment prior to CO_2 activation, and the third method involved treatment of the wood with phosphoric acid (H_3PO_4).

Char production and CO_2 activation were carried out in a vertical packed bed reactor setup described previously [22]. Approximately 7.5 g of the dried and crushed wood chips (particle diameter 0.3–2.0 mm and particle length 0.3–5.0 mm) were loaded into the reactor (bed volume 35 cm^3). The reactor was heated at 4 °C·min^{-1} under N_2 flow (space velocity 1 min^{-1}) to 600 °C and pyrolyzed at this temperature and gas flow for 30 min to produce char (named Char-600). Bio-oil was collected in a flask attached to the outlet of the reactor. For the preparation of activated carbon in this study, the activation temperature was fixed at 800 °C, while the degree of activation and porosity were varied by activation time. Approximately 1.5 g of Char-600 was loaded into the reactor (bed volume 10 cm^3), purged with N_2 for 1 h and then heated at 10 °C·min^{-1} under this N_2 flow to 800 °C. After 800 °C was reached, N_2 was switched to CO_2 (space velocity 10 min^{-1}) for activation. The activation time with CO_2 varied from 0.3 to 3.6 h. The char sample (Char-800) without CO_2 activation was prepared by heating Char-600 in N_2 flow for 1 h at 800 °C. All samples were crushed and sieved (US mesh 100) to obtain particle sizes less than 150 μm before further treatment, adsorption tests and/or characterization.

To reduce the amount of ash and surface oxygen groups, some of the samples were washed with hydrochloric acid and then heat treated in N_2 flow at 900 °C (maximum safe temperature for operation of the setup). To remove acid soluble ash components, approximately 2 g of sample was soaked in 0.1 M hydrochloric acid for 18 h, washed with deionized water until pH 5–5.5 (*i.e.*, pH of deionized water) and then dried at 105 °C for 24 h. To remove surface oxygen groups, the dried sample (~1.5 g) was loaded into a quartz boat, which was then placed into a horizontal furnace, purged with N_2 (100 cm^3·min^{-1}) for 1 h and then heated at 10 °C·min^{-1} under N_2 flow to 900 °C and held at this temperature for 1 h.

Wood impregnated with bio-oil generated in a previous pyrolysis run was also used as a precursor of activated carbon. Two methods were adopted from our previous work [23] to produce char from the mixture of biomass and bio-oil; namely, KOH addition and air pretreatment prior to the pyrolysis. For KOH addition, 0.068 g of KOH (85% pure, Alfa Aesar, Thermo Fisher Scientific Inc., Waltham, MA, USA) was dissolved in 0.1 g of deionized water followed by addition of 4.0 g of bio-oil produced from Char-600 pyrolysis (due to mass loss during transfer, the actual mass of bio-oil added to the wood was ~3.8 g and the resulting KOH content was 0.48% for the feed). The mixture was added to 7.5 g of wood, and the impregnated wood was loaded into the reactor, heated to 150 °C for 1 h in N_2, to remove the water and light organic compounds, and then heated in N_2 at the conditions used for the preparation of Char-600. The char sample was then activated with CO_2 for 1 h at 800 °C following the procedure used for the activation of Char-600.

For experiments with air pretreatment, 3.8 g of bio-oil and 7.5 g of wood were mixed in a beaker and then loaded into the reactor. The reactor was heated at 4 °C· min^{-1} in air (space velocity 1 min^{-1}) first to 150 °C for 1 h, to evaporate water and light organic compounds, and then to 220 °C for 3 h. After the air pretreatment, samples were purged with N_2 (space velocity 1 min^{-1}) for 0.5 h and heated in N_2 at the conditions used for the preparation of Char-600.

To prepare H_3PO_4 activated carbon, dried wood chips were used as received. Approximately 60 g of wood chips were placed into a 1 dm^3 beaker to which 300 cm^3 of an aqueous solution of H_3PO_4 (85% pure, BDH Aristar, VWR International, Radnor, PA, USA) were added. The volume of solution was kept constant while the concentration of H_3PO_4 was varied to target H_3PO_4:wood ratios of 1:2, 1:1 and 2:1 by weight. The mixture was soaked for 1 d at ambient temperature (23 °C) and then water was evaporated from the surface of wood chips by heating at 100 °C on a hot plate. The prepared mixture of H_3PO_4 and wood was left for two more days at ambient temperature for impregnation and then dried in air at 110 °C overnight. Activation was carried out in a horizontal alumina reactor with an inner diameter of 8 cm. Approximately 30 g of the H_3PO_4:wood mixture was loaded into the alumina reactor and heated at a constant rate (4 °C· min^{-1}) to 500 °C under N_2 flow (space velocity 1 min^{-1}). The temperature was maintained at 500 °C for 0.5 h and then the reactor was cooled under N_2 to 70 °C for sample removal. The activated carbon samples were washed with deionized water until pH of 5–5.5, dried overnight at 110 °C and then crushed and sieved (US mesh 100) to obtain particles less than 150 µm. After washing, the samples were either used or subjected to a further heat treatment at 900 °C under N_2 for 1 h.

A commercial activated carbon, ColorSorb G5 prepared by steam activation of wood was provided by Jacobi Carbons AB (Kalmar, Sweden).

3.2. Characterization of Char and Activated Carbon

Yields of char and CO_2 activated carbon were calculated as mass of carbon after pyrolysis and activation, respectively, per mass of dried wood. The yield of H_3PO_4 activated carbon was calculated as mass of dried carbon, after H_3PO_4 removal by water washing, per mass of dried wood.

Activated carbon samples were characterized with N_2 and CO_2 adsorption measured at −196 °C and 0 °C, respectively (Tristar 3000, Micromeritics Instrument Co., Norcross, GA, USA). Surface areas were calculated from N_2 adsorption isotherms using the BET method [35]. Samples (~0.05 g) were degassed at 300 °C under vacuum (100 mTorr) for 3 h before adsorption. The surface areas were determined from linear BET plots at the relative pressure ranges between 0.02 and 0.3, assuming a value of 0.164 nm^2 for the cross-section of the N_2 molecule. Micropore volumes were obtained from N_2 and CO_2 adsorption isotherms using the t-plot with Carbon black STSA reference isotherm and Dubinin-Radushkevich methods [36,37], respectively. Total pore volumes were determined by N_2 adsorption at relative pressures of 0.96–0.97. The meso/macropore volume was calculated as the difference between total pore volume and micropore volume determined by N_2 adsorption. The non-local density functional theory (NLDFT, Autosorp 1C, Quantachrome Instruments, Boynton Beach, FL, USA) [38] was used to calculate the pore size distributions of the activated carbon.

The surface oxygen groups were quantified by temperature-programmed decomposition. Approximately 10 mg of sample was placed in a thermogravimetric analyzer (Cahn Thermax 500 apparatus, Thermo Fisher Scientific Inc., Waltham, MA, USA) and heated in N_2 at 5 $°C \cdot min^{-1}$ to 1000 $°C$. The evolution of CO_2 and CO during heating was detected with an infrared gas analyzer (Uras 26, ABB AO2020, ABB Ltd., Zurich, Switzerland) and recorded.

3.3. Total Organic Carbon Removal from Water

Steam assisted gravity drainage (SAGD) water was collected from the process stream of an industrial SAGD facility (Athabasca region, AB, Canada) immediately following the water softening stage but prior to entering the boiler. This 80 $°C$ water had a pH of 9.5–9.8 and total organic carbon (TOC) content ranging from 910 to 1162 $mg \cdot dm^{-3}$, depending on the sample. After cooling, the samples were shipped to the University of Calgary and stored at 4 $°C$.

The removal of TOC from SAGD water by activated carbon was evaluated by batch adsorption. In a glass vial, 0.1 g of activated carbon was mixed with 10 cm^3 of SAGD water reheated to 80 $°C$. To measure the initial TOC content in the SAGD water, vials with 10 cm^3 of the reheated water without carbon addition were prepared. The mixtures and pure SAGD water were shaken at 80 $°C$ (typical temperature for field operations) and 175 rpm for 18 h before analysis. A series of kinetic measurements have been carried out, and 18 h treatment was sufficient to attain adsorption equilibrium on the carbon samples. After settling for 2–3 min, the supernatant (~9 cm^3) was obtained with a syringe and filtered through a 0.45 μm nylon membrane filter (VWR International, Radnor, PA, USA). The cooled solutions were analyzed by a TOC analyzer (TOC-VCPN, Shimadzu Corp., Kyoto, Japan). To avoid premature precipitation of the organic acids in the SAGD water (Figure S3), acidification (by HCl addition) was done directly in the TOC analyzer. Three replicates with each adsorbent were prepared and the results are reported as averages ± standard deviation of three measurements. The statistical significance of differences between obtained results was determined using a two-tailed *t*-test with a confidence interval of 95%.

After TOC removal by the carbon samples, the pH of water was measured using a Lab 850 pH meter (Schott Instruments GmbH/SI Analytics GmbH, Mainz, Germany) with BlueLine 56 pH electrode (SI Analytics GmbH, Mainz, Germany).

4. Conclusions

Conversion of wood to activated carbon for industrial water treatment was investigated. Both chemically (H_3PO_4) and physically (CO_2) activated carbon samples removed similar percentages of total organic carbon from SAGD water. As CO_2 activated carbon had lower yields compared to carbon prepared with H_3PO_4, a method to increase the yield, and the conversion efficiency of wood by CO_2 activation was investigated. Specifically, bio-oil was recycled and added to the wood feed before CO_2 activation. Air pretreatment of the bio-oil and biomass mixture (220 $°C$ for 3 h) was required but increased the yield of as prepared activated carbon by 1.3 times and total organic carbon uptake from water per mass of utilized wood by 1.2 times.

Acknowledgments: The authors thank Climate Change and Emissions Management Corporation (CCEMC) and the Natural Science and Engineering Research Council (NSERC) of Canada for funding this project. The authors also acknowledge Alberta-Pacific Forest Industries Inc. for providing wood chips and Jacobi Carbons AB for providing ColorSorb G5. The authors would like to acknowledge laboratory help provided by Waheed Zaman and Joseph Kimetu.

Author Contributions: Andrei Veksha and Josephine M. Hill contributed to design of experiments and composed the manuscript. Andrei Veksha and Tazul I. Bhuiyan performed experiments and data analysis. Josephine M. Hill led the project.

Conflicts of Interest: The authors declare no conflict of interest. The funding sponsors had no role in the design of the study, in the collection, analyses, or interpretation of data, in the writing of the manuscript, and in the decision to publish the results.

References

1. Moreno-Castilla, C. Adsorption of organic molecules from aqueous solutions on carbon materials. *Carbon* **2004**, *42*, 83–94. [CrossRef]

2. Li, L.; Quinlivan, P.A.; Knappe, D.R.U. Effects of activated carbon surface chemistry and pore structure on the adsorption of organic contaminants from aqueous solution. *Carbon* **2002**, *40*, 2085–2100. [CrossRef]

3. Dastgheib, S.A.; Karanfil, T.; Cheng, W. Tailoring activated carbon for enhanced removal of natural organic matter from natural waters. *Carbon* **2004**, *42*, 547–557. [CrossRef]

4. Dias, J.M.; Alvim-Ferraz, M.C.M.; Almeida, M.F.; Rivera-Utrilla, J.; Sanchez-Polo, M. Waste materials for activated carbon preparation and its use in aqueous-phase treatment: A review. *J. Environ. Manag.* **2007**, *85*, 833–846. [CrossRef] [PubMed]

5. Okada, K.; Yamamoto, N.; Kameshima, Y.; Yasumori, A. Adsorption properties of activated carbon from waste newspaper prepared by chemical and physical activation. *J. Colloid Interface Sci.* **2003**, *262*, 194–199. [CrossRef]

6. Girgis, B.S.; Yunis, S.S.; Soliman, A.M. Characteristics of activated carbon from peanut hulls in relation to conditions of preparation. *Mater. Lett.* **2002**, *57*, 164–172. [CrossRef]

7. Iranmanesh, S.; Harding, T.; Abedi, J.; Seyedeyn-Azad, F.; Layzell, D.B. Adsorption of naphthenic acids on high surface area activated carbon. *J. Environ. Sci. Health A* **2014**, *49*, 913–922. [CrossRef] [PubMed]

8. Mestre, A.S.; Pires, R.A.; Aroso, I.; Fernandes, E.M.; Pinto, M.L.; Reis, R.L.; Andrade, M.A.; Pires, J.; Silva, S.P.; Carvalho, A.P. Activated carbon prepared from industrial pre-treated cork: Sustainable adsorbents for pharmaceutical compounds removal. *Chem. Eng. J.* **2014**, *253*, 408–417. [CrossRef]

9. Toles, C.A.; Marshall, W.E.; Johns, M.M.; Wartelle, L.H.; McAloon, A. Acid-activated carbon from almond shells: Physical, chemical and adsorptive properties and estimated cost of production. *Bioresour. Technol.* **2000**, *71*, 87–92. [CrossRef]

10. Toles, C.A.; Marshall, W.E.; Wartelle, L.H.; McAloon, A. Steam- or carbon dioxide-activated carbon from almond shells: Physical, chemical and adsorptive properties and estimated cost of production. *Bioresour. Technol.* **2000**, *75*, 197–203. [CrossRef]

11. Ng, C.; Marshall, W.E.; Rao, R.M.; Bansode, R.R.; Losso, J.N. Activated carbon from pecan shell: Process description and economic analysis. *Ind. Crops Prod.* **2003**, *17*, 209–217. [CrossRef]

12. Altenor, S.; Ncibi, M.C.; Brehm, N.; Emmanuel, E.; Gaspard, S. Pilot-scale synthesis of activated carbon from vetiver roots and sugar cane bagasse. *Waste Biomass Valor.* **2013**, *4*, 485–495. [CrossRef]

13. Lee, Y.; Park, J.; Ryu, C.; Gang, K.S.; Yang, W.; Park, Y.K.; Jinho, J.; Hyun, S. Comparison of biochar properties from biomass residues produced by slow pyrolysis at 500 °C. *Bioresour. Technol.* **2013**, *148*, 196–201. [CrossRef] [PubMed]

14. Antal, M.J., Jr.; Gronli, M. The art, science and technology of charcoal production. *Ind. Eng. Chem. Res.* **2003**, *42*, 1619–1640. [CrossRef]

15. Dai, X.; Antal, M.J., Jr. Synthesis of a high-yield activated carbon by air gasification of macadamia nut shell charcoal. *Ind. Eng. Chem. Res.* **1999**, *38*, 3386–3395. [CrossRef]

16. Tam, M.S.; Antal, M.J., Jr.; Jakab, E.; Varhegyi, G. Activated carbon from macadamia nut shell by air oxidation in boiling water. *Ind. Eng. Chem. Res.* **2001**, *40*, 578–588. [CrossRef]

17. Py, X.; Guillot, A.; Cagnon, B. Activated carbon porosity tailoring by cyclic sorption/decomposition of molecular oxygen. *Carbon* **2003**, *41*, 1533–1543. [CrossRef]

18. Kimetu, J.M.; Hill, J.M.; Husein, M.; Bergerson, J.; Layzell, D.B. Using activated biochar for greenhouse gas mitigation and industrial water treatment. *Mitig. Adapt. Strateg. Glob. Change* **2014**. [CrossRef]

19. Oasmaa, A.; Czernik, S. Fuel oil quality of biomass pyrolysis oils—State of the art for the end users. *Energy Fuels* **1999**, *13*, 914–921. [CrossRef]

20. Huang, Y.; Kudo, S.; Norinaga, K.; Amaike, M.; Hayashi, J. Selective production of light oil by biomass pyrolysis with feedstock-mediated recycling of heavy oil. *Energy Fuels* **2012**, *26*, 256–264. [CrossRef]

21. Huang, Y.; Kudo, S.; Masek, O.; Norinaga, K.; Hayashi, J. Simultaneous maximization of the char yield and volatility of oil from biomass pyrolysis. *Energy Fuels* **2013**, *27*, 247–254. [CrossRef]

22. Veksha, A.; McLaughlin, H.; Layzell, D.B.; Hill, J.M. Pyrolysis of wood to biochar: Increasing yield while maintaining microporosity. *Bioresour. Technol.* **2014**, *153*, 173–179. [CrossRef] [PubMed]

23. Veksha, A.; Zaman, W.; Layzell, D.B.; Hill, J.M. Enhancing biochar yield by co-pyrolysis of bio-oil with biomass: Impacts of potassium hydroxide addition and air pretreatment prior to co-pyrolysis. *Bioresour. Technol.* **2014**, *171*, 88–94. [CrossRef] [PubMed]

24. Kawaguchi, H.; Li, Z.; Masuda, Y.; Sato, K.; Nakagawa, H. Dissolved organic compounds in reused process water for steam-assisted gravity drainage oil sands extraction. *Water Res.* **2012**, *46*, 5566–5574. [CrossRef] [PubMed]

25. Veksha, A.; Pandya, P.; Hill, J.M. The removal of methyl orange from aqueous solution by biochar and activated carbon under microwave irradiation and in the presence of hydrogen peroxide. *J. Environ. Chem. Eng.* **2015**, *3*, 1452–1458. [CrossRef]

26. Petersen, M.A.; Grade, H. Analysis of steam assisted gravity drainage produced water using two-dimensional gas chromatography with time-of-flight mass spectrometry. *Ind. Eng. Chem. Res.* **2011**, *50*, 12217–12224. [CrossRef]

27. Di Blasi, C. Modeling chemical and physical processes of wood and biomass pyrolysis. *Prog. Energ. Combust. Sci.* **2008**, *34*, 47–90. [CrossRef]

28. Katyal, S.; Thambimuthu, K.; Valix, M. Carbonisation of bagasse in a fixed bed reactor: Influence of process variables on char yield and characteristics. *Renew. Energy* **2003**, *28*, 713–725. [CrossRef]

29. Jagtoyen, M.; Derbyshire, F. Activated carbon from yellow poplar and white oak by H_3PO_4 activation. *Carbon* **1998**, *36*, 1085–1097. [CrossRef]

30. Sams, D.A.; Shadman, F. Catalytic effect of potassium on the rate of char-CO_2 gasification. *Fuel* **1983**, *62*, 880–882. [CrossRef]

31. Wigmans, T.; Hoogland, A.; Tromp, P.; Moulijn, J.A. The influence of potassium carbonate on surface area development and reactivity during gasification of activated carbon by carbon dioxide. *Carbon* **1983**, *21*, 13–22. [CrossRef]

32. Figueiredo, J.L.; Pereira, M.F.R.; Freitas, M.M.A.; Orfao, J.J.M. Modification of the surface chemistry of activated carbons. *Carbon* **1999**, *37*, 1379–1389. [CrossRef]

33. Figueiredo, J.L.; Pereira, M.F.R.; Freitas, M.M.A.; Orfao, J.J.M. Characterization of active sites on carbon catalysts. *Ind. Eng. Chem. Res.* **2007**, *46*, 4110–4115. [CrossRef]

34. Zielke, U.; Hüttinger, K.J.; Hoffman, W.P. Surface-oxidized carbon fibers: I. Surface structure and chemistry. *Carbon* **1996**, *34*, 983–998. [CrossRef]

35. Brunauer, S.; Emmett, P.H.; Teller, E. Adsorption of gases in multimolecular layers. *J. Am. Chem. Soc.* **1938**, *60*, 309–319. [CrossRef]

36. Lippens, B.C.; de Boer, J.H. Studies on pore systems in catalysts: V. the t method. *J. Catal.* **1968**, *4*, 319–323. [CrossRef]

37. Dubinin, M.M. Physical adsorption of gases and vapors in micropores. *Prog. Surf. Membr. Sci.* **1975**, *9*, 1–70.

38. Lastoskie, C.; Gubbins, K.E.; Quirkeft, N. Pore size distribution analysis of microporous carbons: A density functional theory approach. *J. Phys. Chem.* **1993**, *97*, 4786–4796. [CrossRef]

6

Cutting Modeling of Hybrid CFRP/Ti Composite with Induced Damage Analysis

Jinyang Xu * and Mohamed El Mansori

Academic Editor: Sanjay Mathur

MSMP—EA 7350 Laboratoire, Arts et Métiers ParisTech, Rue Saint Dominique B.P. 508, 51006 Châlons-en-Champagne, France; mohamed.elmansori@ensam.eu
* Correspondence: jinyang.xu@ensam.eu

Abstract: In hybrid carbon fiber reinforced polymer (CFRP)/Ti machining, the bi-material interface is the weakest region vulnerable to severe damage formation when the tool cutting from one phase to another phase and *vice versa*. The interface delamination as well as the composite-phase damage is the most serious failure dominating the bi-material machining. In this paper, an original finite element (FE) model was developed to inspect the key mechanisms governing the induced damage formation when cutting this multi-phase material. The hybrid composite model was constructed by establishing three disparate physical constituents, *i.e.*, the Ti phase, the interface, and the CFRP phase. Different constitutive laws and damage criteria were implemented to build up the entire cutting behavior of the bi-material system. The developed orthogonal cutting (OC) model aims to characterize the dynamic mechanisms of interface delamination formation and the affected interface zone (AIZ). Special focus was made on the quantitative analyses of the parametric effects on the interface delamination and composite-phase damage. The numerical results highlighted the pivotal role of AIZ in affecting the formation of interface delamination, and the significant impacts of feed rate and cutting speed on delamination extent and fiber/matrix failure.

Keywords: hybrid composite; FE modeling; orthogonal cutting; induced damage; interface delamination; fiber/matrix failure

1. Introduction

Hybrid composites, especially those carbon fiber reinforced polymer (CFRP)/Ti stacks, have been identified as an innovative structural configuration in the modern aerospace industry. The enhanced mechanical properties and improved structural functions have given the material a high demand for manufacturing key aircraft structures subjected to high thermo-mechanical stresses. A typical application is the use of wing-fuselage connections in the new-generation Boeing 787 Dreamliner. The CFRP-to-Ti coupling typically provides the best combination of metallurgical and physical properties including high strength-to-weight ratio, low density, and superior corrosion/erosion resistance that favor energy saving in industrial applications [1–3]. Generally, the CFRP/Ti composite exhibits a high strength-to-weight ratio with yield strength as high as 830 MPa and a density of roughly $4\,\text{g/cm}^3$ [4].

Prior to their post applications, structural components made of hybrid CFRP/Ti composite are mostly manufactured in near-net-shape in order to achieve dimensional tolerance and assembly requirement. However, due to the disparate machinability behaviors of each stacked constituent, the manufacturing hybrid composite exhibits the most challenging task in industrial sectors. For instance, the titanium phase exhibits high mechanical properties, low thermal conductivity, and strong chemical affinity to tool materials, which commonly results in high force/heat generation, serious tool wear (abrasive wear and adhesion wear), and short tool life [5–7]. In contrast, the CFRP phase shows anisotropic behavior, abrasive nature, and low thermal conductivity, which leads to severe subsurface damage, poor heat dissipation, and excessive tool wear [8–10].

In hybrid CFRP/Ti machining, typically three cutting stages are involved, *i.e.*, the Ti-phase cutting, interface cutting, and CFRP-phase cutting. Among them, the interface region (also refer to the "Ti-to-CFRP" contact boundary) represents the most difficult-to-cut zone vulnerable to severe damage formation when the tool edges are cutting from one phase (Ti phase) to another phase (CFRP phase) and *vice versa*. The interface region is usually characterized as a physically intermediate transition zone that really exists in the bi-material machining process. During interface cutting, the interface area usually suffers changeable chip-separation modes and experiences severe mechanical/physical phenomena transition exerted at the bi-material contact boundary. In such circumstances, the interface region becomes the most challenging cutting zone as compared to absolute Ti-phase cutting and absolute CFRP-phase cutting while machining the hybrid composite. The discontinuity of the tool-work interaction governing interface cutting commonly makes the machining behavior more complicated and interrelated. Inspections of interface damage in CFRP/Ti cutting via the experimental method have been shown to be very challenging and highly difficult [11]. Despite the fact that several experimental investigations [1,2,12] have been well performed, some key issues have still not been clearly addressed: (*i*) the key mechanisms and physical phenomena controlling the CFRP/Ti interface cutting, (*ii*) the parametric effects on the interface machining and subsequently induced damage extent, and (*iii*) the machinability classification of hybrid CFRP/Ti machining, *i.e.*, which region of cutting actually reflects the machinability of pure Ti-phase cutting (M_{Ti}), stacked material cutting ($M_{CFRP/Ti}$), and pure CFRP-phase cutting (M_{CFRP}), respectively. In addition, the conventional experimental method is cost-prohibitive and time-consuming. In contrast, the numerical approach should be a qualified tool helpful to enable feasible inspections of the damage formation when cutting this bi-material. Furthermore, although a considerable amount of scientific work has dealt with single Ti-cutting modeling and single CFRP-cutting modeling, comprehensive numerical studies concerning hybrid CFRP/Ti machining have still been only rarely reported.

These are the key incentives that motivated the current work to develop an original finite element (FE) model to address the mentioned topics. To inspect the fundamental mechanisms controlling the bi-material machining, the simplified orthogonal cutting configuration (OCC) was adopted. The OCC represents a convenient way to reveal the most fundamental machining physics governing the various actual manufacturing operations of hybrid CFRP/Ti composite, e.g., drilling, grinding, *etc*. The key objective of this investigation aims to establish an FE model for damage predictions and failure analyses when orthogonally cutting hybrid CFRP/Ti composite. The established numerical model incorporated three physical constituents, *i.e.*, the Ti phase, CFRP phase, and interface layer. The CFRP/Ti model was rigorously validated prior to its post-application. The multiple aspects of machining responses including cutting process, interface delamination, and subsurface damage formation were precisely investigated via finite element (FE) analyses. A particular concentration was made to characterize the dynamic process of delamination formation and affected interface zone (AIZ). The numerical results highlighted the significant role of AIZ and bi-material interface consumption (BIC) in controlling the induced interface damage formation.

2. Orthogonal Cutting (OC) Model for Hybrid CFRP/Ti Cutting

2.1. Numerical Setup of the OC Model

In the current work, a 2D orthogonal cutting model was developed by using the commercial software Abaqus/Explicit code (Version 6.11, Dassault Simulia, Paris, France). To simulate the hybrid cutting operation, the machining process was specified as shown schematically in Figure 1. The basic geometries of the tool-workpiece couple and boundary condition are illustrated in Figure 1. The established orthogonal cutting (OC) model is comprised of four basic phases, *i.e.*, the tool part,

Ti part, interface part, and CFRP part with total dimensions of 2 mm × 1 mm ($L \times H$) for the workpiece material.

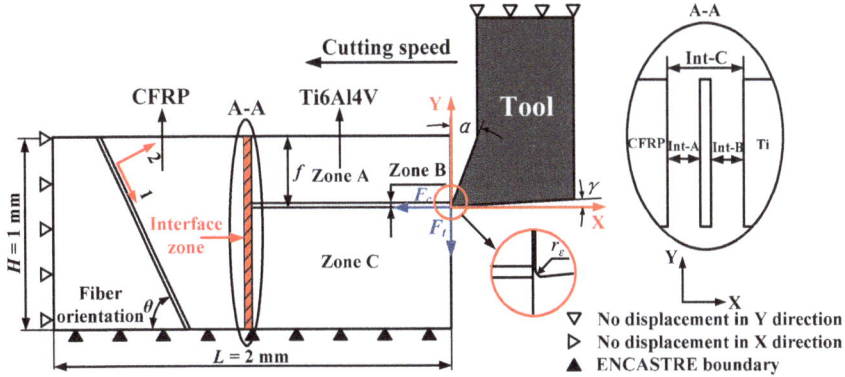

Figure 1. Scheme of the established orthogonal cutting (OC) model for hybrid CFRP/Ti machining ($\alpha = 12°$ and $\gamma = 7°$): (1,2) represents the material coordinate system where 1 → fiber direction, 2 → transverse direction, F_c indicates the cutting force and F_t signifies the thrust force (ENCASTRE boundary indicates the fully fixed end). (see Abbreviations Section).

The cutting tool was modeled as a rigid body and imposed by a cutting velocity on its reference node toward the horizontal direction (*i.e.*, negative *X*-axis) to finalize the cutting simulation. The tool was configured by defined geometries of rake and clearance angles ($\alpha = 12°$ and $\gamma = 7°$) as depicted in Figure 1. The center of the tool tip was placed exactly at the feed-rate distance from the upper surface. Fixed displacements were applied on both the bottom and left edges of the stack model. The bottom edge of the OC model was restrained in all directions (ENCASTRE) while the left edge was constrained to move along the horizontal direction (*X* direction), as shown in Figure 1.

The Ti phase was modeled as a fully isotropic and homogeneous material. A four-node plane-strain thermally coupled quadrilateral element type CPE4RT, which has better convergence properties was utilized for a coupled temperature-displacement analysis and enhanced hourglass control was selected for the whole set of the Ti elements. The entire Ti phase was divided into three physical zones: (*i*) Zone A denoted the separated chip layer, (*ii*) Zone B signified the predefined separation path (joint layer) and (*iii*) Zone C represented the machined Ti surface, where mesh generations exhibited different characteristics. Both Zones A and B were defined by very fine mesh density, whereas Zone C was constructed by coarse mesh element density far away from the vicinity of the tool-work contact region. The surface-to-surface contact algorithm was used to model the interaction between the cutting tool, Zone A and Zone B. The kinematic contact algorithm was assigned to the contact pairs in order to avoid element penetrations. Friction in the orthogonal cutting commonly occurs at the contact surfaces of tool and workpiece, rake face and chip surface. The frictional shearing stress is the average of the shearing stress at the tool and chip interface. In the present work, the Coulomb's friction law was utilized to describe the contact behavior.

For the CFRP part, the composite in reality consists of two distinct phases (fiber and matrix) and globally exhibits anisotropic properties. However, for simulations in a macro-mechanical model, the reinforced laminate has commonly been assumed as an equivalent homogeneous material (EHM) by most numerical study cases [13–16]. In the present model, the CFRP layer was modeled as EHM by using four-node plane-stress linearly interpolated elements (CPS4R) with reduced integration and automatic hourglass control. It should be noted that a plane strain analysis, which is used typically for metal cutting, was not appropriate for machining CFRP laminates due to the extent of out of plane material displacement observed in the cutting experiments [17]. The interaction between the CFRP phase and tool was regulated by the algorithm surface-node-surface contact available in the Abaqus/Explicit code.

To link the Ti phase and CFRP phase together, an interface layer was introduced in the FE model and simulated as a quick transition zone by using a cohesive element. It should be stressed that the use of the interface layer here serves as a technical control for the "Ti-to-CFRP" contact management during simulation. A triangular traction-separation cohesive formulation with linear softening was used to represent its mechanical response. The assembly of the interface layer with both the Ti phase and CFRP phase was carried out by setting a constraint type join (tie constraint). Furthermore, two contact pairs (Int-A and Int-B) with specifications of penalty contact algorithm and rough friction formulation were established in the interface zone, Ti phase and CFRP phase as shown in Figure 1, which made the interface zone a slave surface. Moreover, an additional contact pair referring to the Int-C (as shown in Figure 1) was also assigned between the Ti phase and CFRP phase in order to avoid them penetrating each other when the cohesive elements were eroded.

2.2. Ti-phase Model

The material behavior of the Ti phase was assumed to be isotropic and elastic-plastic with thermal softening by using the isotropic plasticity model available in the Abaqus/Explicit code. The material properties for the Ti phase (Ti6Al4V) are summarized in Table 1 [18]. The Young's modulus (E), thermal expansion coefficient (α_T), thermal conductivity (λ) and specific heat (c_p) were considered to be temperature-dependent in order to accurately represent the property variation of titanium phase *versus* thermal influence during the cutting process.

Table 1. Mechanical properties of the Ti6Al4V phase [18].

Physical Parameter	Ti6Al4V
Density (ρ)	4430 (kg/m^3)
Young's modulus (E)	$E = 0.7412T + 113.375$ (GPa)
Poisson's ratio (v)	0.342
Thermal expansion coefficient (α_T)	$\alpha_T = 2 \times 10^{-9} \times T + 9 \times 10^{-6}$ (°C^{-1})
Melting temperature (T_m)	1680 (°C)
Room temperature (T_r)	25 (°C)
Thermal conductivity (λ)	$\lambda = 7.039e^{0.0011T}$ (W/(m·°C))
Specific heat (c_p)	$c_p = 2.24e^{0.0007T} \times 10^6/\rho$ (J/(kg·°C))

Note: the term "T" indicates the cutting temperature generated inside the Ti6Al4V alloy during the machining process.

In FE modeling, accurate material flow stress models are very much required to capture the constitutive behavior of the work material under high strain/strain rate/temperature encountered in machining. The constitutive model proposed by Johnson-Cook (JC) [19,20] was applied in this investigation, which offers a satisfactory description of ductile material behavior by considering large strains, high strain rates, and temperature-dependent visco-plasticity encountered in machining. The JC material model also takes into account the effects of strain hardening, strain rate sensitivity, and thermal softening behavior as illustrated in Equation (1).

$$\bar{\sigma} = \underbrace{(A + B\bar{\varepsilon}^n)}_{Strain\,hardening} \underbrace{\left(1 + C\ln\frac{\dot{\bar{\varepsilon}}}{\dot{\bar{\varepsilon}}_0}\right)}_{Strain\,rate\,sensitivity} \underbrace{\left[1 - \left(\frac{T - T_r}{T_m - T_r}\right)^m\right]}_{Thermal\,softening\,behavior} \tag{1}$$

Where $\bar{\sigma}$ denotes the equivalent flow stress, $\bar{\varepsilon}$ is the equivalent plastic strain, $\dot{\bar{\varepsilon}}$ is the equivalent plastic strain rate, $\dot{\bar{\varepsilon}}_0$ is the reference equivalent plastic strain rate, T is the cutting temperature of Ti6Al4V alloy during the machining process, T_m is the material melting temperature, and T_r is the room temperature. A, B, C, m and n are material constants, which are usually determined by fitting the strain-stress curves obtained by split-Hopkinson bar. In this simulation, the JC material constants are selected carefully from the open literature [21,22] as shown in Table 2.

In order to simulate the chip separation process, an energy-based ductile failure criterion was applied in the FE computation. The failure damage criteria consist of two-stage laws, *i.e.*, failure initiation law and failure evolution law, to describe the failure responses of ductile material as shown in Figure 2 [23]. The failure formation includes two steps described below.

Table 2. Input parameters for Johnson-Cook (JC) constitutive model and JC damage law [21,22].

JC Model Type	JC Model Parameter				
JC constitutive model	A (MPa)	B (MPa)	C	n	m
	1098	1092	0.014	0.93	1.1
JC damage law	D_1	D_2	D_3	D_4	D_5
	−0.09	0.25	−0.5	0.014	3.87

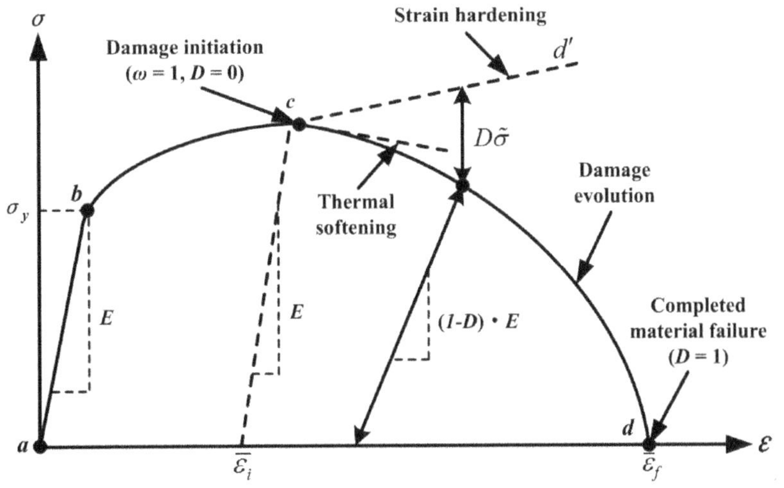

Figure 2. Typical uniaxial strain–stress (ε–σ) responses of ductile material failure process [23]. (Note: σ_y is the yield stress, $\tilde{\sigma}$ is the effective (or undamaged) flow stress, $\bar{\varepsilon}_i$ implies the equivalent plastic strain at damage initiation and $\bar{\varepsilon}_f$ indicates the equivalent plastic strain at failure).

Step 1: Damage Initiation. The JC failure model was used as a damage initiation criterion, which contains five failure parameters that need to be determined (D_1–D_5) as presented in Equation (2). In the JC failure model, damage initiation is assumed to happen when a scalar damage parameter (ω) reaches 1. The ω parameter is defined based on a cumulative law as described in Equation (3).

$$\bar{\varepsilon}_i = \left[D_1 + D_2\exp\left(D_3\frac{P}{\bar{\sigma}}\right)\right]\left(1 + D_4\ln\frac{\dot{\bar{\varepsilon}}}{\dot{\bar{\varepsilon}}_0}\right)\left[1 + D_5\left(\frac{T - T_r}{T_m - T_r}\right)\right] \tag{2}$$

$$\omega = \sum\frac{\Delta\bar{\varepsilon}}{\bar{\varepsilon}_i} \tag{3}$$

In the above equations, $\bar{\varepsilon}_i$ is the equivalent plastic strain at damage initiation, P is the hydrostatic pressure, $P/\bar{\sigma}$ is the stress triaxiality, D_1–D_5 are JC damage parameters, ω is the scalar damage parameter and $\Delta\bar{\varepsilon}$ is the equivalent plastic strain increment. The following parameters summarized in Table 2 [21] were adopted for D_1–D_5 in the cutting simulation.

Step 2: Damage Evolution. When the ductile material damage is initiated, the strain-stress relationship no longer accurately represents the real material behavior. Based on this, Hillerborg's fracture energy proposal [24] was used to reduce mesh dependency by creating a displacement-stress response after damage initiation. Hillerborg defines the energy required to open a unit area of crack (G_f) as a material parameter, and the fracture energy is represented as follows:

$$G_f = \int_{\overline{\varepsilon}_i}^{\overline{\varepsilon}_f} L\sigma_y d\overline{\varepsilon} = \int_0^{\overline{u}_f} \sigma_y d\overline{u} \qquad (4)$$

In the equation, $\overline{\varepsilon}_i$ is the equivalent plastic strain at damage initiation, $\overline{\varepsilon}_f$ is the equivalent plastic strain at failure, L denotes the characteristic length, σ_y signifies the yield stress, $\overline{\varepsilon}$ is the equivalent plastic strain, $\overline{u}_f$ is the equivalent plastic displacement at failure, and $\overline{u}$ is the equivalent plastic displacement.

The FE model applied the planar quadrilateral continuum element (CPE4RT), and then characteristic length (L) was defined as a half typical length of a line across a second order element. As the direction in which fracture occurs was not known in advance, so the definition of characteristic length was used.

A linear damage parameter (D_l) was used for the joint layer according to the following equation:

$$D_l = \frac{L\overline{\varepsilon}}{\overline{u}_f} = \frac{\overline{u}}{\overline{u}_f} \qquad (5)$$

Where the equivalent plastic displacement at failure ($\overline{u}_f$) was computed as follows:

$$\overline{u}_f = \frac{2G_f}{\sigma_y} \qquad (6)$$

In contrast, an exponential damage parameter (D_e) was used for the chip layer according to the following equation:

$$D_e = 1 - \exp\left(-\int_0^{\overline{u}} \frac{\overline{\sigma}}{G_f} d\overline{u}\right) \qquad (7)$$

At any given time during the FE calculation, the equivalent flow stress in the material is given by the following equation:

$$\overline{\sigma} = (1 - D)\, \tilde{\sigma} \qquad (8)$$

Where $\tilde{\sigma}$ denotes the effective (or undamaged) equivalent flow stress computed in the current increment and D represents the damage parameter (D_l or D_e).

In this study, G_f is provided as an input parameter and theoretically is a function of Poisson's ratio (v), Young's modulus (E), and fracture toughness (K_C) as shown in Equation (9). Considering the different fracture mechanics [25] occurring on the chip-separation process, two different values of fracture energy were utilized as input data in the Abaqus/Explicit code: $(G_f)_I$ for the joint layer (Zone B) and $(G_f)_{II}$ for the chip layer (Zone A). The $(G_f)_I$ denotes the fracture energy of mode I which is a tensile mode (opening mode normal to the plane of the fracture) whereas $(G_f)_{II}$ signifies the fracture energy of mode II which is a shearing one (sliding mode acting parallel to the plane of the fracture).

$$\left(G_f\right)_{I,II} = \left(\frac{1 - v^2}{E}\right)\left(K_C^2\right)_{I,II} \qquad (9)$$

2.3. CFRP-Phase Model

For CFRP phase, the simulated material was unidirectional carbon/epoxy T300/914 laminate (T300/914 represents the standard specification of the used CFRP material) and its mechanical/physical properties are summarized in Table 3 [26–28]. The definition of composite fiber orientation (θ) was made based on the introduction of the material coordinate system into the CFRP phase as shown schematically in Figure 1. To replicate the rupture of the fiber/matrix system, Hashin damage criteria [29] were adopted for the numerical computation. The Hashin damage criteria take into account four fundamental failure modes commonly occurring in composite machining, *i.e.*, fiber-tensile failure, fiber-compression failure, matrix-tensile failure, and matrix-compression failure, as summarized in Table 4. In Table 4, σ_{11} signifies the stress in the fiber direction, σ_{22} denotes the stress in the transverse direction, and σ_{12} represents the in-plane shear stress.

Table 3. Material properties of T300/914 CFRP used in simulation [26–28].

Material Properties	CFRP
Longitudinal modulus, E_1 (GPa)	136.6
Transverse modulus, E_2 (GPa)	9.6
In-plane shear modulus, G_{12} (GPa)	5.2
Major Poisson's ratio, v_{12}	0.29
Longitudinal tensile strength, X_T (MPa)	1500
Longitudinal compressive strength, X_C (MPa)	900
Transverse tensile strength, Y_T (MPa)	27
Transverse compressive strength, Y_C (MPa)	200
In-plane shear strength, S_{12} (MPa)	80
Longitudinal shear strength, S_L (MPa)	80
Transverse shear strength, S_T (MPa)	60

The element erosion of the CFRP phase is conducted through the concept of stiffness degradation., *i.e.*, when one type of the fiber/matrix failure occurs, the relative material properties will be degraded as shown in Table 4. In the present analysis, the material property degradation depends on four associated defined variables, as listed below: (*i*) the first variable noted HSNFTCRT represents the fiber-tensile failure mode; (*ii*) the second HSNFCCRT represents the fiber-compression failure mode; (*iii*) the third HSNMTCRT represents the matrix-tensile failure made and (*iv*) the fourth HSNMCCRT represents the matrix-compression failure mode.

Table 4. General formulation of 2D Hashin damage criteria for the CFRP phase [29].

Failure Criteria	Failure Mode	Associated Defined Variable	Reduced Material Properties
Fiber-tensile failure ($\sigma_{11} \geqslant 0$)	$D_{ft}^2 = \left(\dfrac{\sigma_{11}}{X_T}\right)^2 + \left(\dfrac{\sigma_{12}}{S_L}\right)^2$	HSNFTCRT	$E_1, E_2, G_{12}, v_{12} \to 0$
Fiber-compression failure ($\sigma_{11} < 0$)	$D_{fc}^2 = \left(\dfrac{\sigma_{11}}{X_C}\right)^2$	HSNFCCRT	$E_1, E_2, G_{12}, v_{12} \to 0$
Matrix-tensile failure ($\sigma_{22} \geqslant 0$)	$D_{mt}^2 = \left(\dfrac{\sigma_{22}}{Y_T}\right)^2 + \left(\dfrac{\sigma_{12}}{S_L}\right)^2$	HSNMTCRT	$E_2, G_{12} \to 0$
Matrix-compression failure ($\sigma_{22} < 0$)	$D_{mc}^2 = \left(\dfrac{\sigma_{22}}{2S_T}\right)^2 + \left[\left(\dfrac{Y_C}{2S_T}\right)^2 - 1\right]\dfrac{\sigma_{22}}{Y_C} + \left(\dfrac{\sigma_{12}}{S_L}\right)^2$	HSNMCCRT	$E_2, G_{12} \to 0$

During the FE computation, the material properties at each integration point were evaluated and degraded depending on which set of failure mode was used. If any failure index reached unity, the relevant material properties were automatically reduced to zero according to the implemented stiffness degradation scheme. The procedure was repeated until the occurrence of complete chip formation.

2.4. Interface Model

The interface model used here aims to serve as a technical control for the "Ti-to-CFRP" contact management and to facilitate the characterization of interface damage formation during the simulation. It should be noted that in real CFRP/Ti configurations, some of them do not exist with such a third layer and only combine together for machining. The interface layer was modeled by using cohesive interaction allowing interfacial-damage propagation between the two joint phases as a fracture mechanics phenomenon with a very small thickness (probably 5 μm). Note that the use of a small interface thickness aims to minimize its influence on some other machining responses such as Ti/CFRP chip separation modes, force generation, *etc.* The surface-based traction-separation law with linear softening was adopted to produce the mechanical responses of the cohesive interaction. The failure initiation law required to motivate damage among the interface layer is based on the quadratic stress criterion as illustrated in the following equation.

$$\left(\frac{\sigma_{33}}{t_n^f}\right)^2 + \left(\frac{\sigma_{13}}{t_s^f}\right)^2 + \left(\frac{\sigma_{23}}{t_t^f}\right)^2 = 1 \tag{10}$$

In which, σ_{33}, σ_{13} and σ_{23} represent the normal traction stress, and shear traction stresses in two directions, respectively; t_n^f, t_s^f and t_t^f denote the peak normal failure strength and peak shear failure strengths in two directions, respectively.

Once the damage onset was satisfied, the Benzeggagh-Kenane (BK) damage criteria [30] and potential law [31] were utilized to simulate damage evolution dominating the cohesive interaction, as presented in Equations (11) and (12). BK criteria are based on the energy dissipated due to failure considering traction-separation responses characterized in terms of released rate energies in the normal and two shear directions (G_n, G_s and G_t).

$$G_n^C + \left(G_s^C - G_n^C\right) \left(\frac{G_s + G_t}{G_n + G_s + G_t}\right)^\eta = G^C \tag{11}$$

$$\left(\frac{G_n}{G_n^C}\right)^\beta + \left(\frac{G_s}{G_s^C}\right)^\beta + \left(\frac{G_t}{G_t^C}\right)^\beta = 1 \tag{12}$$

In which, G_n, G_s and G_t are the released rate energies in the normal and two shear directions respectively; G_n^C, G_s^C and G_t^C are the critical values of released rate energies, η and β are the parameters of the laws.

In addition, the traction-separation law controlling the responses of the interface zone is specified by means of the stiffness in the normal and in the two shear directions (K_{nn}, K_{ss} and K_{tt}), the interface resistance in each direction (t_n^f, t_s^f and t_t^f) and the damage evolution through the critical released rate energy (G_n^C, G_s^C and G_t^C). The input parameters for the interface zone were adopted rigorously based on the comprehensive selection of relevant research works [32,33], as summarized in Table 5.

Table 5. Material properties of the interface zone.

Parameter	υ	K_{nn}	$K_{ss} = K_{tt}$	t_n^f	$t_s^f = t_t^f$	G_n^C	$G_s^C = G_t^C$
Value	0.33	2.0 GPa	1.5 GPa	60 MPa	80 MPa	0.78 N/mm	1.36 N/mm

3. Experimental Validation of the OC Model

Due to the significant lack of experimental studies concerning orthogonal cutting of hybrid CFRP/Ti composite in the open literature, the stack model was validated separately in terms of each constituent with experimental results from the literature. Moreover, since the CFRP/Ti interface was considered as a quick transition zone and a very small thickness was defined, its influence on some other machining-induced responses (e.g., Ti/CFRP chip formation mode, force generation, strain/stress) could be ignored. Besides, the input parameters for interface zone were also selected carefully from the literature, in which it had already been validated with experimental results and indicated good suitability for multi-material modeling. Therefore, the validation work was performed solely focused on Ti phase verification and CFRP phase verification by referring to the open literature. Each model was improved and refined carefully until it was capable to replicate consistent results with the experimental observations. For validation purposes, all the numerical simulations were run under the same cutting conditions as used in the literature.

The Ti-phase model was validated rigorously by means of force generation and chip morphology with experimental data from the literature [34–36], which were the commonly-used metrics for validations of metal cutting modeling. Figure 3 and Table 6 show the comparison between simulated and experimentally measured force magnitudes [34,35], and the calculated average errors among them, respectively. Note that the force magnitudes (in N/mm) were normalized as the ratio between the average level of force generation and the workpiece thickness. It was apparent that the simulated force generation yielded strong agreement with the experimental measurements for various cutting speeds and feed rates as depicted in Figure 3. Globally, the average errors between the simulated and experimental results were controlled below 10% as summarized in Table 6. Furthermore, comparisons between simulated and experimental chip morphologies were also performed as shown in Figure 4. The serrated chip morphology was validated by three parameters, *i.e.*, valley, peak, and pitch, as compared with the experimental results from the literature [36]. As shown in Table 7, the dimensions of the simulated chip morphologies matched well with the experimental ones. The above validations confirmed the credibility of the developed Ti constituent for Ti-phase modeling.

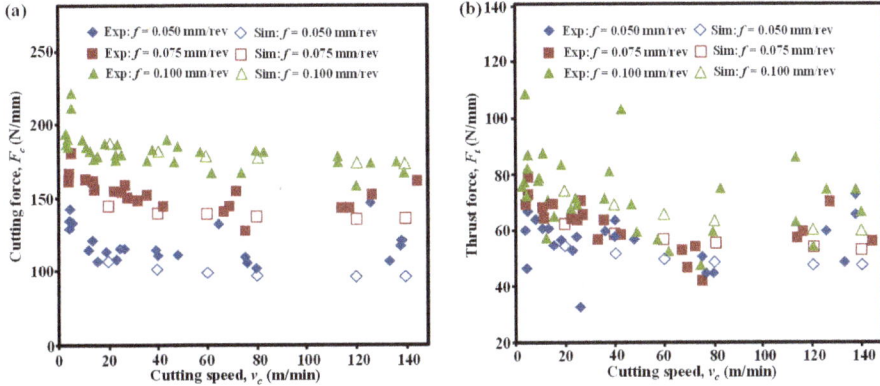

Figure 3. Comparison of the simulated (Sim.) and experimental (Exp.) force generation in Ti-phase cutting modeling [34,35] for different cutting speeds at feeds of 0.05, 0.075 and 0.100 mm/rev: (**a**) cutting force; (**b**) thrust force.

Table 6. Average error between the simulated and experimental force generations in Figure 3.

Test Condition	Average Error for Cutting Force (%)	Average Error For Thrust Force (%)
$f = 0.050$ mm/rev	−9.31	−8.32
$f = 0.075$ mm/rev	−6.86	−2.15
$f = 0.100$ mm/rev	+1.73	+1.36

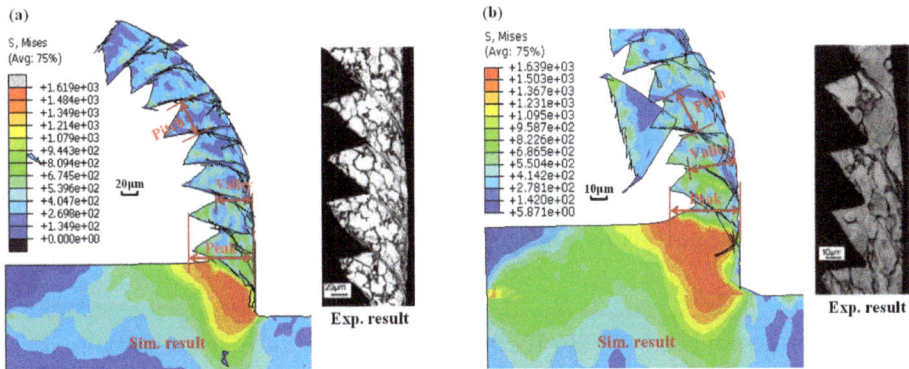

Figure 4. Comparison of the simulated (Sim.) and experimental (Exp.) chip morphologies in Ti-phase cutting modeling [36]: (**a**) test condition 1 ($v_c = 1200$ m/min and $f = 70$ μm/rev); (**b**) test condition 2 ($v_c = 4800$ m/min and $f = 35$ μm/rev) (Note: the symbol "S" in the figure represents the von Mises stress and the unit is MPa).

Table 7. Comparison between simulated (Sim.) and experimental (Exp.) chip geometries in Figure 4.

Sim./Exp.	Cutting Condition	Chip Morphology (μm)		
		Average Valley	Average Peak	Average Pitch
Simulation		47.2	95.6	52.3
Experiment	Test condition 1	50.3	105.7	68.2
Error		6.16%	9.56%	23.31%
Simulation		25.4	41.7	27.2
Experiment	Test condition 2	23.6	45.6	36.6
Error		−7.63%	8.56%	25.68%

Moreover, the CFRP-phase model was validated through the simplest manner of force generation (cutting force and thrust force) comparison. Figure 5 presents the comparative results of predicted and experimentally measured forces *versus* fiber orientation (θ). It was noticeable that the predicted force magnitudes of both cutting force and thrust force yielded a strong correlation and a consistent variation trend with the experimental results gained by Iliescu *et al.* [28], which confirmed sufficient credibility of the proposed CFRP-phase model.

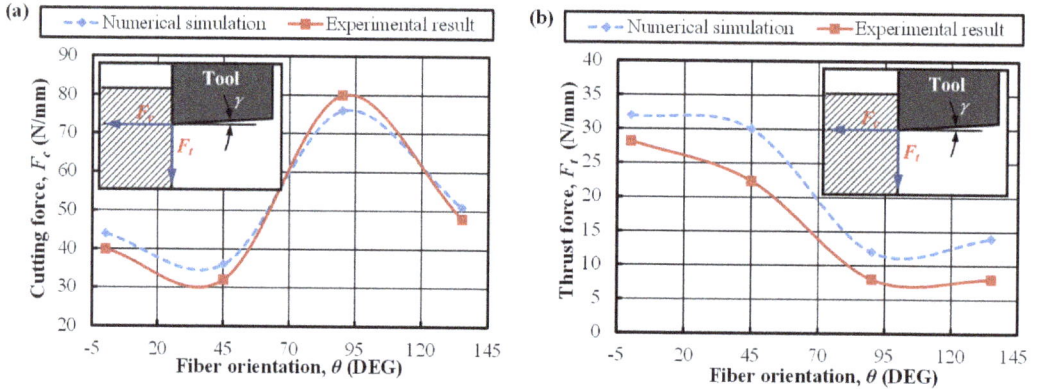

Figure 5. Comparison of the simulated and experimental force generations in CFRP-phase cutting modeling [28]: (a) cutting force; (b) thrust force (cutting condition: v_c = 6 m/min, f = 0.2 mm, α = 0°).

4. Numerical Results and Discussion

In this Section, numerical analyses concerning hybrid CFRP/Ti cutting are presented with the aim of better machining comprehension. Since Ti phase cutting and CFRP phase cutting require different optimal cutting parameters (v_c and f), the cutting conditions used in the FE simulations were selected based on the compromise selection of the optimal parametric range for both the two phases of machining. Special focus was made on the analyses of the cutting process, interface damage formation, and parametric effects on fiber/matrix damage extent.

4.1. Cutting Process Investigation

The cutting process of hybrid CFRP/Ti exhibits quite differently from the single-composite and single-metal cutting cases due to the multi-tool-work interaction domains. The disparate natures of each constituent make the chip separation modes more interrelated and coupled governing the bi-material interface consumption (BIC) [37,38]. The interface cutting commonly experienced changeable chip-separation modes and severe transitions of thermo/mechanical responses (e.g., force generation, cutting temperature, strain/stress flow). To reveal the key phenomena controlling CFRP/Ti cutting, the evolution of force generation *versus* cutting time and also the chip formation progression under fixed cutting conditions of v_c = 40 m/min, f = 0.2 mm/rev and θ = 0° are presented in Figures 6 and 7 respectively. The force generations in CFRP/Ti cutting were split into two components, *i.e.*, the cutting-force component (F_c) and the thrust-force component (F_t), which signify the tribological interactions between tool rake and chip surfaces, together with tool flank and machined surfaces, respectively. It was noticeable that typically three cutting stages referring to the Ti-phase cutting, interface cutting, and CFRP-phase cutting could be seen from the force signal variation and the chip-morphology evolution.

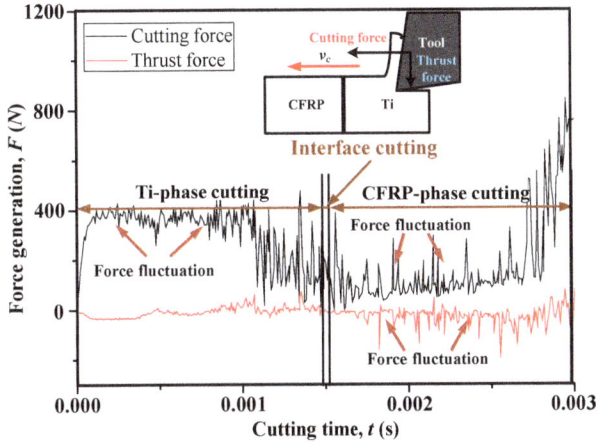

Figure 6. Evolution of the force generation *versus* cutting time when cutting hybrid CFRP/Ti composite (v_c = 40 m/min, f = 0.2 mm/rev and θ = 0°).

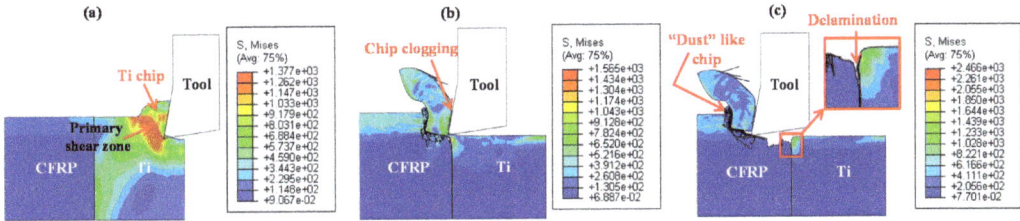

Figure 7. Chip-morphology evolution during CFRP/Ti cutting: (a) Ti-phase cutting; (b) interface cutting; (c) CFRP-phase cutting (v_c = 40 m/min, f = 0.2 mm/rev and θ = 0°) (Note: the symbol "S" in the figure represents the von Mises stress and the unit is MPa).

As depicted in Figures 6 and 7 when the tool edges initially cut into the Ti phase, material separation occurred through the elastic-plastic deformation mode that controlled the tool-Ti interaction area. The shearing actions arising from the thermo-mechanical coupling effects produced "continuous" chip morphology that flowed on the tool rake face. It should be noted that the Ti-chip morphology exhibited strong sensitivity to the input parameters (cutting parameters and tool geometries) during machining. Since very low cutting speed and positive tool rake angle were used in the current simulation case, the resected chip shape exhibited a more "continuous" rather than a "serrated" appearance. Under the fixed cutting conditions, the machining operation gradually achieved a steady state for which the forces generation approximately attained a stable variation condition. Besides, despite reaching a steady cutting process, high-frequency force fluctuation was also pronounced in Ti-phase cutting as described in Figure 6. The force-generation usually signifies the mechanical-energy consumption of tool-work interactions in cutting and presents a close relation with the inherent properties of the studied workpiece. Since the Ti alloy exhibited ductile behavior and low thermal conductivity, the chip separation typically involved serious thermo-plastic instability and shear localization in the primary cutting zone. Such a phenomenon would promote the quick occurrence of crack initiation and progression dominating the active cutting zones and result in the instability of the tool-work interaction controlling the chip removal process, as reflected in the cyclic force fluctuations. In addition, when the cutting time (t) approximately exceeded 1.0 ms, the cutting-force component was observed to suffer a gradual reduction. The physical phenomenon could be explained by the decreased uncut Ti-chip thickness contributing to the reduction of force resistance when the tool tended to finalize the Ti-phase cutting. With tool advancement, especially when the cutting edges cut into the interface region, the previously-resected Ti chip adhered on the tool rake face and replaced the tool edges for further chip separation. Due to the transition from tool-Ti interaction to tool-CFRP interaction, the cutting force magnitudes underwent quick drop throughout the interface cutting. In such circumstances, the chip separation mode shifted gradually from plastic-defamation into brittle-fracture, which inevitably resulted in the serious transfer of mechanical/physical loads exerted on the tool-work system. The harsh cutting conditions dominating the CFRP/Ti interface cutting was the key contributor promoting the severe damage formation in the bi-material interface.

With further cutting progression, the tool tip cut across the interface region completely and induced a large extent of delamination damage focused on the interface area as shown in the magnified view of Figure 7c. When the tool penetrated into the CFRP phase, material removal took place through successive ruptures aided by the diverse nature and uneven load sharing among the fiber/matrix systems. Since brittle fracture operated as the predominant cutting mode, the resected chip morphology was produced in the form of "discontinuous" shape (especially "dust" like appearance), as illustrated in Figure 7c. Concerning force generation, both the cutting force and thrust force signals were predicted to undergo severe fluctuation and high-frequency variation. However, the mechanisms governing the physical phenomena were disparate. For the cutting force variation, it was induced due to the crack initiation and evolution governing the chip separation process. In contrast, the thrust force fluctuation was mainly attributed to the intense bouncing-back effects on the tool flank surface arising from released carbon fibers in the machined surface [39].

Moreover, when the cutting tool approached the finish of the cutting process, a significant increase of cutting force generation became pronounced as portrayed in Figure 6. This abnormal phenomenon could be explained as follows. When the tool cuts into the CFRP phase, the Ti chip adhered on the tool rake face and replaced the cutting edges for further chip separation. In addition, the produced "dust" like composite chips also caused serious clogging on the head of the Ti chip. Due to these phenomena, the accumulated chip volume inevitably led to a dramatic rise of the cutting resistance, and hence the higher cutting force generation.

4.2. Inspection of Interface Damage Formation

In CFRP/Ti cutting, the bi-material interface consumption (BIC) signifies the tremendous mechanical/thermal energy transfer arising from disparate phase cutting, the interrelated chip separation modes, and mixed machining responses controlling the hybrid cutting process. BIC plays a significant role in affecting the final surface quality and subsurface damage formation. The interface cutting can be identified as the weakest region vulnerable to serious damage formation. Figure 8 presents the FE observation of interface damage morphology with OC of CFRP/Ti composite under the cutting conditions of v_c = 40 m/min, f = 0.2 mm/rev and θ = 0°. It was noticeable that the key characteristic of the interface damage was a "V-shape" like notch damage focused on the CFRP-Ti contact boundary. A large extent of delamination damage became pronounced in both tool-cutting direction (X-direction) and through-thickness direction (Y-direction) as depicted in Figure 8. However, the key manifestation was the severe tearing in the CFRP-phase boundary deviating far from the Ti-phase boundary. The physical phenomena demonstrated that the crack trajectory would experience a quick damage evolution concerning the tool cutting direction during the machining process. The key mechanisms controlling the crack trajectory and evolution in the bi-material interface were strongly influenced by two important factors. The first was the relative fracture toughness arising from the stacked constituents of the bi-material system. It is believed that the crack path prefers to take place nearer the more brittle constituent characterized by lower fracture toughness since it needs less energy to open and propagate the crack damage. In such a case, the preferred crack path occurring in the CFRP/Ti interface should exhibit more closely to the CFRP-phase boundary due to its relatively lower fracture toughness as compared to its counterpart. Another factor was the specific fracture mode encountered along the bi-material interface. Figure 9 shows the schematization of the crack path and fracture modes I and II in cutting. It could be seen that when the tool edges completely passed through the Ti-phase boundary, the load path exerting on the bi-material system was uniquely applied on the uncut CFRP and interface layers while the machined Ti-phase surface was deprived of load occupation. As such, the cutting load parallel to the tool cutting direction together with the thrust load perpendicular to the cutting direction would produce mixed damage modes (fracture modes I and II) on the interface region. Consequently, it resulted in a sole path evolution of crack trajectory approaching the cutting direction. This phenomenon inevitably gave rise to the so-called "delamination" damage.

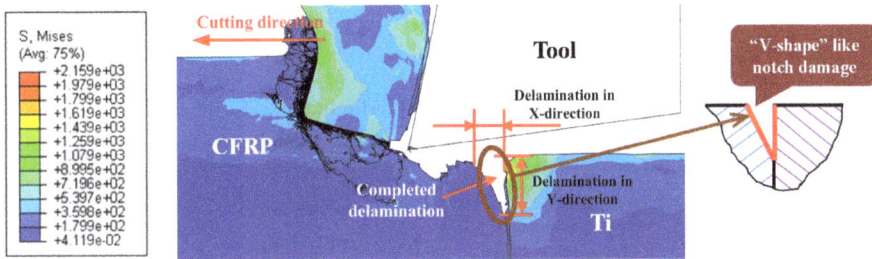

Figure 8. FE observation of interface damage when cutting hybrid CFRP/Ti composite ($v_c = 40$ m/min, $f = 0.2$ mm/rev and $\theta = 0°$) (Note: the symbol "S" in the figure represents the von Mises stress and the unit is MPa).

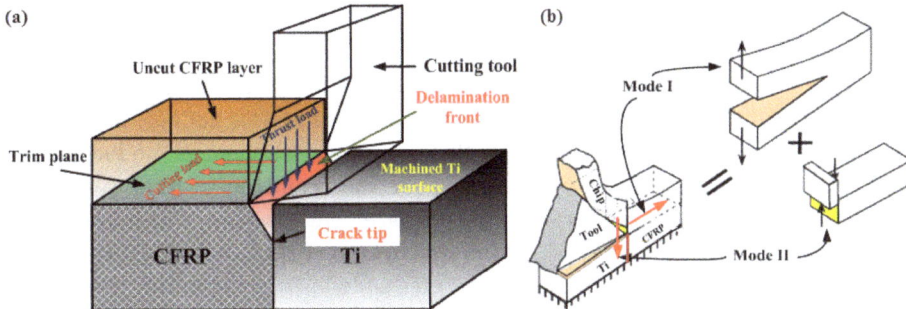

Figure 9. Scheme of (a) crack path in interface region and (b) fracture modes I and II in cutting.

To specify the dynamic process of delamination formation and also the affected interface zone (AIZ), nine interface nodes beneath the trimmed plane were selected to characterize their X-direction displacement *versus* cutting time (*t*) as shown in Figure 10. Table 8 summarizes the depth beneath the trim plane of the selected interface nodes. The AIZ studied here was devoted to clarifying the actual cutting time that influenced the interface damage formation during the orthogonal cutting process.

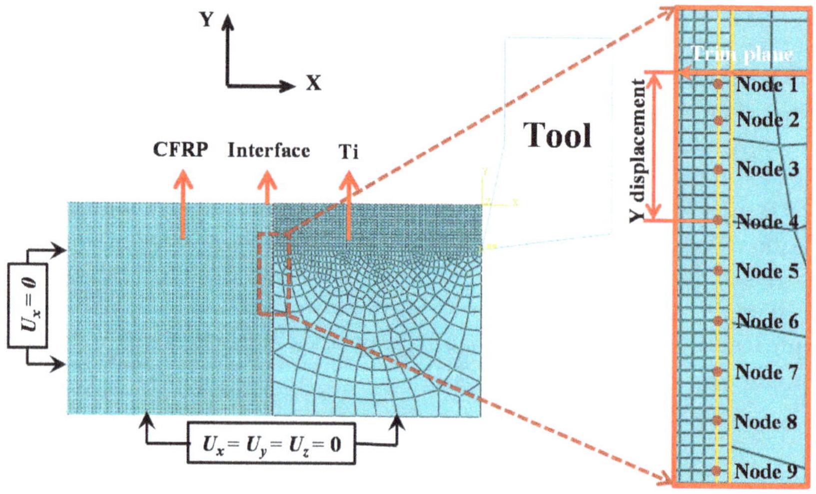

Figure 10. Scheme of selected interface nodes in the CFRP/Ti model.

Table 8. Y-displacement from the trim plane of selected nodes in the interface zone.

Number of Node	Depth beneath the Trim Plane (μm)	Number of Node	Depth beneath the Trim Plane (μm)
1	5	2	20
3	40	4	60
5	80	6	100
7	120	8	140
9	160	-	

Figure 11 then illustrates the X displacement evolution of the selected interface nodes with respect to the cutting time (t) under the fixed cutting conditions of v_c = 40 m/min, f = 0.2 mm/rev, and θ = 0°. It was apparent that the X displacements of the selected interface nodes generally underwent three variation stages, i.e., the initial variation stage, rapid variation stage, and steady variation stage during the total cutting duration (t = 3.0 ms). At the beginning of cutting, all the selected interface nodes were predicted to suffer slow-rate displacement variation, which indicated that the chip removal process initially exhibited slight/minor effects on the interface region. However, with the cutting progression, the machining operation gradually exerted significant influences on the output responses of the interface region with evidence that all the selected interface nodes began to experience high-extent displacement fluctuations and quickly enter into the rapid variation stage. The physical phenomena were predicted to take place in the Ti-phase cutting period approximately at a cutting time of 1.0 ms. Such evidence signified the advent of the affected interface zone (AIZ) prior to the interface-cutting period as illustrated in Figure 11. The occurrence of AIZ commonly implied that the CFRP/Ti interface had suffered dramatic influences arising from the Ti-chip removal process. As depicted in Figure 11a, the produced Ti chip caused severe bending and inclination effects on the interface zone and also the uncut CFRP-chip layer. The pronounced displacement variation in AIZ strongly demonstrated the appearance of serious detaching and separation concerning the interface zone. In addition, the interface nodes located near the trim plane typically underwent larger X displacement compared to their counterparts far away from the trim plane. The reason might be due to the different levels of influences arising from tool-work interactions. When the cutting time approximately reached 2.0 ms, the X displacements of the selected interface nodes gradually achieved the steady variation state, indicating the diminishing trend of the cutting influences on the interface region. For instance, as shown in Figure 11c, both the deamination lengths in X and Y directions reached their stable values. In such a case, the AIZ could be defined probably at a duration of $t \in$ [1.0, 2.0 ms]. With regard to the severe displacement fluctuation occurring in the steady variation state, the reason might be induced as being due to the bouncing-back effects from adjacent CFRP elements still affected by the local cutting operation. At the end of the cutting process, all the selected interface nodes approached attainment of constant displacements in the X direction except the interface node 1. The dramatic displacement increase of interface node 1 demonstrated that the node had been deleted completely from the interface zone.

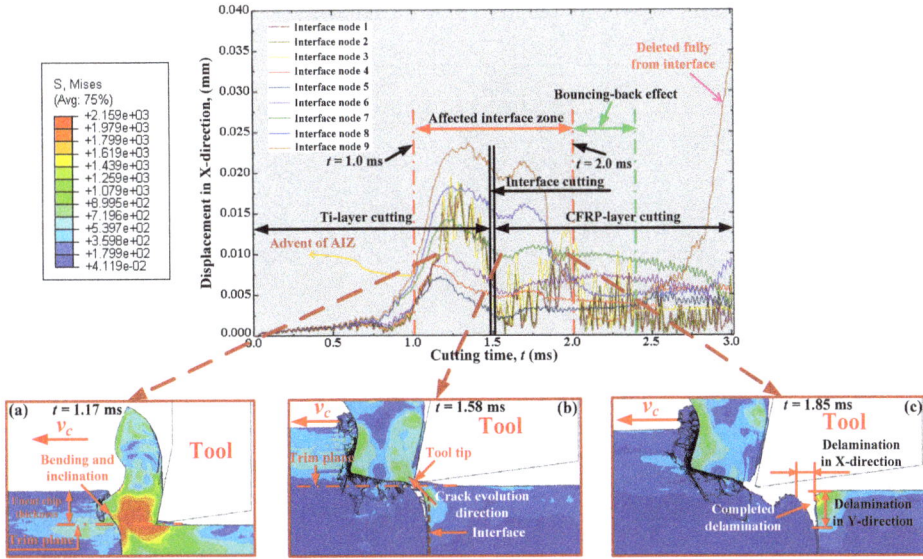

Figure 11. Displacement in X-direction of selected interface nodes *versus* cutting time (t): (**a**) cutting process at $t = 1.17$ ms; (**b**) cutting process at $t = 1.58$ ms and (**c**) cutting process at $t = 1.85$ ms.

To quantify the delamination extent occurring in the interface zone, three indicators, *i.e.*, D_X, D_Y, and S_D were introduced for evaluation in this study, where D_X signified the delamination length in the tool-cutting direction (X-direction), D_Y denoted the delamination length in the through-thickness direction (Y-direction) and S_D implied the delamination area of the damage zone as shown schematically in Figure 12. The delamination area (S_D) was defined as illustrated in Equation (13).

$$S_D = 0.5 \times D_X \times D_Y \tag{13}$$

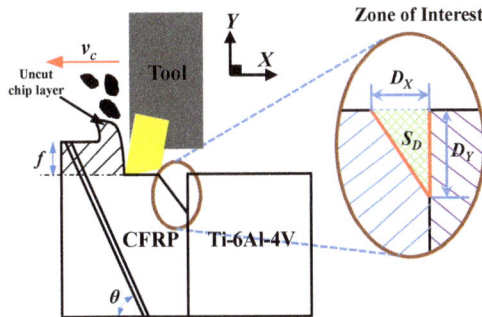

Figure 12. Scheme of delamination damage evaluation in CFRP/Ti cutting.

Figure 13 presents the evolution of interface delamination damage *versus* cutting time (t). The results confirmed that all the delamination indicators (D_X, D_Y and S_D) exhibited similar evolution trends as the X displacements of the selected interface nodes *versus* cutting time (t) in Figure 11. The comparison mutually demonstrated the reliability of the two numerical observations. Moreover, as depicted in Figure 13, from the geometrical consideration, the entire hybrid CFRP/Ti cutting can be divided into three stages referring to (*i*) Ti-phase cutting, (*ii*) interface cutting, and (*iii*) CFRP-phase cutting in terms of their respective cutting lengths. However, such a category actually couldn't reflect the real machining behavior of the bi-material system due to ignorance of the interactive cutting influences arising from each-phase cutting. In reality, the entire hybrid cutting process could be divided into three basic zones based on the interactive influences of each phase machining.

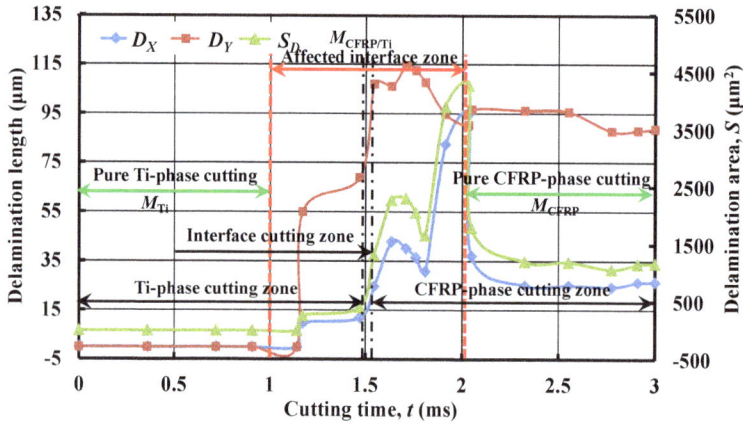

Figure 13. Evolution of delamination damage (D_X, D_Y and S_D) as function of cutting time (t).

(*i*) Pure Ti-phase cutting $t \in$ [0, 1.0 ms] signifies the cutting period concerning the absolute Ti-phase machining with minimum influence affecting the CFRP/Ti interface and the CFRP zone. Such a zone represents the machinability of the single Ti-alloy phase (M_{Ti}).

(*ii*) AIZ cutting $t \in$ [1.0, 2.0 ms] denotes the cutting duration concerning the machining variation of the bi-material interface arising from Ti-phase cutting and CFRP-phase cutting. Moreover, the AIZ cutting zone in reality represents the real machinability of the stacked CFRP/Ti material ($M_{CFRP/Ti}$).

(*iii*) Pure CFRP-phase cutting $t \in$ [2.0, 3.0 ms] implies the cutting time concerning the absolute CFRP-phase machining and reflects the machinability of the standard CFRP phase (M_{CFRP}).

During the pure Ti-phase cutting zone, all D_X, D_Y and S_D nearly remained constant values (probably zero values), indicating that no delamination occurred in the interface zone. With the continuation of the cutting process, especially when t reached the AIZ, the three delamination indicators totally suffered a dramatic increase with elevated cutting time. Such a phenomenon signified that the delamination damage was principally formed at the AIZ cutting period. Besides, the subsequent reduction of the delamination indicators in the AIZ might be caused by the bouncing-back effects from adjacent CFRP elements affected by the cutting operation. Moreover, with further increased t, especially in the pure CFRP-phase cutting zone, D_X, D_Y and S_D gradually reached their steady values, which meant that the later cutting process (post CFRP-phase cutting) generated less effect on the interface delamination formation.

Moreover, to inspect the effects of input variables on the AIZ and also on the interface delamination extent, a parametric study was performed by considering various cutting speeds and feed rates. Figures 14 and 15 present the acquired results. Note that in Figure 14, the symbol "I" indicates the cutting duration of AIZ and the interface delamination extent (D) was measured based on the D_Y length (delamination length in the through-thickness direction (Y-direction)). As can be seen in Figure 14, both v_c and f were found to have significant effects on the AIZ cutting time. Specifically, v_c exhibited a negative impact on the AIZ cutting duration while the impact of f was positive. A parametric combination of high v_c and low f commonly resulted in the lowest AIZ cutting duration. Moreover, as is evident in Figure 15, all the examined parametric variables showed pronounced influences on the interface delamination extent. The impact of v_c was found to be negative, i.e., an increase of v_c resulted in the reduction of D, while the impact of f was positive. The mechanisms controlling the variation phenomena could be associated with their specific influences on AIZ. This was because when v_c was elevated, typically a reduced cutting time of AIZ was achieved as illustrated in Figure 14. As such, the decreased AIZ inevitably alleviated the damage extent of induced interface delamination. In contrast, an increase of f usually led to an increased AIZ (as depicted in Figure 14), and hence gave rise to a higher extent of interface delamination as shown in Figure 15b. Therefore, from the above analyses, for minimizing the extent of interface delamination, a high cutting speed and a low feed rate are strongly preferred.

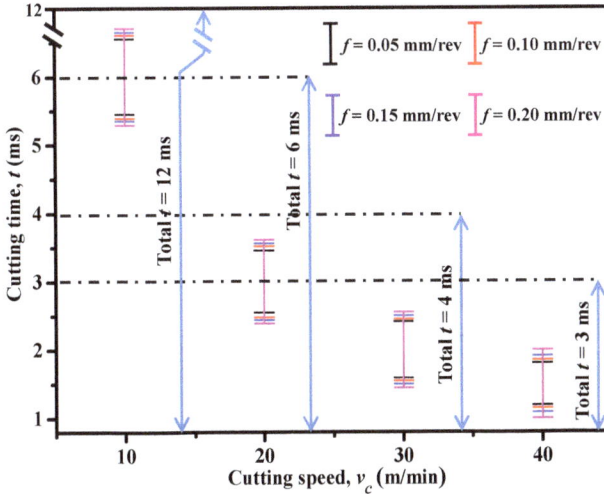

Figure 14. Effects of cutting speed (v_c) and feed rate (f) on cutting duration of AIZ (Note: "I" indicates the cutting duration of AIZ) ($\theta = 0°$).

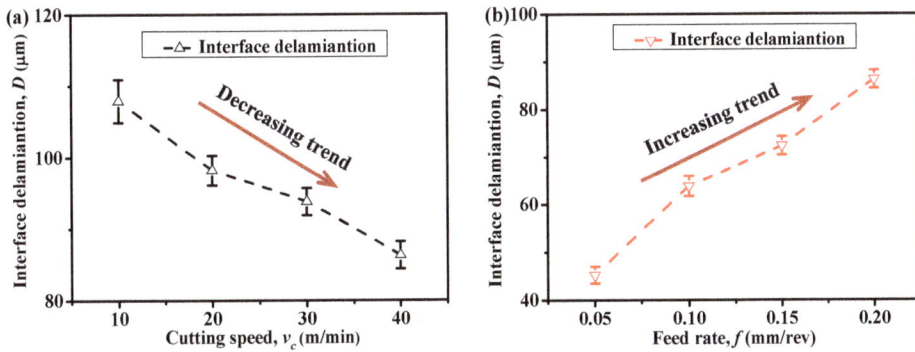

Figure 15. Interface delamination extent (D) in function of: (**a**) cutting speed (v_c) ($f = 0.2$ mm/rev and $\theta = 0°$) and (**b**) feed rate (f) ($v_c = 40$ m/min and $\theta = 0°$).

4.3. Subsurface Damage Study

Apart from the severe interface imperfection, the subsurface damage occurring in the composite phase is also a particular concern when machining hybrid CFRP/Ti composite. As discussed above, the composite-phase damage takes place following four types of fiber/matrix failure, i.e., fiber compression damage, fiber tensile damage, matrix compression damage, and matrix tensile damage. To inspect the subsurface composite damage resulting from the hybrid cutting operation, a parametric analysis was performed. The fiber/matrix failure was measured as the largest distance from the machined CFRP surface to the deepest fully damaged area, as shown schematically in Figure 16. The fiber/matrix failure extent was predicted and studied *versus* the fiber orientation (θ) under the fixed cutting conditions of v_c = 40 m/min and f = 0.2 mm/rev as depicted in Figure 17. It was noticeable that the θ exhibited significant influence on the evolution of the various types of fiber/matrix damage during the chip removal process. The subsurface damage extent increased with elevated θ, especially the elevation became pronounced when θ was above 60° as depicted in Figure 17. All the damage types tended to reach their maximum extent at θ = 90°. With θ's further increase, the subsurface damage appeared to suffer a slight decrease. Such findings of θ's effect on composite damage agreed well with the observations of Arola *et al.* [17], and Wang and Zhang [39] when cutting CFRP laminates. The mechanisms controlling the mentioned phenomena could be attributed to the change of failure modes from bending and crushing in case of θ = 0° to a fracture by compression and interfacial-shearing mode for positive fiber orientation (0° < θ ⩽ 90°) [40], which inevitably led to in-depth fiber/matrix damage. For θ above 90°, chip-separation modes were shifted to be dominated by pressing, inter-laminar shearing, matrix cracking, and fiber/matrix interface debonding. This situation would result in less energy-consumption for chip separation and lower cutting resistance for machining. As a result, less extent of damage formation was promoted.

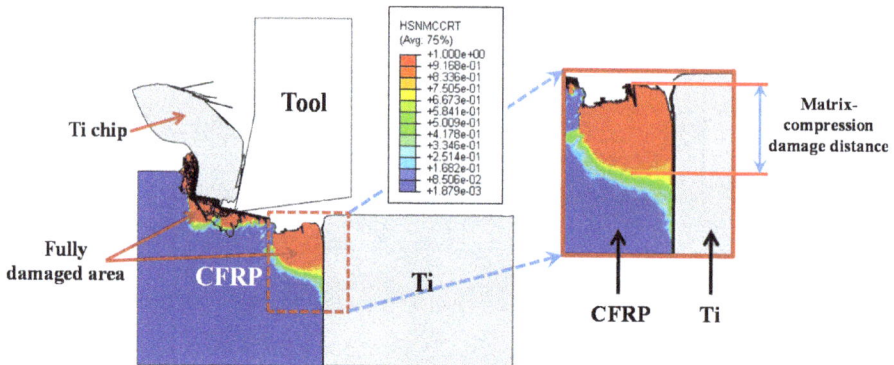

Figure 16. Scheme of damage measurement in the CFRP phase (Note: HSNMCCRT represents the matrix-compression failure mode).

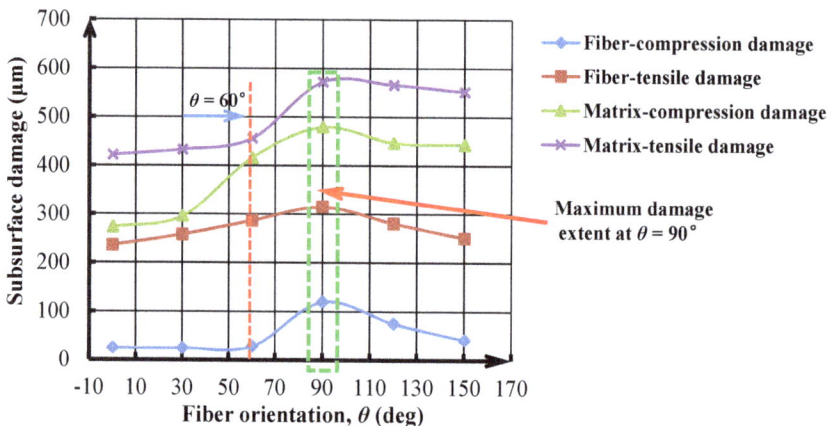

Figure 17. Fiber/matrix damage extent *versus* fiber orientation (θ) (v_c = 40 m/min and f = 0.2 mm/rev).

In addition, Figure 18 also shows the parametric effects of v_c and f on the cutting-induced composite-phase damage. Note that the composite-phase damage (D_{CFRP}) was evaluated based on the type of fiber/matrix failure that caused the largest extent of damage (e.g., as shown in

Figure 17, matrix-compression damage caused the maximum damage extent). Afterward, D_{CFRP} was measured as the largest distance from the machined CFRP surface to the deepest fully damaged area. Each measurement was repeated three times in order to ensure sufficient credibility for the acquired results. As shown in Figure 18, it was noticeable that the f had remarkable effects on the composite-phase damage in such a manner that a small increase of f resulted in a dramatically elevated D_{CFRP}. In contrast, an increase of v_c typically led to a direct reduction of D_{CFRP}. The phenomena implied that the use of high-speed cutting (HSC) might benefit the reduction of composite-phase damage formation when machining hybrid CFRP/Ti composite. Therefore, based on the above analyses, the optimum cutting parameters for composite damage minimization should consist of small fiber orientation, high cutting speed, and low feed rate when machining this bi-material system.

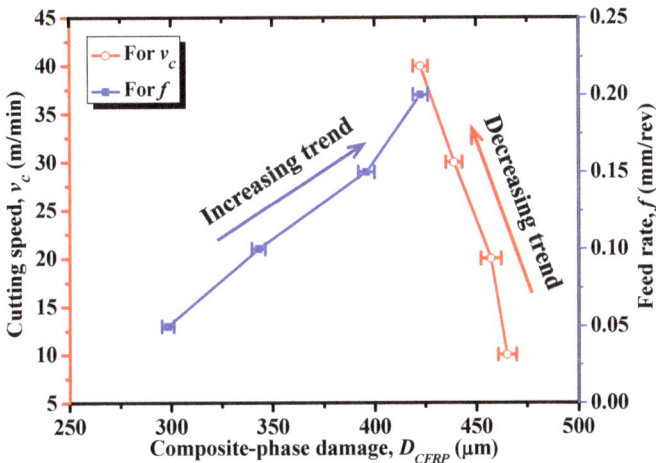

Figure 18. Effects of cutting speed (v_c) and feed rate (f) on composite-phase damage extent (D_{CFRP}) ($\theta = 0°$).

5. Conclusions

In this paper, a finite element model was developed to simulate the chip formation process with orthogonal cutting of hybrid CFRP/Ti composite. The proposed OC model attempted to address the fundamental mechanisms dominating hybrid composite machining. The numerical studies provided a better comprehension of various cutting phenomena induced in the hybrid cutting process. Based on the results acquired, key conclusions can be drawn as follows.

(1) The disparate natures of each stacked constituent resulted in the hybrid cutting process exhibiting three distinct cutting stages, *i.e.*, Ti-phase cutting, interface cutting, and CFRP-phase cutting. The chip separation modes controlling the bi-material cutting comprise both elastic-plastic deformation and brittle fracture. The changeable chip removal mechanisms typically resulted in extremely harsh conditions dominating the interface cutting. Severe force fluctuation was observed as the main characteristic governing the hybrid CFRP/Ti cutting. Serious chip adhesion on the tool rake face was found to be the key factor contributing to the dramatic increase of cutting force generation in post CFRP-phase machining.

(2) Numerical results highlighted the occurrence of severe delamination damage concerning the CFRP/Ti interface. FE analyses revealed that the initiation and evolution of interface delamination primarily took place at the cutting period of AIZ. The formation of CFRP/Ti interface damage involved a series of detaching and tearing variations governing the AIZ cutting time. The key mechanisms governing the interface damage formation could be attributed to the coupling effects of fracture modes I and II arising from the unique mechanical loads exerted on the bi-material interface region. The key

morphological characteristic of the interface delamination was a "V-shape" like notch focused on the CFRP-Ti contact boundary.

(3) Parametric studies confirmed the significant role of the cutting speed and feed rate in affecting the AIZ cutting duration. The input variables (v_c and f) indeed exhibited pronounced effects on the interface delamination extent via the manner of their influences on the AIZ. For minimizing severe interface damage formation, a parametric combination of high cutting speed and low feed rate is strongly preferred.

(4) The fiber/matrix damage promoted in hybrid CFRP/Ti cutting was carefully studied *versus* input variables. The numerical results pointed out the crucial role of fiber orientation, feed rate, and cutting speed in affecting induced composite-phase damage. For composite-phase damage minimization, the optimum cutting parameters should comprise small fiber orientation, high cutting speed, and low feed rate when machining hybrid CFRP/Ti composite.

(5) Another key finding revealed in this paper showed that the entire hybrid cutting could be physically divided into three cutting zones from the viewpoint of the interrelated influences arising from each-phase machining: (*i*) pure Ti-phase cutting, (*ii*) AIZ cutting and (*iii*) pure CFRP-phase cutting. Also, among them, the AIZ cutting in reality reflects the real machinability of the stacked CFRP/Ti material ($M_{CFRP/Ti}$), which can be considered as the most difficult-to-cut zone due to the occurrence of interrelated cutting influences and the existence of the weakest interface zone. In such a case, to fundamentally improve the machinability of the hybrid CFRP/Ti composite, special focus should be made concerning the in-depth mechanism investigation and cutting-parameter optimization of the AIZ cutting zone. In the future, more systematic studies should be performed to address precisely the aforementioned issues.

Acknowledgments: The authors gratefully acknowledge the financial support of China Scholarship Council (CSC) (Contract No. 201306230091).

Author Contributions: Jinyang Xu developed the finite element (FE) model and wrote the paper; Mohamed El Mansori proposed the idea of numerical analysis and made the corrections of the paper.

Conflicts of Interest: The authors declare no conflict of interest.

Abbreviations

The following abbreviations are used in this manuscript:

AIZ	Affected interface zone
BIC	Bi-material interface consumption
CFRP	Carbon Fiber Reinforced Polymer
CZ	Cohesive zone
D	Interface delamination extent
D_{CFRP}	Composite-phase damage
D_X	Delamination length in the tool-cutting direction (X direction)
D_Y	Delamination length in the through-thickness direction (Y direction)
EHM	Equivalent homogeneous material
f	Feed rate
F_c	Cutting force
FE	Finite element
F_t	Thrust force
JC	Johnson-Cook
M_{CFRP}	Machinability of pure CFRP-phase cutting
$M_{CFRP/Ti}$	Machinability of stacked CFRP/Ti composite
M_{Ti}	Machinability of pure Ti-phase cutting
OC	Orthogonal cutting

S_D	Delamination area of the damage zone
α	Tool rake angle
θ	Fiber orientation
γ	Tool clearance angle
t	Cutting time
v_c	Cutting speed

References

1. Ramulu, M.; Branson, T.; Kim, D. A study on the drilling of composite and titanium stacks. *Compos. Struct.* **2001**, *54*, 67–77. [CrossRef]
2. Brinksmeier, E.; Janssen, R. Drilling of multi-layer composite materials consisting of carbon fiber reinforced plastics (CFRP), titanium and aluminum alloys. *CIRP Ann. Manuf. Technol.* **2002**, *51*, 87–90. [CrossRef]
3. Kim, D.; Ramulu, M. Study on the drilling of titanium/graphite hybrid composites. *J. Eng. Mater. Technol. Trans. ASME* **2007**, *129*, 390–396. [CrossRef]
4. Park, K.H.; Beal, A.; Kim, D.; Kwon, P.; Lantrip, J. A comparative study of carbide tools in drilling of CFRP and CFRP-Ti stacks. *J. Manuf. Sci. Eng. Trans. ASME* **2014**, *136*. [CrossRef]
5. Liu, Z.; Xu, J.; Han, S.; Chen, M. A coupling method of response surfaces (CRSM) for cutting parameters optimization in machining titanium alloy under minimum quantity lubrication (MQL) condition. *Int. J. Precis. Eng. Manuf.* **2013**, *14*, 693–702. [CrossRef]
6. Nurul Amin, A.K.M.; Ismail, A.F.; Nor Khairusshima, M.K. Effectiveness of uncoated WC–Co and PCD inserts in end milling of titanium alloy—Ti–6Al–4V. *J. Mater. Process. Technol.* **2007**, *192*, 147–158. [CrossRef]
7. Ezugwu, E.O.; Bonney, J.; Da Silva, R.B.; Cakir, O. Surface integrity of finished turned Ti–6Al–4V alloy with PCD tools using conventional and high pressure coolant supplies. *Int. J. Mach. Tools Manuf.* **2007**, *47*, 884–891. [CrossRef]
8. Xu, J.; An, Q.; Cai, X.; Chen, M. Drilling machinability evaluation on new developed high-strength T800S/250F CFRP laminates. *Int. J. Precis. Eng. Manuf.* **2013**, *14*, 1687–1696. [CrossRef]
9. Wei, Y.; An, Q.; Cai, X.; Chen, M.; Ming, W. Influence of fiber orientation on single-point cutting fracture behavior of carbon-fiber/epoxy prepreg sheets. *Materials* **2015**, *8*, 6738–6751. [CrossRef]
10. Xu, J.; An, Q.; Chen, M. A comparative evaluation of polycrystalline diamond drills in drilling high-strength T800S/250F CFRP. *Compos. Struct.* **2014**, *117*, 71–82. [CrossRef]
11. Xu, J.; Mkaddem, A.; El Mansori, M. Recent advances in drilling hybrid FRP/Ti composite: A state-of-the-art review. *Compos. Struct.* **2016**, *135*, 316–338. [CrossRef]
12. Isbilir, O.; Ghassemieh, E. Comparative study of tool life and hole quality in drilling of CFRP/titanium stack using coated carbide drill. *Mach. Sci. Technol.* **2013**, *17*, 380–409. [CrossRef]
13. Arola, D.; Ramulu, M. Orthogonal cutting of fiber-reinforced composites: a finite element analysis. *Int. J. Mech. Sci.* **1997**, *39*, 597–613. [CrossRef]
14. Ramesh, M.V.; Seetharamu, K.N.; Ganesan, N.; Sivakumar, M.S. Analysis of machining of FRPs using FEM. *Int. J. Mach. Tools Manuf.* **1998**, *38*, 1531–1549. [CrossRef]
15. Mahdi, M.; Zhang, L. A finite element model for the orthogonal cutting of fiber-reinforced composite materials. *J. Mater. Process. Technol.* **2001**, *113*, 373–377. [CrossRef]
16. Mkaddem, A.; Demirci, I.; El Mansori, M. A micro-macro combined approach using FEM for modelling of machining of FRP composites: Cutting forces analysis. *Compos. Sci. Technol.* **2008**, *68*, 3123–3127. [CrossRef]
17. Arola, D.; Sultan, M.B.; Ramulu, M. Finite element modeling of edge trimming fiber reinforced plastics. *J. Manuf. Sci. Eng. Trans. ASME* **2002**, *124*, 32–41. [CrossRef]
18. Sima, M.; Özel, T. Modified material constitutive models for serrated chip formation simulations and experimental validation in machining of titanium alloy Ti-6Al-4V. *Int. J. Mach. Tools Manuf.* **2010**, *50*, 943–960. [CrossRef]
19. Johnson, G.R.; Cook, W.H. A constitutive model and data for metals subjected to large strains, high strain rates and high temperatures. In Proceedings of the 7th International Symposium on Ballistics, the Hague, The Netherlands, 19–21 April 1983.

20. Johnson, G.R.; Cook, W.H. Fracture characteristics of three metals subjected to various strains, strain rates, temperatures and pressures. *Eng. Fract. Mech.* **1985**, *21*, 31–48. [CrossRef]

21. Lesuer, D. Experimental Investigation of Material Models for Ti-6Al-4V and 2024-T3. Avaliable online: https://e-reports-ext.llnl.gov/pdf/236167.pdf (accessed on 29 December 2015).

22. Xi, Y.; Bermingham, M.; Wang, G.; Dargusch, M. Finite element modeling of cutting force and chip formation during thermally assisted machining of Ti6Al4V alloy. *J. Manuf. Sci. Eng. Trans. ASME* **2013**, *135*. [CrossRef]

23. Abaqus Version 6.7. Abaqus Theory Manuals. Avalibale online: http://www.egr.msu.edu/software/abaqus/Documentation/docs/v6.7/books/stm/default.htm?startat=ch01s01ath01.html (accessed on 30 December 2015).

24. Hillerborg, A.; Modeer, M.; Petersson, P.E. Analysis of crack formation and crack growth in concrete by means of fracture mechanics and finite elements. *Cem. Concr. Res.* **1976**, *6*, 773–781. [CrossRef]

25. William, J.; Callister, D. *Materials Science and Engineering: An Introduction*; Wiley: New York, NY, USA, 1994.

26. Iliescu, D.; Gehin, D.; Nouari, M.; Girot, F. Damage modes of the aeronautic multidirectional carbon/epoxy composite T300/914 in machining. *Int. J. Mater. Prod. Technol.* **2008**, *32*, 118–135. [CrossRef]

27. Iliescu, D. Approches Experimentale et Numerique de L'usinage a sec des Composites Carbone-Epoxy. Ph.D. Thesis, Arts et Métiers ParisTech, Paris, French, 2008. (In French).

28. Iliescu, D.; Gehin, D.; Iordanoff, I.; Girot, F.; Gutiérrez, M.E. A discrete element method for the simulation of CFRP cutting. *Compos. Sci. Technol.* **2010**, *70*, 73–80. [CrossRef]

29. Hashin, Z.; Rotem, A. A fatigue failure criterion for fiber reinforced materials. *J. Compos. Mater.* **1973**, *7*, 448–464. [CrossRef]

30. Benzeggagh, M.L.; Kenane, M. Measurement of mixed-mode delamination fracture toughness of unidirectional glass/epoxy composites with mixed-mode bending apparatus. *Compos. Sci. Technol.* **1996**, *56*, 439–449. [CrossRef]

31. Lapczyk, I.; Hurtado, J.A. Progressive damage modeling in fiber-reinforced materials. *Compos. Part A Appl. Sci. Manuf.* **2007**, *38*, 2333–2341. [CrossRef]

32. Aymerich, F.; Dore, F.; Priolo, P. Prediction of impact-induced delamination in cross-ply composite laminates using cohesive interface elements. *Compos. Sci. Technol.* **2008**, *68*, 2383–2390. [CrossRef]

33. Savani, E.; Pirondi, A.; Carta1, F.; Nogueira, A.C.; Hombergsmeier, E. Modeling delamination of Ti-CFRP interfaces. In Proceedings of the 15th European Conference on Composite Materials, Venice, Italy, 24–28 June 2012.

34. Cotterell, M.; Byrne, G. Dynamics of chip formation during orthogonal cutting of titanium alloy Ti-6Al-4V. *CIRP Ann. Manuf. Technol.* **2008**, *57*, 93–96. [CrossRef]

35. Liu, R.; Melkote, S.; Pucha, R.; Morehouse, J.; Man, X.; Marusich, T. An enhanced constitutive material model for machining of Ti-6Al-4V alloy. *J. Mater. Process. Technol.* **2013**, *213*, 2238–2246. [CrossRef]

36. Gentel, A.; Hoffmeister, H.W. Chip formation in machining Ti6A14V at extremely high cutting speeds. *CIRP Ann. Manuf. Technol.* **2001**, *50*, 14–17.

37. Xu, J.; El Mansori, M. Cutting modeling using cohesive zone concept of titanium/CFRP composite stacks. *Int. J. Precis. Eng. Manuf.* **2015**, *16*, 2091–2100. [CrossRef]

38. Xu, J.; El Mansori, M. Numerical modeling of stacked composite CFRP/Ti machining under different cutting sequence strategies. *Int. J. Precis. Eng. Manuf.* **2016**, *17*, 99–107.

39. Wang, X.M.; Zhang, L.C. An experimental investigation into the orthogonal cutting of unidirectional fibre reinforced plastics. *Int. J. Mach. Tools Manuf.* **2003**, *43*, 1015–1022. [CrossRef]

40. Rahman, M.; Ramakrishna, S.; Prakash, J.; Tan, D. Machinability study of carbon fiber reinforced composite. *J. Mater. Process. Technol.* **1999**, *89–90*, 292–297. [CrossRef]

7

Preparation and Characterization of Inorganic PCM Microcapsules by Fluidized Bed Method

Svetlana Ushak [1,2,†,*], M. Judith Cruz [1,†], Luisa F. Cabeza [3,†] and Mario Grágeda [1,2,†]

Academic Editor: A. Inés Fernández

[1] Department of Chemical Engineering and Mineral Processing and Center for Advanced Study of Lithium and Industrial Minerals (CELiMIN), University of Antofagasta, Av. Universidad de Antofagasta 02800, Campus Coloso, Antofagasta 127300, Chile; mjudith.cruz@uantof.cl (M.J.C.); mario.grageda@uantof.cl (M.G.)

[2] Solar Energy Research Center (SERC-Chile), Av Tupper 2007, Piso 4, Santiago 8370451, Chile

[3] GREA Innovació Concurrent, Edifici CREA, Universitat de Lleida, Pere de Cabrera s/n, Lleida 25001, Spain; lcabeza@diei.udl.cat

* Correspondence: svetlana.ushak@uantof.cl

† These authors contributed equally to this work.

Abstract: The literature shows that inorganic phase change materials (PCM) have been very seldom microencapsulated, so this study aims to contribute to filling this research gap. Bischofite, a by-product from the non-metallic industry identified as having good potential to be used as inorganic PCM, was microencapsulated by means of a fluidized bed method with acrylic as polymer and chloroform as solvent, after compatibility studies of both several solvents and several polymers. The formation of bischofite and pure $MgCl_2 \cdot 6H_2O$ microcapsules was investigated and analyzed. Results showed an efficiency in microencapsulation of 95% could be achieved when using 2 min of fluidization time and 2 kg/h of atomization flow. The final microcapsules had excellent melting temperatures and enthalpy compared to the original PCM, 104.6 °C and 95 J/g for bischofite, and 95.3 and 118.3 for $MgCl_2 \cdot 6H_2O$.

Keywords: phase change material; inorganic; microencapsulation; fluidization; bischofite; $MgCl_2 \cdot 6H_2O$

1. Introduction

In recent years, the use of thermal energy storage (TES) with latent heat storage has become a very popular topic within researchers. The main advantage of latent heat storage is the high storage density in small temperature intervals, showing very big potential to be used in building applications [1]. However, in most cases, the materials used in latent heat storage, known as phase change materials (PCM), need to be encapsulated to avoid leakage when it is in the liquid phase. There are three means of encapsulation: micro-encapsulation, macro-encapsulation and shape-stabilization [2], although recently nano-encapsulation has also grown in interest [3,4].

Microencapsulation is the encapsulation in particles smaller than 1 mm in diameter, known as microcapsules, microparticles, microspheres [5]. Microencapsulation serves several purposes, such as holding the liquid PCM and preventing changes of its composition through contact with the environment; improving material compatibility with the surrounding, through building a barrier; improving handling of the PCM in a production; reducing external volume changes, which is usually also a positive effect for an application; improving heat transfer to the surrounding through its large surface to volume ratio; and improving cycling stability since phase separation is restricted to microscopic distances.

Microencapsulation processes can be categorized into two groups: physical processes and chemical processes. Physical methods include spray cooling, spray drying, and fluidized bed processes; chemical processes include *in-situ* polymerization (interfacial polycondensation, suspension polymerization, and emulsion polymerization), complex coacervation, sol-gel method, and solvent extraction/evaporation method. Physical methods are limited by their granulated sizes thus making them useful for producing microencapsulated PCM particles, and chemical methods can produce much smaller encapsulated PCM particles [3,5–7]. Hawlader *et al.* [8] reported a substantial drop in heat storage capacity with the physical methods as compared to that of chemical methods.

In 2011, Cabeza *et al.* [1] claimed that only hydrophobic PCM could be microencapsulated. In 2015, Su *et al.* [3] claimed that inorganic PCM micro-/nano-encapsulation is limited to the solvent extraction/evaporation method, probably based in the existence of the study from Salaun *et al.* [9]. In 2015, Khadiran *et al.* [5] and Giro-Paloma *et al.* [6] reviewed only the encapsulation techniques of organic PCM. Therefore, there is a research gap on finding ways to encapsulate inorganic PCM.

At the time of writing this paper, microencapsulation of inorganic PCM can be found in very few papers. For example, Salaun *et al.* [9] microencapsulated sodium phosphate dodecahydrate (DSP) by solvent evaporation-precipitation method using various organic solvents and cellulose acetated butyrate (CAB) crosslinked by methylene di-isocyanate (MDI) as coating polymer. Those authors identified that the nature of the solvent was one of the most influencing parameters in the final surface morphology of the microcapsule. Similarly, Huang *et al.* [10] microencapsulated disodium hydrogen phosphate heptahydrate ($Na_2HPO_4 \cdot 7H_2O$) by means of the suspension copolymerization-solvent volatile method with modified PMMA as coating polymer. Hassabo *et al.* [11] microencapsulated six different salt hydrates (calcium nitrate tetrahydrate, calcium chloride hexahydrate, sodium sulphate decahydrate, disodium hydrogen phosphate dodecahydrate, ferric nitrate nonahydrate, and manganese (II) nitrate hexahydrate) by polycondensation of tetraethoxysilane.

Moreover, microencapsulated PCM are composed of two main parts (Figure 1), the core (the PCM) and the shell (usually a polymer). However, the process of microencapsulation always involves two solvents, that should not be miscible between them (Figure 2 shows the process of microsuspension polymerization as example). So the problem of microencapsulating inorganic PCM is that water is always used as solvent in microencapsulation processes and salt hydrates are soluble in water.

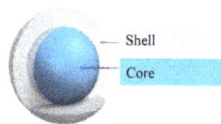

Figure 1. Structure of a microencapsulated phase change materials (PCM) (adapted from [5]).

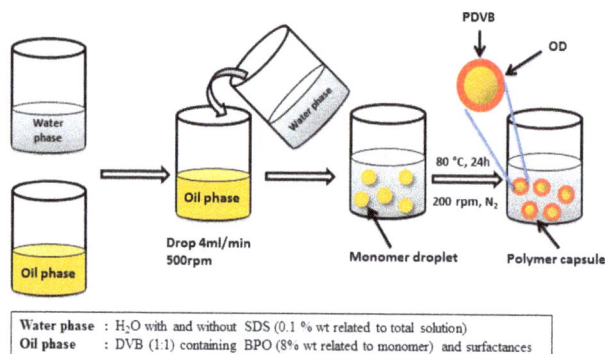

Figure 2. Microsuspension polymerization process [12].

The aim of this paper was to microencapsulate inorganic PCM. To achieve this objective, an encapsulation method had to be selected taking into consideration that not only the PCM or the shell material (polymer) would influence the process, but also the solvent to be used during the encapsulation process (Figure 3); therefore, between the available methods, a fluidized bed process was selected.

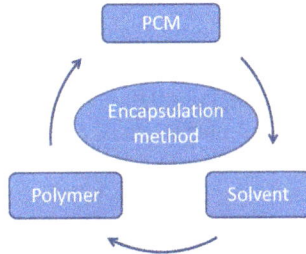

Figure 3. Material selection influencing an encapsulation method.

Encapsulation using fluidized bed has been used extensively in areas such as food [13–15] and agriculture [16] but to the author's knowledge it has never been used for microencapsulated PCM. Notoriously, fluidization of PCM was used as early as 1988 by Sozen *et al.* [17] to increase the heat storage efficiency of Glauber salt, an inorganic PCM. In this study, fluidization provided enhanced heat transfer to or from the storage medium and resulted in a steady-state heat storage efficiency of about 60% after repeated heating and cooling cycles. However, this technology was not used again until 2013 when Izquierdo-Barrientos *et al.* started a series of papers on the study of thermal energy storage in a fluidized bed of PCM [18–20]. The results showed that fluidized PCM can increase the efficiency of the system.

2. Results

2.1. Compatibility Studies

The results on the solubility of the considered polymers with the solvents are presented in Table 1. Results show that polypropylene is not soluble in any tested solvents, nor in bar form neither in prill, probably due to the reticulation within the polymer; therefore this polymer was disregarded. Polystyrene is soluble in the four considered organic solvents, requiring solvent volumes over 60%. Acrylic was partially soluble in chloroform and slightly soluble in THR. Finally, the resin epoxy was non-soluble in the four solvents tested. From these results, it could be concluded that polystyrene and acrylic are the best polymers to encapsulate PCM, both using chloroform as solvent in the percentages shown in Table 2.

Table 1. Solubility of polymers into solvents.

Polymer		% Polymer	Chloroform	THF	Acetone	Xylene
Polypropylene	Bar	10	Non-soluble	Non-soluble	Non-soluble	Non-soluble
		40	Non-soluble	Non-soluble	Non-soluble	Non-soluble
	Prill	10	Non-soluble	Non-soluble	Non-soluble	Non-soluble
		40	Non-soluble	Non-soluble	Non-soluble	Non-soluble
Polystyrene		10	Soluble	Soluble	Soluble	Soluble
		40	Soluble	Soluble	Soluble	Soluble
Acrylic		10	Soluble	Slightly soluble	Non-soluble	Non-soluble
		40	Partially soluble	Slightly soluble	Non-soluble	Non-soluble
Resin epoxy		10	Non-soluble	Non-soluble	Non-soluble	Non-soluble
		40	Non-soluble	Non-soluble	Non-soluble	Non-soluble

Table 2. Percentage of polymers acrylic and polystyrene to be used in chloroform.

Solute	Solvent	% Solute	% Solvent
Polystyrene	Chloroform	40	60
Acrylic	Chloroform	10	90

Figure 4 shows the DSC (Differential Scanning Calorimetry) analysis of the considered polymers, polystyrene and acrylic. In this analysis, polystyrene became malleable at 70 °C, contrary to that found in the literature that indicates that the thermal degradation of this polymer in contact with air starts at 200 °C [21]. On the other hand, acrylic polymer starts its transition at 140 °C, so this polymer would not be adequate to be used as coating of salts having a melting temperature below 140 °C.

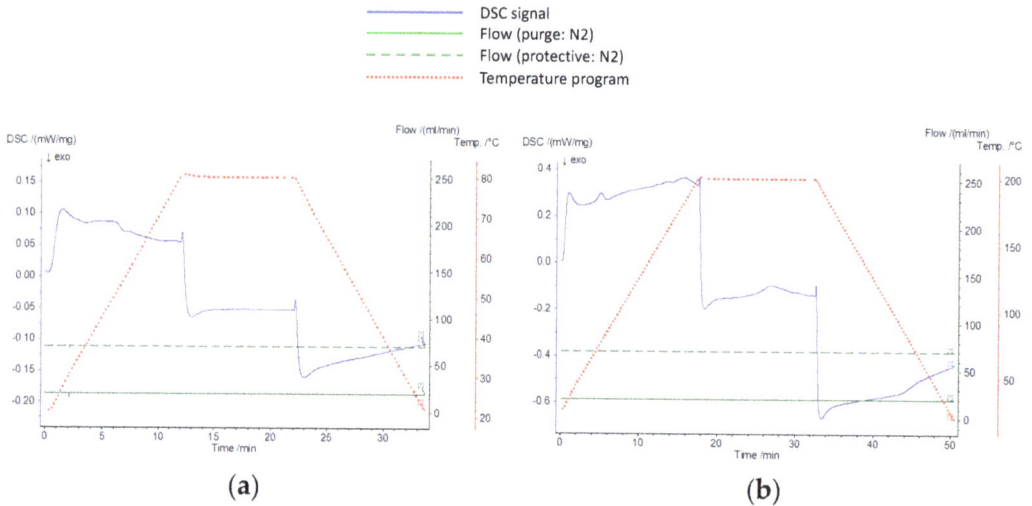

Figure 4. Melting and solidification curve of polymers tested. (**a**): polystyrene; (**b**): acrylic.

The results of the interaction between the studied PCM and the considered organic solvents are presented in Table 3. Taking into consideration that, as explained above, the PCM should not dissolve in to the solvent, these results show that all solvents would be adequate. Therefore, considering the solubility of the polymers and the PCM into the considered solvents (Tables 1 and 2), the best results were obtained with chloroform.

Table 3. PCM-solvent interaction.

Solvent	$MgCl_2 \cdot 6H_2O$	Bischofite
Acetone	Non-soluble	Non-soluble
Chloroform	Non-soluble	Non-soluble
Xylene	Non-soluble	Non-soluble
Tetrahydrofuran	Non-soluble	Non-soluble

The results of the stability of the polymers in contact with the considered PCM melted are shown in percentage of mass loss. Figure 5 shows that acrylic lost 0.87% of its initial mass in 30 days when in contact with $MgCl_2 \cdot 6H_2O$ and 3.43% when in contact with bischofite. However, after 30 days there was no mass loss in any polymer. Therefore, acrylic is physically and thermally stable in contact with the studied PCM.

Moreover the acrylic samples immersed in the PCM did show adhesion of the salts on the polymers (Figure 6), which would be beneficial in the encapsulation of the PCM with the polymer.

Magnesium chloride hexahydrate could be cleaned easily while bischofite required a more aggressive cleaning method to be removed completely. Moreover, the samples of acrylic that were in contact with bischofite became yellowish.

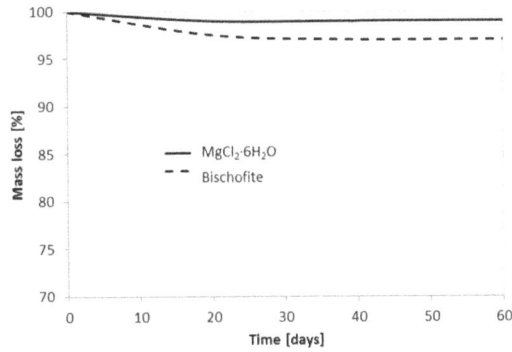

Figure 5. Mass loss of acrylic in contact with PCM.

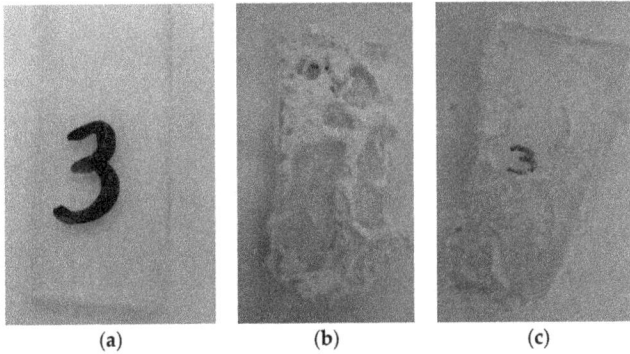

(a)	(b)	(c)

Figure 6. Acrylic samples in contact with PCM. (a) Initial; (b) immersed in $MgCl_2 \cdot H_2O$ during 60 days; (c) immersed in bischofite during 60 days.

2.2. Caracterization of Microencapsulated Particles

2.2.1. Fluidization Production Yield

The production yield is presented in Table 4. The highest production yield for bischofite was obtained with a polymer atomization flow of 2 kg/h and a fluidization time of 2 min, when 72.5% of the initial mass was encapsulated. The best production yield for $MgCl_2 \cdot 6H_2O$ was obtained with the same fluidization conditions. Bischofite has always a lower production yield than $MgCl_2 \cdot 6H_2O$, mainly due to the hygroscopy of bischofite.

Table 4. Fluidization production yield.

Material	Fluidization Time (s)	Atomization Flow (kg/h)	Yield (%)
Bischofite	60	2	53.8
		4	41.4
	120	2	**72.5**
		4	31.2
$MgCl_2 \cdot 6H_2O$	60	2	57.7
		4	43.9
	120	2	**84.9**
		4	46.8

2.2.2. Thermal Characterization

The results of the thermal characterization of microencapsulated PCM are presented in Table 5. The results are an average of eight samples cycled three times each one. The highest melting and crystallization enthalpy was obtained for an atomization flow of 2 kg/h and fluidization time of 2 min. As expected, the melting and crystallization enthalpy of the microencapsulated PCM were lower than that of the pure PCM.

Table 5. Thermal characterization of microencapsulated PCM.

Material	Fluidization Time (s)	Atomization Flow (kg/h)	Melting Temperature (°C)	Solidification Temperature (°C)	Melting Enthalpy (J/g)	Solidifiation Enthalpy (J/g)
Bischofite	–	–	108.5	88.5	104.5	103.1
Microencapsulated bischofite	60	2	79.6	71.5	70.2	64.6
	60	4	78.6	65.2	51.1	50.3
	120	2	104.6	85.4	95.0	104.8
	120	4	80.3	65.4	53.8	55.1
$MgCl_2 \cdot 6H_2O$	–	–	117.1	83.7	127.2	125.8
Microencapsulated $MgCl_2 \cdot 6H_2O$	60	2	96.1	61.5	89.6	85.6
	60	4	97.8	62.5	57.0	58.3
	120	2	95.3	61.0	118.3	119.2
	120	4	95.3	78.2	39.1	41.3

Salunkhe [7] stated that high encapsulation efficiency (E) is desirable, since it will result in microcapsules with higher mechanical strength and leak proof characteristics, and that the phase change enthalpy of the encapsulated PCM is a strong function of the encapsulation ratio and encapsulation efficiency. The encapsulation efficiency is defined with the following equation:

$$E = \frac{(\Delta H)_{\text{fusion,PCMencaps}} + (\Delta H)_{\text{solidif,PCMencaps}}}{(\Delta H)_{\text{fusion,PCM}} + (\Delta H)_{\text{solidif,PCM}}} \times 100 \qquad (1)$$

Table 6 presents the encapsulation efficiency of the carried out processes. Once more, the best results were obtained for an atomization flow of 2 kg/h and fluidization time of 2 min. Under these conditions, the obtained encapsulation efficiency was 87% for bischofite and 92% for $MgCl_2 \cdot 6H_2O$. These efficiencies are similar to the best found in the literature but better than most of those. For example, Alkan and Sari [22] reported an efficiency of 80% when encapsulating fatty acids with PMMA via in-situ polymerization; Fang et al. [23] obtained an efficiency of 60% when encapsulating n-tetradecane with UREA/formaldehyde using in-situ polymerization; Ma et al. [24] claimed an encapsulation efficiency of 48%–68% when encapsulating paraffin with an acrylic-based polymer using suspension-like polymerization as claimed by the authors; Alay et al. [25] reported an efficiency of 29%–61% when encapsulating n-hexadecane with PMMA using emulsion polymerization; and Fei et al. [26] obtained an efficiency of around 49% when encapsulating paraffin RT-27 with LDPE/EVA or polystyrene by spray-drying.

Table 6. Encapsulation efficiency.

Material	Fluidization Time (s)	Atomization Flow (kg/h)	Encapsulation Efficiency (%)
Bischofite	60	2	58.61
	60	4	44.28
	120	2	87.02
	120	4	46.99
$MgCl_2 \cdot 6H_2O$	60	2	66.16
	60	4	44.30
	120	2	92.22
	120	4	30.85

2.2.3. Morphological Characteristics of the Microencapsulated PCM Particles

The morphology of the microencapsulated inorganic PCM was done by visual observation and with an optical microscope. The results are presented in Table 7 for a given experiment of each PCM as example. Images from $MgCl_2 \cdot 6H_2O$ have been selected to show one of the best results, where the encapsulation efficiency was very high (therefore, the sample is nearly completely red); and images for bischofite have been selected to show one of the worst results, where blue and white parts are seen, showing the encapsulated crystals and the non-encapsulated ones. This characterization was used to corroborate the previously shown examples.

Table 7. View of the materials before encapsulation (**left**), after encapsulation (**middle**) and after encapsulation with a microscope (**right**).

Material	PCM	Encapsulated PCM	Microscopic View of Encapsulated PCM-X10
$MgCl_2 \cdot 6H_2O$			
Bischofite			

3. Discussion

Although there is a need for microencapsulation of PCM and organic PCM have been widely microencapsulated by many different methods, microencapsulation of inorganic PCM has been not studied adequately. This paper shows that fluidization is a good method to do so, since this study shows that efficiencies of around 90% were achieved microencapsulting pure $MgCl_2 \cdot 6H_2O$ and the by-product bischofite, with 95% $MgCl_2 \cdot 6H_2O$. To do so, not only the fluidization parameters such as fluidization time and atomization flow need to be selected, 2 min and 2 kg/h were used respectively, but also the compatibility between the three materials involved in the process: PCM, polymer and solvent. Results show that for the PCM studied, $MgCl_2 \cdot 6H_2O$ and bischofite, the solvent to be used should be chloroform and the polymer acrylic as shell material. The final microcapsules had excellent melting temperatures and enthalpy compared to the original PCM, 104.6 °C and 95 J/g for bischofite, and 95.3 °C and 118.3 J/g for $MgCl_2 \cdot 6H_2O$.

4. Materials and Methods

4.1. Materials

As material to be encapsulated, two salt hydrates were used as PCM, magnesium chloride hexahydrate (99% Merck S.A, Santiago, Chile) and bischofite (Salmag, Antofagasta, Chile). Bischofite is a mineral that precipitates in the evaporation ponds during potassium chloride production process in the Salar de Atacama (Chile) [27]. Bischofite is a by-product with a chemical composition of at least 95% $MgCl_2 \cdot 6H_2O$ that melts at 101 °C with a heat of fusion of 116.2 J/g.

For a material to be a good encapsulating material it needs to be compatible with the PCM that will encapsulated, needs to be thermally and physically stable during the melting and solidification cycles, should have low density and be non-corrosive, and should be easy to produce.

The materials chosen in the paper that fulfil the requirements listed above were high density polyethylene (HDPE), resin epoxy (both from Plastigen S.A., Antofagasta, Chile), polystyrene and acrylic (from Norglass, Santiago, Chile). The HDPE was used both as a bar and in granular form (prill or perl). The solvents used were acetone, chloroform, xylene, and tetrahydrofuran (THF), all >99.7% from Norglass.

4.2. Compatibility Studies

Since the polymers will be solubilized in the PCM in the fluidized bed, first of all, the compatibility between polymers and solvents were carried out. Between 1 and 50 mL of solvent were mixed with 1 to 5 g of polymer and the mixture was agitated constantly during 24 h. The polymer with the best solubility with the organic solvents was chosen.

When the mixture polymer-solvent covers the PCM, this could be partially solubilized by the organic solvent, which would mean PCM losses; therefore, the solubility between the PCM and the considered solvents was determined. For this study, two PCM-solvent ratios were used, 60:40 and 90:10. The variables measured were the initial mass and the final mass of PCM after the interaction with the solvent. The optimal mixture was that where there was not mass difference.

The compatibility between the PCM and the polymer was also studied to ensure the stability of the final microcapsules. The experiments were carried out immersing four sheets of each polymer, having previously been weight, in melted PCM at, approximately, 10 °C over the melting temperature of the PCM inside an oven during 60 days. On sheet was evaluated after 15, 30, 45 and 60 days of experimentation. The difference between the initial and the final mass of the polymer was used to determine the possible degradation of the polymer.

4.3. Microencapsulation via Fluidization

The used equipment was a glass fluidization chamber, an air source with a mess distributor to distribute the air flow evenly in the chamber, and a spray system. Based upon the literature [28], it seemed necessary to use a pressure nozzle to achieve the desired droplets size.

The studied variables were:

- The polymer concentration was established at 60% and 90%. Concentrations lower than 60% were tried but no good results could be achieved.
- The polymer atomization flow was set at 2 kg/h and 4 kg/h.
- The fluidization time was selected to be 60 s and 120 s.
- The PCM mass was fixed at 100 g of salt hydrate.

The fluidization method used was particles fluidization, where the crystals are suspended in an air flow as shown in Figure 7. The dimensions of the fluidization chamber are summarized in Table 8. As uniformization section, a cone of glass with the dimensions shown in Table 9 was used. The distributor was PVC mesh with a diameter of 200 μm.

The particles properties studied were that required for fluidization method: sphericity, diameter and density. The characteristics of $MgCl_2 \cdot 6H_2O$ particles are presented in Table 10 [29]. The same values were considered for bischofite.

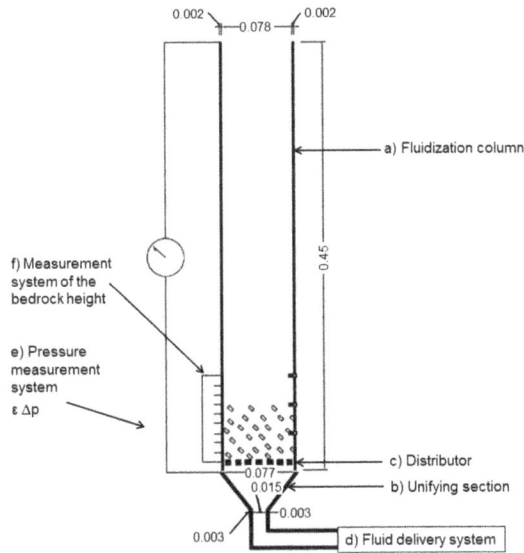

Figure 7. Fluidization column used in this study (units: m).

Table 8. Dimensions of the fluidization chamber.

Dimensions	Value (m)
Tube external diameter	0.08
Tube internal diameter	0.077
Thickness	0.0015
Height	0.45

Table 9. Dimensions of the uniformization section of the fluidization chamber.

Dimensions	Value (m)
Internal diameter of the tube in its upper part	0.077
Internal diameter of the tube in its lower part	0.015
Thickness	0.0015
Height	0.03

Table 10. $MgCl_2 \cdot 6H_2O$ particles properties [29].

Parameter	Value
Sphericity, $\varnothing$	0.86
Average particle diameter, d_p	500 μm
Density	1570 kg/m^3

4.4. Chemical Analysis

Thermophysical properties of the encapsulated PCM were analyzed by differential scanning calorimetry (DSC) with a Foenix F 201 (NETZSCH Group, Santiago, Chile). Measurements were done with 40 μL micro-crucibles hermetically closed.

Morphological characteristics of the salts hydrates and the microcapsules were determined with an optical phase contrast microscopy Olympus with a mechanism coaxial coarse and fine focus adjustment objectives 4×, 10×, 40×, and 60× connected to a precision chamber. Bischofite polymer was dyed with a blue pigment and $MgCl_2 \cdot 6H_2O$ polymer was dyed with a red pigment, so microcapsules could be differentiated from the non-encapsulated PCM.

5. Conclusions

In this study, the microencapsulation of two inorganic PCM was done by means of a fluidized bed method. Bischofite, a by-product from the non-metallic industry identified as having good potential to be used as inorganic PCM, was microencapsulated with acrylic as shell polymer and chloroform as solvent, after compatibility studies of both several solvents and several polymers. The formation of bischofite and pure $MgCl_2 \cdot 6H_2O$ microcapsules was investigated and analyzed. Results showed that efficiency in microencapsulation of 95% could be achieved when using 2 min of fluidization time and 2 kg/h of atomization flow. The final microcapsules had excellent melting temperatures and enthalpy compared to the original PCM, 104.6 °C and 95 J/g for bischofite, and 95.3 and 118.3 for $MgCl_2 \cdot 6H_2O$.

Acknowledgments: The authors would like to acknowledge the collaboration of the company SALMAG. The authors acknowledge to FONDECYT No. 1120422, CONICYT/FONDAP No. 15110019, and the Education Ministry of Chile Grant PMI ANT 1201 for the financial support. Luisa F. Cabeza would like to acknowledge the Generalitat de Catalunya for the quality recognition 2014-SGR-123. This project has received funding from the European Commission Seventh Framework Programme (FP/2007-2013) under Grant agreement N PIRSES-GA-2013-610692 (INNOSTORAGE) and from the European Union's Horizon 2020 research and innovation programme under grant agreement No 657466 (INPATH-TES).

Author Contributions: Svetlana Ushak conceived the experiments and analyzed the data; Judith M. Cruz performed the experiments and analyzed the data; Luisa F. Cabeza analyzed the data and wrote the paper; Mario Grágeda contributed in planning experiments, and in the design of fluidized bed.

Conflicts of Interest: The authors declare no conflict of interest. The founding sponsors had no role in the design of the study; in the collection, analyses, or interpretation of data; in the writing of the manuscript, and in the decision to publish the results.

References

1. Cabeza, L.F.; Castell, A.; Barreneche, C.; de Gracia, A.; Fernández, A.I. Materials used as PCM in thermal energy storage in buildings: A review. *Renew. Sustain. Energy Rev.* **2011**, *15*, 1675–1695. [CrossRef]
2. Khudhair, A.M.; Farid, M.M. A review on energy conservation in building applications with thermal storage by latent heat using phase change materials. *Energy Convers. Manag.* **2004**, *45*, 263–275. [CrossRef]
3. Su, W.; Darkwa, J.; Kokogiannakis, G. Review of solid–liquid phase change materials and their encapsulation technologies. *Renew. Sustain. Energy Rev.* **2015**, *48*, 373–391. [CrossRef]
4. Liu, C.; Rao, Z.; Zhao, J.; Huo, Y.; Li, Y. Review on nanoencapsulated phase change materials: Preparation, characterization and heat transfer enhancement. *Nano Energy* **2015**, *13*, 814–826. [CrossRef]
5. Khadiran, T.; Hussein, M.Z.; Zainal, Z.; Rusli, R. Encapsulation techniques for organic phase change materials as thermal energy storage medium: A review. *Sol. Energy Mater. Sol. Cells* **2015**, *143*, 78–98. [CrossRef]
6. Giro-Paloma, J.; Martinez, M.; Cabeza, L.F.; Fernandez, A.I. Types, methods, techniques, and applications for Microencapsulated Phase Change Materials (MPCM): A review. *Renew. Sustain. Energy Rev.* **2015**. in press. [CrossRef]
7. Salunkhe, P.B.; Shembekar, P.S. A review on effect of phase change material encapsulation on the thermal performance of a system. *Renew. Sustain. Energy Rev.* **2012**, *16*, 5603–5616. [CrossRef]
8. Hawlader, M.N.A.; Uddin, M.S.; Khin, M.M. Microencapsulated PCM thermal-energy storage system. *Appl. Energy* **2003**, *74*, 195–202. [CrossRef]
9. Salaun, F.; Devaux, E.; Bourbigot, S.; Rumeau, P. Influence of the solvent on the microencapsulation of an hydrated salt. *Carbohydr. Polym.* **2010**, *79*, 964–974. [CrossRef]
10. Huang, J.; Wang, T.; Zhu, P.; Ziao, J. Preparation, characterization, and thermal properties of the microencapsulation of a hydrated salt as phase change energy storage materials. *Thermochim. Acta* **2013**, *557*, 1–6. [CrossRef]
11. Hassabo, A.G.; Mohamed, A.L.; Wang, H.; Popescu, C.; Moller, M. Metal salts rented in silica microcapsules as inorganic phase change materials for textile usage. *ICAIJ* **2015**, *10*, 59–65.
12. Chaiyasat, P.; Chaiyasat, A. Preparation of poly(divinylbenzene) microencapsulated octadecane by microsuspension polymerization: Oil droplets generated by phase inversion emulsification. *RSC Adv.* **2013**, *3*, 10202–10207. [CrossRef]

13. Desai, K.G.H.; Park, H.J. Recent developments in microencapsulation of food ingredients. *Dry. Technol.* **2005**, *23*, 1361–1394. [CrossRef]
14. Shahidi, F.; Han, X.Q. Encapsulation of food ingredients. *Crit. Rev. Food Sci. Nutr.* **1993**, *33*, 501–547. [CrossRef] [PubMed]
15. Feng, T.; Xiao, Z.; Tian, H. Recent patents in flavor microencapsulation. *Recent Pat. Food Nutr. Agric.* **2009**, *1*, 193–202. [CrossRef] [PubMed]
16. Schoebitz, M.; López, M.D.; Roldán, A. Bioencapsulation of microbial inoculants for better soil-plant fertilization. A review. *Agron. Sustain. Dev.* **2013**, *33*, 751–765. [CrossRef]
17. Sozen, Z.Z.; Grace, J.R.; Pinder, K.L. Thermal energy storage by agitated capsules of phase change material. 1. Pilot scale experiments. *Ind. Eng. Chem. Res.* **1988**, *27*, 679–684. [CrossRef]
18. Izquierdo-Barrientos, M.A.; Sobrino, C.; Almendros-Ibáñez, J.A. Thermal energy storage in a fluidized bed of PCM. *Chem. Eng. J.* **2013**, *230*, 573–583. [CrossRef]
19. Izquierdo-Barrientos, M.A.; Sobrino, C.; Almendros-Ibáñez, J.A. Experimental heat transfer coefficients between a surface and fixed and fluidized beds with PCM. *Appl. Therm. Eng.* **2015**, *78*, 373–379. [CrossRef]
20. Izquierdo-Barrientos, M.A.; Sobrino, C.; Almendros-Ibáñez, J.A. Energy storage with PCM in fluidized beds: Modeling and experiments. *Chem. Eng. J.* **2015**, *264*, 497–505. [CrossRef]
21. Peterson, J.D.; Vyazovkin, S.; Wight, C.A. Kinetics of the thermal and thermo-oxidative degradation of polystyrene, polyethylene and poly(propylene). *Macromol. Chem. Phys.* **2001**, *2020*, 775–784. [CrossRef]
22. Alkan, C.; Sari, A. Fatty acid/poly(methyl methacrylate) (PMMA) blends as form-stable phase change materials for latent heat thermal energy storage. *Sol. Energy* **2008**, *82*, 118–124. [CrossRef]
23. Fang, G.; Li, H.; Yang, F.; Liu, X.; Wu, S. Preparation and characterization of nanoencapsulated n-tetradecane as phase change material for thermal energy storage. *Chem. Eng. J.* **2009**, *153*, 217–221. [CrossRef]
24. Ma, Y.; Chu, X.; Tang, G.; Yao, Y. Synthesis and thermal properties of acrylate-based polymer shell microcapsules with binary core as phase change materials. *Mater Lett.* **2013**, *91*, 133–135. [CrossRef]
25. Alay, S.; Alkan, C.; Göde, F. Synthesis and characterization of poly(methylmethacrylate)/n-hexadecane microcapsules using different cross-linkers and their application to some fabrics. *Thermochim. Acta* **2011**, *518*, 1–8. [CrossRef]
26. Fei, B.; Lu, H.; Qi, K.; Shi, H.; Liu, T.; Li, X.; Xin, J.H. Multi-functional microcapsules produced by aerosol reaction. *J. Aerosol. Sci.* **2008**, *39*, 1089–1098. [CrossRef]
27. Ushak, S.; Gutierrez, A.; Galleguillos, H.; Fernandez, A.G.; Cabeza, L.F.; Grágeda, M. Thermophysical characterization of a by-product from the non-metallic industry as inorganic PCM. *Sol. Energy Mater. Sol. Cells* **2015**, *132*, 385–391. [CrossRef]
28. Mujumdar, A.S. *Handbook of Industrial Drying*; Marcel Dekker, Inc.: New York, NY, USA, 1995; pp. 263–309.
29. Gómez, F.A. Almacenamiento de Energía Solar Térmica Usando Cloruro de Magnesio Hexahidratado. Master's Thesis, The Universidad Nacional de Colombia, Medellin, Colombia, 2007.

8

A Comparison of Simple Methods to Incorporate Material Temperature Dependency in the Green's Function Method for Estimating Transient Thermal Stresses in Thick-Walled Power Plant Components

James Rouse *,† and Christopher Hyde †

Academic Editor: Robert Lancaster

Department of Mechanical, Materials and Manufacturing Engineering, University of Nottingham, Nottingham, Nottinghamshire NG7 2RD, UK; Christopher.Hyde@nottingham.ac.uk
* Correspondence: James.Rouse@nottingham.ac.uk
† These authors contributed equally to this work.

Abstract: The threat of thermal fatigue is an increasing concern for thermal power plant operators due to the increasing tendency to adopt "two-shifting" operating procedures. Thermal plants are likely to remain part of the energy portfolio for the foreseeable future and are under societal pressures to generate in a highly flexible and efficient manner. The Green's function method offers a flexible approach to determine reference elastic solutions for transient thermal stress problems. In order to simplify integration, it is often assumed that Green's functions (derived from finite element unit temperature step solutions) are temperature independent (this is not the case due to the temperature dependency of material parameters). The present work offers a simple method to approximate a material's temperature dependency using multiple reference unit solutions and an interpolation procedure. Thermal stress histories are predicted and compared for realistic temperature cycles using distinct techniques. The proposed interpolation method generally performs as well as (if not better) than the optimum single Green's function or the previously-suggested weighting function technique (particularly for large temperature increments). Coefficients of determination are typically above 0.96, and peak stress differences between true and predicted datasets are always less than 10 MPa.

Keywords: thermal fatigue; header; Green's function method; power plant; P91

1. Introduction

There is a clear need in many industries to be able to predict the long-term behaviour of components operating in demanding environments in order to prevent/understand material failure. The aim is that with a greater understanding of how a component reacts due to a particular loading pattern, remnant life can be quantified. With this confidence, plant efficiency and longevity could be maximised safely. In particular, pressure is mounting on thermal power plant operators to generate electricity in an efficient and economical manner. Unit loads are expected to fluctuate with higher frequencies and steeper "ramp up and down" rates as drivers attempt to match market demands. Such so-called "two-shifting" or "partial-load" operating conditions have been in use for many years [1]; however, concern over their implementation is mounting as the amount of time a plant has to come on line reduces [2]. Generally, as steam pressures and temperatures vary with time, potentially large thermal stresses will develop in thick-walled components, such as steam headers. The fluctuation of total stress (mechanical and thermal) in components makes fatigue an important structural integrity concern in power plant components; a problem that is significantly complicated by the transient nature of thermal stresses. The present work looks to establish a technique based on the Green's function method that estimates transient thermal stresses while accounting for temperature-dependent material properties.

Many novel monitoring systems have been developed for assessing the structural integrity of at-risk power plant components, including "on line" management systems that monitor power station load characteristics (such as main steam temperature and pressure) and estimate component degradation using generalised finite element models and creep/fatigue damage fraction rules [3–6]. An example of one of these products is Areva's fatigue monitoring system FAMOSi (Erlangen, Germany) [7,8], where thermal loads are recorded using on site thermocouples and converted to thermal stresses using FEA (finite element analysis) models at critical points in a system. Alternatively, accurate stress histories in a component may be estimated through bespoke analyses utilising complex visco-plastic material models [9]; however, this is commonly computationally intensive and is typically impractical for on line component assessment.

While these advances have shown some success, established design codes and analysis procedures are still by far the most commonly-used tools in industry for component fitness assessment, along with frequent inspection during outage periods [10]. In the UK, the R5 [11,12] procedure is commonly used for high temperature assessment and the R6 [13] procedure for low temperature fracture assessment of power plant components. These step by step methods usually involve decomposing a loading history into cycles. The likelihood of failure by various mechanisms, such as plastic collapse, creep and fatigue, is calculated by estimating damage accumulation and mechanism interaction factors.

The Green's function method provides a general approach to estimate the transient linear elastic thermal stress responses at a point in a structure by integrating the response due to a unit thermal load change. In the context of steam headers, thermal stress histories may be estimated at a point of interest for any bulk steam temperature history. While limited to linear analysis (due to the inherent summation during integration), the Green's function method is still of use in component failure assessment, particularly where damage is suspected to be localised. The Green's function method (see Section 2.2) has been show to be a useful tool in predicting transient thermal stresses by several authors. In particular, the technique has been applied to fatigue analysis problems in the nuclear power industry [14–16]. It has often been assumed (for simplicity of integration) that Green's functions are temperature independent. In reality, this is not the case due to variations in material properties with temperature. The work of Koo *et al.* suggested the implementation of a temperature-dependent weighting function [17]; however, this neglects second order variations (*i.e.*, it assumes time-independent scaling of Green's function). The present work looks to compare this method to a developed interpolation procedure in order to establish the importance of these second order effects.

2. Background

2.1. The Thermoelastic Problem

The governing equations for a linear coupled thermoelastic problem may be derived from the fundamental principles of mechanics and thermodynamics. When loads applied to a body give rise to variations in strain within the body, variations in temperature are also observed. This causes heat flow and therefore an increase in entropy for the body (this irrecoverable mechanical dissipation is known as thermoelastic dissipation). There is an internal generation of heat due to mechanical deformation that will affect the temperature field within a body in addition to any thermal boundary conditions. Deformation, however, is not only controlled by the application of, say, body forces. Temperature fields cause thermal expansion within elements of the body, generating additional internal surface forces between the elements. There exists therefore a coupling between the solutions for temperature and displacement fields, $T(P, t)$ and $u(P, t)$, respectively (where P is a point within the body specified by coordinates using the coordinate system x_1, x_2, $x_3 = x$ and t is time). For a linear coupled thermoelastic problem, it may be shown that a unique solution may be found (for a given set of initial and boundary conditions) using the heat equation with mechanical coupling (Equation (1)), the equilibrium condition (Equation (2); note the inclusion of an inertia term on the right hand side of the equation), the strain-displacement relations (Equation (3)) and the stress-strain relations (Equation (4)) [18]. Note the use of indicial notation ($\frac{\delta g_i}{\delta x_j} = g_{i,j}$, where g_i is a vector component in the i-th direction and x_j is the basis vector in the j-th direction of the coordinate system, $i, j = 1, 2, 3$) and the Einstein summation convention. Note also that dots are used to denote derivatives with respect to time.

$$kT_{,mm} = \rho C \dot{T} + (3\lambda + 2\mu) \alpha T_0 \dot{\epsilon}_{kk} \tag{1}$$

$$\sigma_{ij,j} + f_i = \rho \ddot{u}_i \tag{2}$$

$$\epsilon_{ij} = \frac{1}{2} \left(u_{i,j} + u_{j,i} \right) \tag{3}$$

$$\sigma_{ij} = \delta_{ij} \lambda \epsilon_{kk} + 2\mu \epsilon_{ij} - \delta_{ij} (3\lambda + 2\mu) \alpha T \tag{4}$$

where ϵ, σ and f are the small strain tensor, the stress tensor and the body force vector field, respectively; T_0 is a reference temperature at which, in the absence of body forces, the material will be in a stress-free state. The material-dependent parameters are thermal conductivity (k), density (ρ), specific heat capacity at constant deformation (C) and the thermal expansion coefficient (α). Lamé's first and second parameter are defined in terms of Young's modulus (E) and Poisson's ratio (ν) in Equation (5). δ_{ij} is the Kronecker delta ($\delta_{ij} = 1$ if $i = j$, else $\delta_{ij} = 0$).

$$\begin{aligned} \lambda &= \frac{E\nu}{(1+\nu)(1-2\nu)} \\ \mu &= \frac{E}{2(1+\nu)} \end{aligned} \tag{5}$$

The existence of the coupling term in the energy equation (Equation (1)) greatly complicates the solution process for the thermoelastic problem (clearly, temperature and displacement field solutions must be found simultaneously to satisfy Equations (1)–(4) and the problem-specific initial and boundary conditions). In general, temperature variations due to mechanical deformations are small (particularly if the small strain theory is implemented). Similarly, differences between heat transfer solutions in deformed and undeformed bodies are also small (deformations from either thermal expansion or external mechanical agencies do not change the dimensions of the structure to such an extent that heat transfer is significantly affected). If thermoelastic dissipation can be neglected (as is almost always the case [18]), an uncoupled formulation may be derived for the thermoelastic problem. In this case, internal heat generation due to deformation is ignored, and temperature fields can be found first by solving the well-known heat equation (Equation (6)), where κ is the thermal diffusivity ($\kappa = k/\rho C$). Once the temperature field has been determined, the corresponding displacement field (dependent on thermal expansion and mechanical body forces) may be found.

If the rate of change of the deformation rates are small (as is the case in many engineering applications), inertia effects may be neglected, and the formulation is termed quasi-static. In this case (with the absence of body forces), the equilibrium equation simplifies to Equation (7). Strain-displacement and stress-strain relations given in Equations (3) and (4), respectively, are still valid in the uncoupled formulation.

$$\kappa T_{,mm} = \dot{T} \tag{6}$$

$$\sigma_{ij,j} = 0 \tag{7}$$

2.2. The Green's Function Method for Predicting Transient Thermal Stresses

The thermoelastic problem has been defined in Section 2.1. Solutions for even the uncoupled formulation with the component geometry used in the present work (power plant steam header) are very complex and generally require numerical methods (such as finite element analysis (FEA)) to estimate a solution. A practical problem with this analysis strategy is that temperature, displacement and, consequently, stress fields would need to be found for each new operating condition. It is not feasible to perform full FEA simulations of the header components for each change to the bulk steam temperature or pressure. A solution to this dilemma however exists through the use of Green's functions.

It can be seen from Section 2.1 that the thermal stress solution is based on the temperature field solution, both of which are unique and dependent on the particular initial and boundary conditions of the problem. This uniqueness allows the use of a Green's function that finds the thermal stresses based on the boundary conditions. It is therefore possible to determine the thermal stress distribution without direct knowledge of the temperature or displacement fields. As the present work is concerned with power plant header applications, the bulk internal steam temperature may be used as a "driving" term for the thermal stress field (it shall be assumed that external surfaces of the header are insulated, and attention is restricted to the uncoupled formulation). Thermal stresses at a point P in the header structure can be found by the integral in Equation (8), where $G(P, t - \tau)$ is Green's function, $\psi(t)$ is the bulk internal steam temperature and τ is the time integration variable.

$$\sigma(P, t) = \int_0^t G(P, t - \tau) \frac{d\psi(\tau)}{d\tau} d\tau \tag{8}$$

Numerical integration of Equation (8) may be accomplished using Equation (9), where $G_{SS}(P) = \lim_{t \to \text{inf}} G(P, t)$. In the present work, the temperature field is allowed to reach equilibrium, and mechanical loads are not considered; therefore, $G_{SS}(P) = 0$.

$$\sigma(P, t) = G_{SS}(P)\psi(\tau) + \sum_{t-t_{CH}}^{t} \bar{G}(P, t - \tau) \Delta\psi(\tau) \tag{9}$$

$\bar{G}(P, t - \tau)$ represents the thermal stress response due to a unit temperature step at the point of interest P (assuming no other loads are present in the structure). Green's function may be represented by a sum of exponential terms (see Equation (10)).

$$\bar{G}(P, t) = \exp\left(\sum_{m=1}^{7} C_m(P) (\ln(t))^{m-1} \right) \tag{10}$$

The work of Koo et al. introduced temperature dependency in the Green's function method using a weighting function dependent on the bulk temperature ($W(\psi)$) [17]; see Equation (11). Note, for simplicity, this method will be refereed to as the "weight function" method for the reminder of the present work. Based on the work of Koo et al., a fourth order polynomial has been assumed for use in a comparison study (see Equation (12)).

$$\sigma(P, t) = \int_0^t G(P, t - \tau) W(\psi) \frac{d\psi(\tau)}{d\tau} d\tau \tag{11}$$

$$W(\psi) = A_0 + A_1\psi + A_2\psi^2 + A_3\psi^3 + A_4\psi^4 \tag{12}$$

3. FEA Header Models and Temperature-Dependent Material Properties

FEA models must be generated in order to determine the coefficients in Equation (10) and, thus, to define Green's functions. FEA has been conducted in the present work using the commercially-available code ABAQUS (Dassault Systèmes, Paris, France). Since the present work looks to establish a method to introduce temperature dependency in the Green's function method, a simplified two-stub penetration model has been used by way of example (see Figure 1a, noting the plane of symmetry assumed between stub penetrations). Despite the simplified geometry, shell and stub dimensions are similar to those found in industry for P91 header components. Uncoupled thermoelastic analysis was conducted by first determining a temperature field from a heat transfer simulation. An insulated exterior boundary condition was assumed ($\dot{q} = 0$) to allow temperature fields in the model to reach equilibrium after the bulk steam temperature experiences a step change. Heat conduction on the inside surface of the header is controlled by convection (see Figure 1b), where the heat transfer coefficient h is taken to be a temperature-independent constant 0.002 W/mm^{2}K. Once the transient temperature field has been determined, it can be used as an input in mechanical analyses to estimate thermal stress histories. Boundary conditions for the mechanical analysis can be seen in Figure 1c. An "equation" type constraint [19] is applied to the upper surface of the model (designated by the label U_Z = Constant in Figure 1c). This enforces equal displacements in the Z direction between all nodes on this plane, thus ensuring it remains planar, and the assumed symmetry holds. Tetrahedral quadratic elements where used, namely DC3D10 (ABAQUS) for thermal analyses and C3D10 (ABAQUS) for mechanical analyses (see Figure 1a for an example mesh) [19]. In has been indicated in the literature that ligament cracking is a potential concern for header components, notably when thermal fatigue is a significant damage mechanism [20]. As discussed previously, the Green's function method allows for estimation of stresses at a singular analysis point only. With these factors in mind and for the illustration of the stress analysis potential of the Green's function method, an analysis point (P; see Figure 1a) is considered in the present work that represents crack initiation at the inner bore [20].

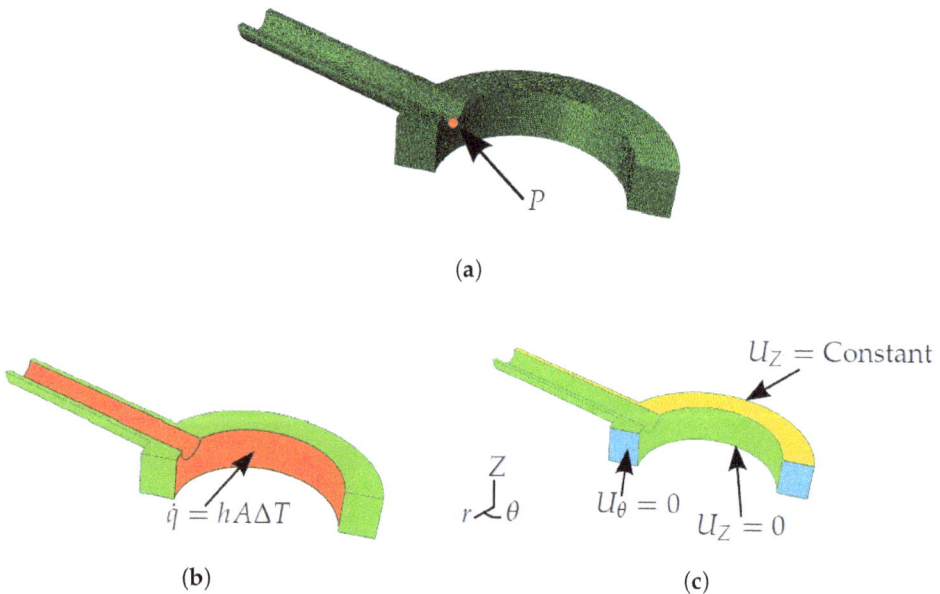

(a)

(b)   **(c)**

Figure 1. Finite element analysis (FEA) models, showing: (**a**) the tetrahedral mesh, exploiting the plane of symmetry between stub penetrations and showing the location of the example point of interest P; (**b**) boundary conditions in the thermal analyses; and (**c**) boundary conditions in the mechanical analyses.

A single material is assumed for the FEA model in the present work (variations in material properties at the stub weld are not considered). Temperature-dependent material parameters are required in order to calculate transient thermal stresses within the header models. Values for Young's modulus (E) and the instantaneous thermal expansion coefficient (α) have been determined from monotonic tests performed on an Instron 8862 thermomechanical fatigue machine (operating under isothermal conditions, Norwood, MA, USA) utilising radio frequency induction heating and using a TA instruments Q400 thermomechanical analyser (New Castle, DE, USA), respectively (see Table 1). Tested temperature ranges were chosen to represent the typical bounds of operation for thermal power plant components. The remainder of the material constants have been taken from the work of Yaghi *et al.* [21] (Table 2). A negligible dependency is assumed in density (ρ) and Poisson's ratio (ν) over the tested temperature range. As such, values for these quantities are taken to be 7.76×10^{-6} kg/mm^3 and 0.3, respectively. Temperature dependent material properties are summarised in Figure 2.

Table 1. A summary of the temperature-dependent material parameters (representative of a P91 chrome steel), determined through experimental analysis, used in the FEA modelling.

Temperature (°C)	Young's Modulus (E) (N/mm^2)	Thermal Expansion Coefficient (α) (1/K)
240	1.88×10^5	1.21×10^{-5}
400	1.77×10^5	1.34×10^{-5}
550	1.51×10^5	1.41×10^{-5}
600	1.30×10^5	1.42×10^{-5}
700	8.58×10^4	1.41×10^{-5}

Table 2. A summary of the temperature-dependent material constants (representative of a P91 chrome steel), taken from the work of Yaghi *et al.* [21], used in the FEA modelling.

Temperature (°C)	Thermal Conductivity (k) (W/mm.K)	Specific Heat Capacity (C) (J/kg.K)
200	0.028	510
250	0.028	530
300	0.028	550
350	0.029	570
375	0.029	585
400	0.029	600
450	0.029	630
500	0.03	660
550	0.03	710
600	0.03	770
650	0.03	860
700	0.0305	942

In order to apply the Green's function method and the temperature interpolation techniques, realistic thermal stress histories must be generated from the FEA models for representative bulk steam temperature profiles (mechanical loading is neglected here, as it is a trivial exercise to scale linear elastic loads based on varying internal pressures). Several "ramp up" temperature profiles have been generated and analysed using the uncoupled thermoelastic procedure (with temperature-dependent material parameters). These are summarised in Table 3 and are representative of the high end limiting bulk steam temperature increments and rates seen in two shifting plant. Plots of the ramp up temperature profiles can be seen in Figure 3 with their respective labels (which are used in the remainder of the present work). An oscillating temperature profile (that represents a control signal correcting bulk steam temperature to a nominal operating temperature of 550 °C) is also considered (see Figure 4), designated Profile "I". This profile was generated using Equation (13), using the parameters $B_1 = 60$, $B_2 = -3.978 \times 10^{-3}$, $B_3 = \pi/50$ and $T_M = 550$ °C.

$$T(t) = B_1 e^{B_2 t} sin(B_3 t) + T_M \tag{13}$$

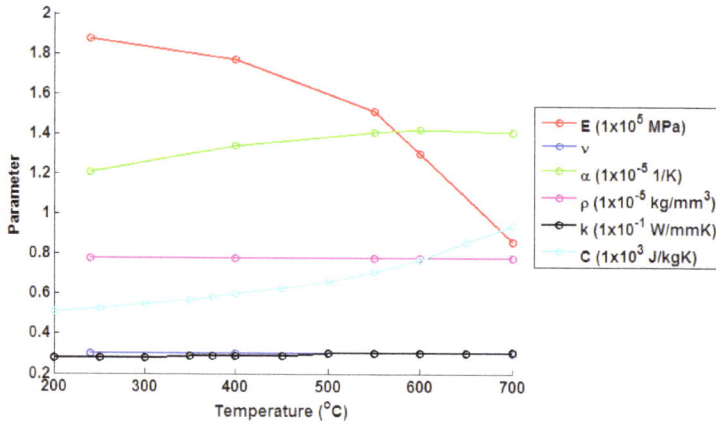

Figure 2. A summary of the temperature-dependent material properties used in the present work to represent a P91 chrome steel.

Table 3. A summary of bulk steam temperature "ramp up" profiles.

Load Case	Start Temperature (°C)	End Temperature (°C)	Temperature Rate (°C/min)
A	250	550	4
B	250	550	8
C	450	550	4
D	450	550	8
E	550	650	4
F	550	650	8
G	250	640	4
H	250	640	8

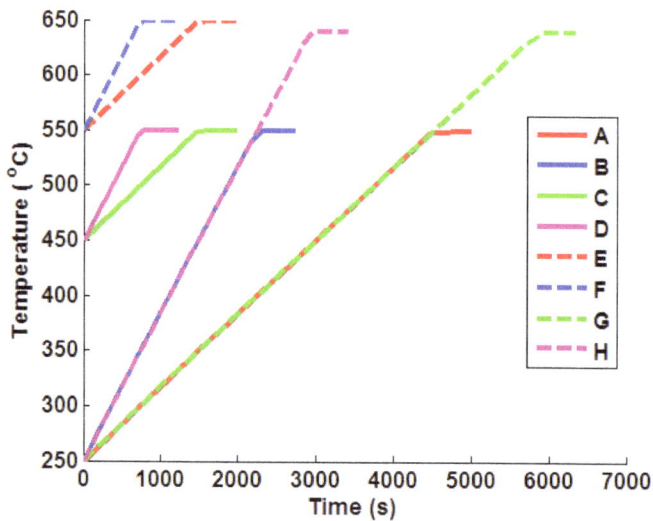

Figure 3. Plots of the representative ramp up temperature profiles used to test the various Green's function implementation techniques.

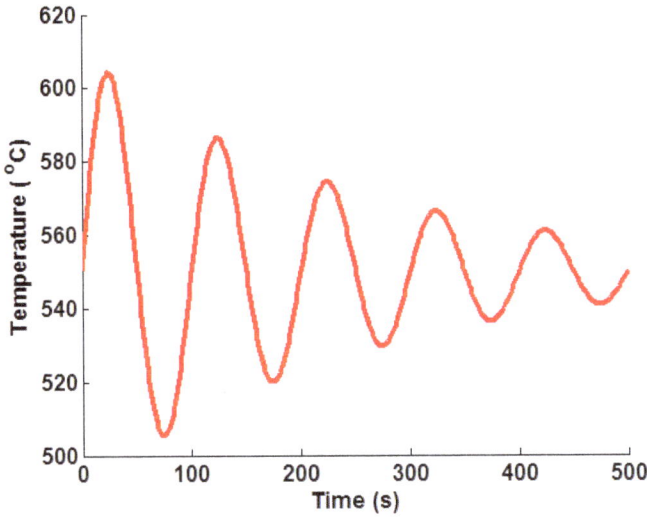

Figure 4. A plot of the representative oscillating (decay) temperature profile used to test the various Green's function implementation techniques.

4. Methodology Overview

Prior to discussing the proposed method to introduce temperature dependency, it is worthwhile briefly discussing the procedure to determine temperature-independent Green's functions. Once the thermally-driven stress profile has been determined from FEA for a unit bulk steam temperature step, the Green's function approximation shown in Equation (10) can by fitted (an example may be seen in Figure 5). This defines the constants C_m, $m = 1...7$. A non-linear least squares optimisation algorithm (the Levenberg—Marquardt algorithm) was used in a MATLAB program (function LSQNONLIN [22], MathWorks, Natick, MA, USA) to optimise the values of Equation (10) in order to fit the FEA solution.

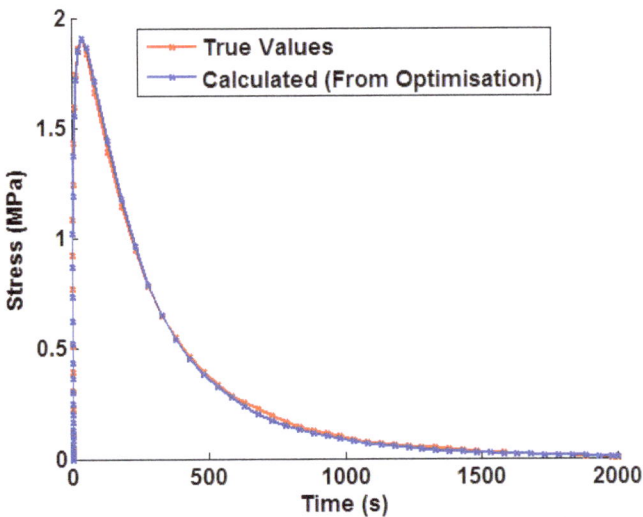

Figure 5. An example of the Green's function approximation (shown in Equation (10)) fitted to the von Mises stress history from an uncoupled thermoelastic FEA simulation (unit temperature step).

A schematic of the proposed interpolation technique is given in Figure 6. The Green's function method fundamentally relies on determining a reference solution for a unit temperature step that may be integrated for a particular thermal loading history. By including temperature dependency, this reference solution will need to be altered over the thermal history. The proposed technique achieves this by interpolating between solutions determined from temperature-independent Green's functions. Green's functions are determined for unit temperature steps, each of which has an associated representative temperature (taken here to be the mean temperature for the unit step). Given some initial conditions ($\sigma = \sigma_{t_i}$, $T = T_{t_i}$), stress increments can be determined for each Green's function $\sigma_{t_f T=T_1}$, $\sigma_{t_f T=T_2}$..., and so on, with the representative temperatures $T_1, T_2, ...$; see Equation (14)). Note that a different (temperature dependent) set of constants ($C_m, m = 1...7$) is used to find each reference solution ($\sigma_{t_f T=T_1}, \sigma_{t_f T=T_2}...$). These constant sets are designated $C_{mT=T_1}, C_{mT=T_2}, ...$ for representative temperatures $T_1, T_2, ...$, respectively, in Equation (14). The relationship between the representative temperatures and reference stress values may then be used to interpolate to the actual instantaneous temperature T_{t_f} and, thus, find the estimated stress increment σ_{t_f}. A suitably high order polynomial may be used to model this relationship. For M reference solutions and representative temperatures, Equation (15) may be used, where $D_i, i = 1...M$ are the coefficients of the polynomial. Generating the reference curves in this way allows the relationship between stress values predicted by particular Green's functions at a time instant to change with time, thus allowing the second order effect ($\frac{\delta^2 \sigma}{\delta t \delta T}$) to be accounted for.

$$\sigma_{t_f T=T_1} = \bar{G}_{T=T_1}(P, t_f) = \exp\left(\sum_{m=1}^{7} C_{mT=T_1}(P) \left(\ln(t_f) \right)^{m-1} \right)$$

$$\sigma_{t_f T=T_2} = \bar{G}_{T=T_2}(P, t_f) = \exp\left(\sum_{m=1}^{7} C_{mT=T_2}(P) \left(\ln(t_f) \right)^{m-1} \right)$$

$$\sigma_{t_f T=T_3} = \bar{G}_{T=T_3}(P, t_f) = \exp\left(\sum_{m=1}^{7} C_{mT=T_3}(P) \left(\ln(t_f) \right)^{m-1} \right) \tag{14}$$

...

$$\sigma_{t_f}(T) = \sum_{i=1}^{M} D_i T^{i-1} \tag{15}$$

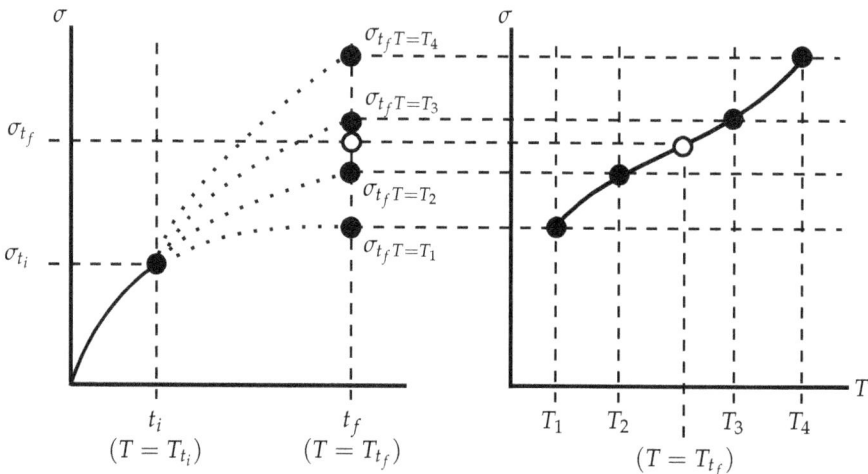

Figure 6. A schematic of the proposed interpolation technique.

5. Results

5.1. Unit Temperature Steps

Thermal stress profiles were generated at the analysis point (P) defined in Figure 1a for unit bulk steam temperature changes using the modelling techniques discussed in Section 3. Figure 7 highlights the difference in the thermal stress responses due to a unit temperature step (as a result of temperature-dependent material properties). For the analysis point considered, the hoop stress is the dominant principal stress, and the other principal stresses are negligible. Equivalent von Mises stresses are presented here, but these (as far as the Green's functions are concerned) may be taken to be the absolute of the maximum principal stress.

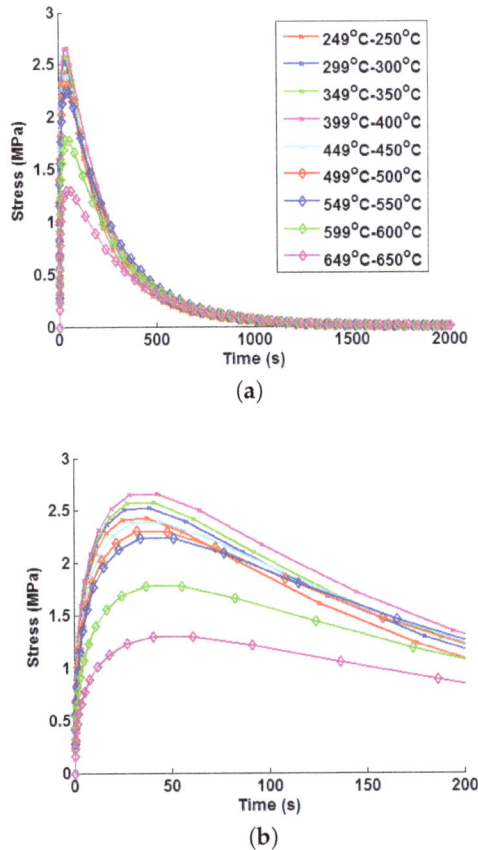

Figure 7. Variations in thermal stress responses due to unit step bulk steam temperature increments, showing: (a) general development and decay behaviour; and (b) magnified thermal stress development behaviour.

Similar general trends are observed for all six components of the symmetric stress tensor (there is development of thermal stress until a maximum value is achieved, after which the thermal stress exponentially decays). Equation (10) may therefore be used in general to describe the unit thermal response of for the six stress components at a specific analysis point. Transient thermal stress behaviour may then be approximated as before using the integration shown in Equation (9). Additionally, the temperature dependency method described in Section 4 may also be implemented for each of the stress components. The present work only considers the von Mises stress. This provides a concise and easy to follow way to compare the applicability of the tested methods. If a user were to determine Green's functions for all six stress components, the von Mises Green's function could be generated with little additional effort and be used as a constraint to limit any interpolation errors accumulated when processing these individual components.

Peak thermal stress values are observed for the 399 °C–400 °C case. Referring to Figure 2, it can be seen that this is due to the significant increase in the thermal expansion coefficient α and the specific heat capacity C. This results in higher internal forces being required for a particular temperature gradient (note that thermal conductivity k is relatively stable over the given temperature range). After 400 °C, a marked reduction in Young's modulus (E) is observed, leading to a loss of stiffness in the material and, hence, a reduction in peak thermal stresses. Green's functions (defined by Equation (10)) have been fitted to each unit step using the method discussed in Equation (4). In all cases, the optimisation terminated due to the change in the sum of squares between iterations falling below a tolerance (set to 1×10^{-7}), suggesting convergence on a local minimum. A summary of the Green's function coefficients can be seen in Table 4. The weighting function described by Equations (11) and (12) has also been fitted to the unit responses using the same optimisation technique and satisfying the same stopping criterion. A summary of these coefficients may be found in Table 5.

Table 4. A summary of Green's function coefficients fitted to the unit thermal stress responses.

Unit Step	C_1	C_2	C_3	C_4	C_5	C_6	C_7
249 °C–250 °C	-7.07×10^{-2}	4.12×10^{-1}	-2.44×10^{-2}	-2.77×10^{-4}	-1.00×10^{-4}	-3.16×10^{-4}	4.07×10^{-6}
299 °C–300 °C	-4.78×10^{-2}	4.14×10^{-1}	-2.40×10^{-2}	-2.99×10^{-4}	-1.00×10^{-4}	-2.97×10^{-4}	2.30×10^{-6}
349 °C–350 °C	-4.19×10^{-2}	4.17×10^{-1}	-2.37×10^{-2}	-3.30×10^{-4}	-1.00×10^{-4}	-2.97×10^{-4}	2.56×10^{-6}
399 °C–400 °C	-3.16×10^{-2}	4.20×10^{-1}	-2.32×10^{-2}	-3.86×10^{-4}	-1.00×10^{-4}	-2.70×10^{-4}	3.89×10^{-8}
449 °C–450 °C	-1.58×10^{-1}	4.21×10^{-1}	-2.29×10^{-2}	-3.65×10^{-4}	-1.00×10^{-4}	-2.52×10^{-4}	-1.24×10^{-6}
499 °C–500 °C	-2.16×10^{-1}	4.25×10^{-1}	-2.24×10^{-2}	-4.11×10^{-4}	-1.00×10^{-4}	-2.48×10^{-4}	-1.23×10^{-6}
549 °C–550 °C	-2.75×10^{-1}	4.28×10^{-1}	-2.18×10^{-2}	-4.47×10^{-4}	-1.00×10^{-4}	-2.17×10^{-4}	-4.01×10^{-6}
599 °C–600 °C	-5.38×10^{-1}	4.31×10^{-1}	-2.13×10^{-2}	-3.35×10^{-4}	-1.00×10^{-4}	-1.97×10^{-4}	-5.30×10^{-6}
649 °C–650 °C	-9.02×10^{-1}	4.35×10^{-1}	-2.11×10^{-2}	-1.56×10^{-4}	-1.00×10^{-4}	-1.69×10^{-4}	-7.19×10^{-6}

Table 5. Coefficient values determined for the weighting function approach.

C_1	-8.04×10^{-1}
C_2	5.13×10^{-1}
C_3	-2.16×10^{-2}
C_4	-7.68×10^{-3}
C_5	-7.19×10^{-6}
C_6	6.08×10^{-5}
C_7	-3.17×10^{-5}
A_0	1.00×10^{-1}
A_1	-1.43×10^{-2}
A_2	4.64×10^{-5}
A_3	-7.71×10^{-8}
A_4	5.45×10^{-11}

5.2. Representative Temperature Profiles

In order to illustrate the importance of considering temperature dependency in the Green's function approach, unit Green's functions have been used individually to predict each of the nine temperature profiles A–I (thus highlighting the potential degree of stress over-/under-estimation). Plots of the true (FEA) and predicted thermal stresses may be seen in Figures 8–10. Similarly, predictions of the thermal stress histories have been made using the temperature dependency techniques discussed (the weighting function and the proposed interpolation technique, shown in Figures 11–13). In order to quantify the relative qualities of fit, coefficients of determination (R^2 [23]) and peak absolute differences between predicted and true stresses ($\Delta\sigma_{VM}$) have been determined for each method. These are summarised in Tables 6 and 7, respectively.

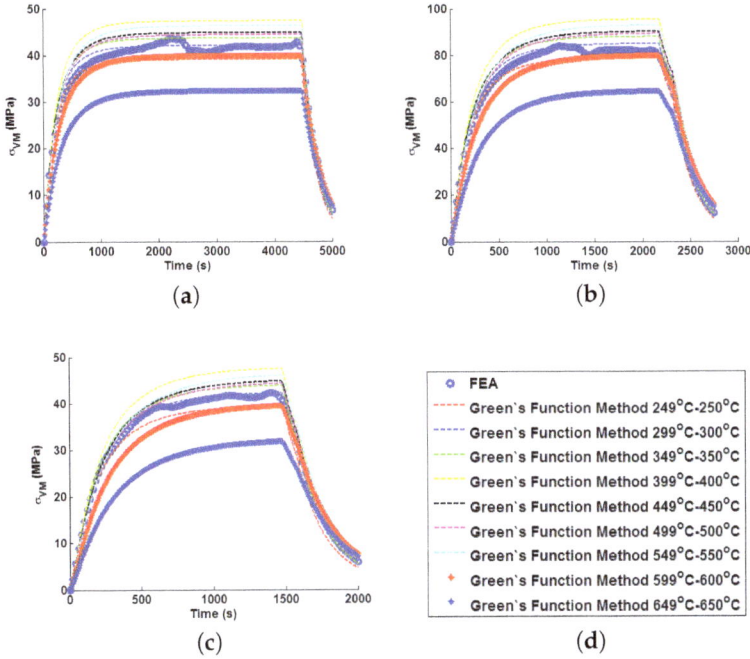

Figure 8. Predictions of thermal stress responses determined using individual Green's functions for: (a) ramp up Profile A; (b) ramp up Profile B; and (c) ramp up Profile C. Sub figure (d) shows the legend used in the plots.

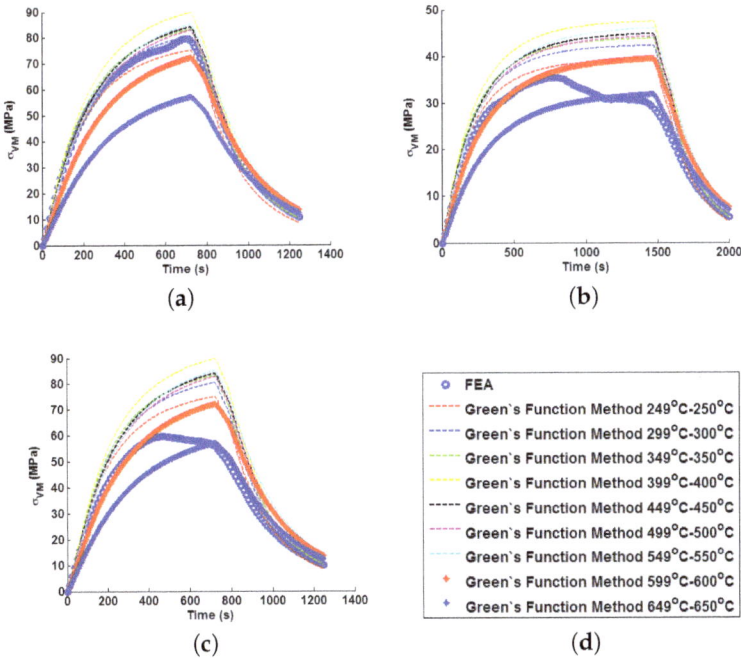

Figure 9. Predictions of thermal stress responses determined using individual Green's functions for: (a) ramp up Profile D; (b) ramp up Profile E; and (c) ramp up Profile F. Sub figure (d) shows the legend used in the plots.

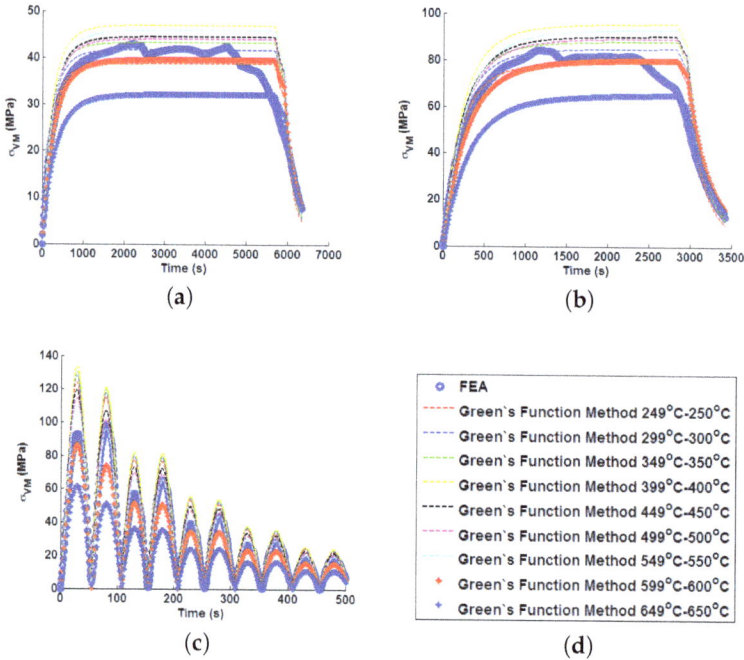

Figure 10. Predictions of thermal stress responses determined using individual Green's functions for: (a) ramp up Profile G; (b) ramp up Profile H; and (c) oscillating Profile I. Sub figure (d) shows the legend used in the plots.

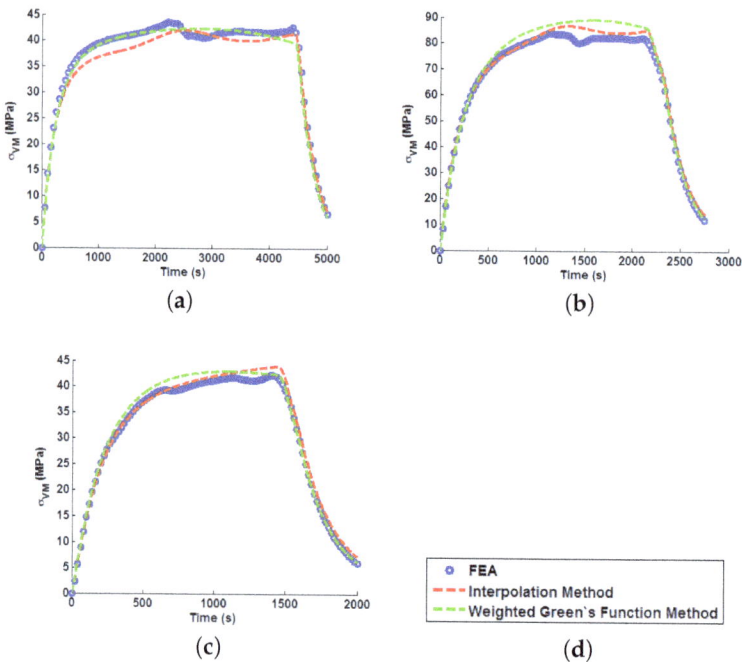

Figure 11. Predictions of thermal stress responses determined using the interpolation and weighted Green's functions methods for: (a) ramp up Profile A; (b) ramp up Profile B; and (c) ramp up Profile C. Sub figure (d) shows the legend used in the plots.

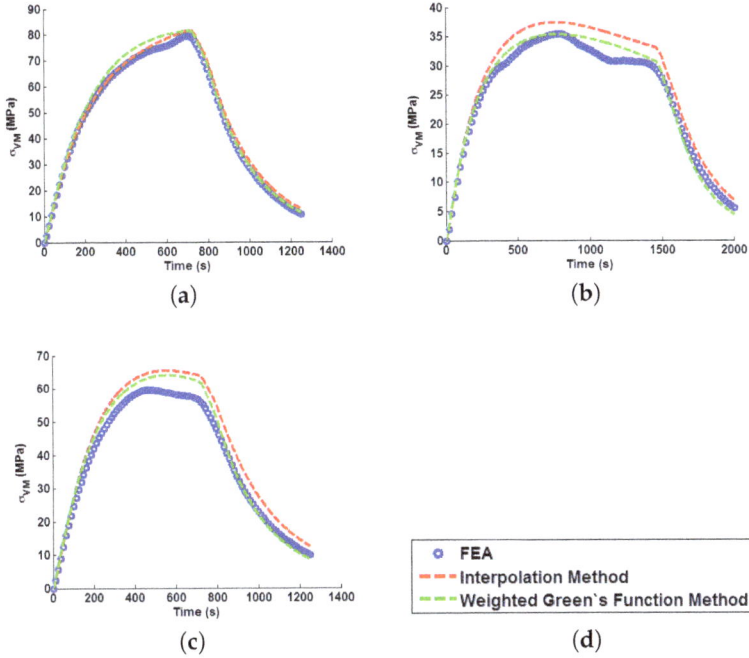

Figure 12. Predictions of thermal stress responses determined using the interpolation and weighted Green's functions methods for: (**a**) ramp up Profile D; (**b**) ramp up Profile E; and (**c**) ramp up Profile F. Sub figure (**d**) shows the legend used in the plots.

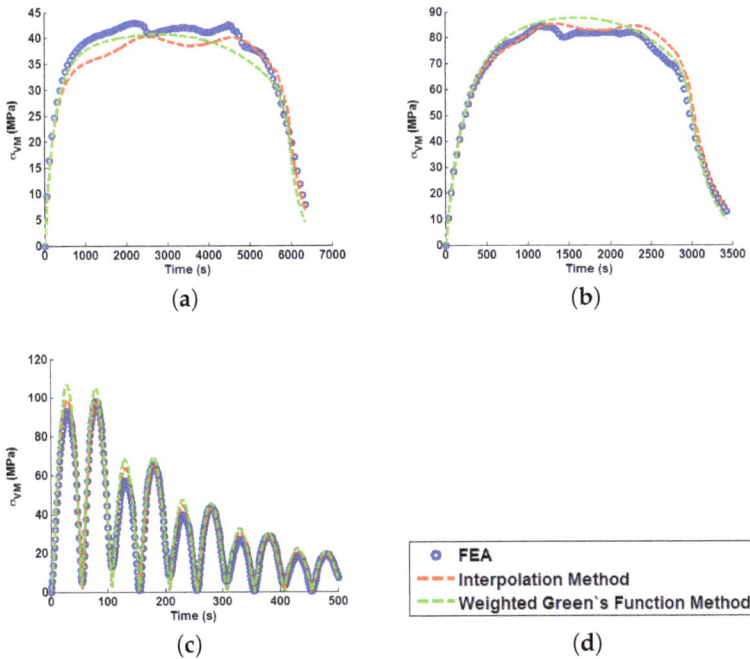

Figure 13. Predictions of thermal stress responses determined using the interpolation and weighted Green's functions methods for: (**a**) ramp up Profile G; (**b**) ramp up Profile H; and (**c**) oscillating Profile I. Sub figure (**d**) shows the legend used in the plots.

Table 6. A summary of coefficients of determination (R^2) for the test temperatures profiles, fitted using the three Green's function implementation methods.

Method	A	B	C	D	E	F	G	H	I
249 °C–250 °C	0.930	0.976	0.966	0.979	0.763	0.759	0.861	0.947	0.827
299 °C–300 °C	0.983	0.988	0.995	0.995	0.508	0.540	0.842	0.897	0.785
349 °C–350 °C	0.949	0.958	0.983	0.986	0.336	0.397	0.753	0.825	0.757
399 °C–400 °C	0.703	0.785	0.885	0.918	-0.185	-0.013	0.320	0.525	0.723
449 °C–450 °C	0.911	0.932	0.974	0.984	0.235	0.357	0.660	0.761	0.905
499 °C–500 °C	0.936	0.950	0.983	0.990	0.307	0.427	0.704	0.793	0.944
549 °C–550 °C	0.826	0.876	0.944	0.966	0.078	0.273	0.495	0.653	0.969
599 °C–600 °C	0.955	0.958	0.935	0.929	0.735	0.802	0.866	0.909	0.918
649 °C–650 °C	0.179	0.452	0.470	0.549	0.790	0.837	0.115	0.471	0.582
Interpolation Method	0.968	0.990	0.991	0.992	0.918	0.921	0.916	0.976	0.983
Weighting Method	0.986	0.967	0.989	0.990	0.979	0.967	0.940	0.976	0.955

Table 7. A summary of peak differences between true (FEA) and predicted stress values ($\Delta\sigma_{VM}$) for the test temperatures profiles, fitted using the three Green's function implementation methods.

Method	A	B	C	D	E	F	G	H	I
249 °C–250 °C	5.09	6.65	3.95	5.90	10.49	19.36	8.94	16.50	31.59
299 °C–300 °C	2.63	5.37	1.68	2.40	13.49	24.79	11.97	22.20	35.63
349 °C–350 °C	3.54	8.47	3.07	5.02	15.02	27.56	13.52	25.10	37.95
399 °C–400 °C	6.73	15.92	6.74	11.06	18.68	34.02	17.26	32.22	41.15
449 °C–450 °C	4.14	10.61	4.03	5.40	15.98	28.36	14.70	27.80	26.72
499 °C–500 °C	3.68	9.64	3.52	4.60	15.45	27.07	14.25	27.12	21.80
549 °C–550 °C	5.59	13.37	5.29	7.63	17.22	29.56	16.19	31.06	17.26
599 °C–600 °C	3.86	8.19	5.28	9.94	10.69	16.58	10.34	19.86	24.43
649 °C–650 °C	11.39	21.86	12.04	23.09	8.01	13.15	10.94	22.38	46.83
Interpolation Method	2.76	6.14	2.53	3.69	4.72	7.85	4.34	9.36	9.75
Weighting Method	3.51	8.69	2.59	4.43	2.92	5.90	5.52	7.38	13.78

6. Discussion and Conclusions

The coefficient of determination (R^2) provides a statistical measure to describe the amount of variance accounted for in a model when predicting some "true" data (taking limiting values of zero if no variance is accounted for and one if all variance is accounted for [23]). The values determined for R^2 shown in Table 6 are plotted in Figure 14 (note the range of values determined when using individual Green's functions is presented). Similarly, a plot of the peak instantaneous stress differences ($\Delta\sigma_{VM}$; see Table 7) is given in Figure 15. This quantity is of interest as, when attempting to quantify the threat of thermal fatigue using a reference elastic solution, stress ranges experienced in a component are one of the most fundamental ways to characterise a particular loading scenario.

In all cases, solutions where temperature dependency was considered in the Green's function method showed an improvement in fitting quality. All results were at least in the 98th percentile of the range predicted when using individual Green's functions (for the majority of cases, R^2 and $\Delta\sigma_{VM}$ values were better than even the optimum individual Green's function solution). Despite the small number of load cases considered, some general comments can be made. R^2 values are comparable for most cases, and in almost all cases, the interpolation technique resulted in lower $\Delta\sigma_{VM}$ values than the weighting function technique (suggesting small stress over-/under-estimations). Anomalies to these observations are found in load Cases E and F, where the weighting function method appears to give superior results, represented by an approximate 2 MPa reduction in stress differences. It is also noted from Figures 11–13 that load cases with large temperature increments (Cases A, B, G and H) do not result in smooth stress development curves that monotonically increase to a maximum value and then decay (i.e., there are "ripples" in the thermal stress histories; see Figure 11b in particular). These features are not seen in more modest temperature step load cases.

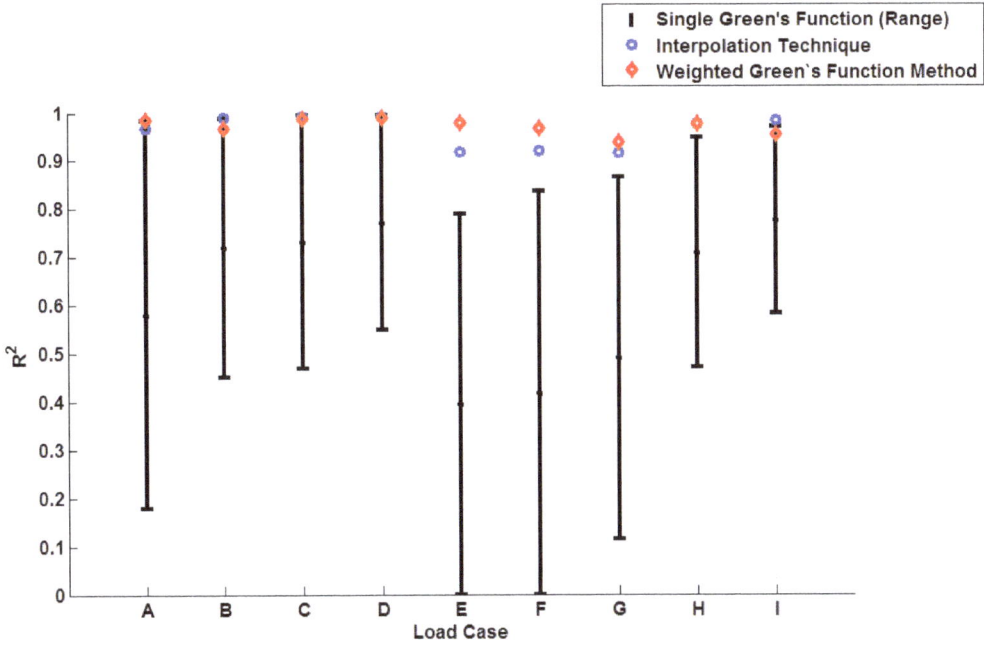

Figure 14. A plot to show the relative performance of the various implementation methods (using R^2).

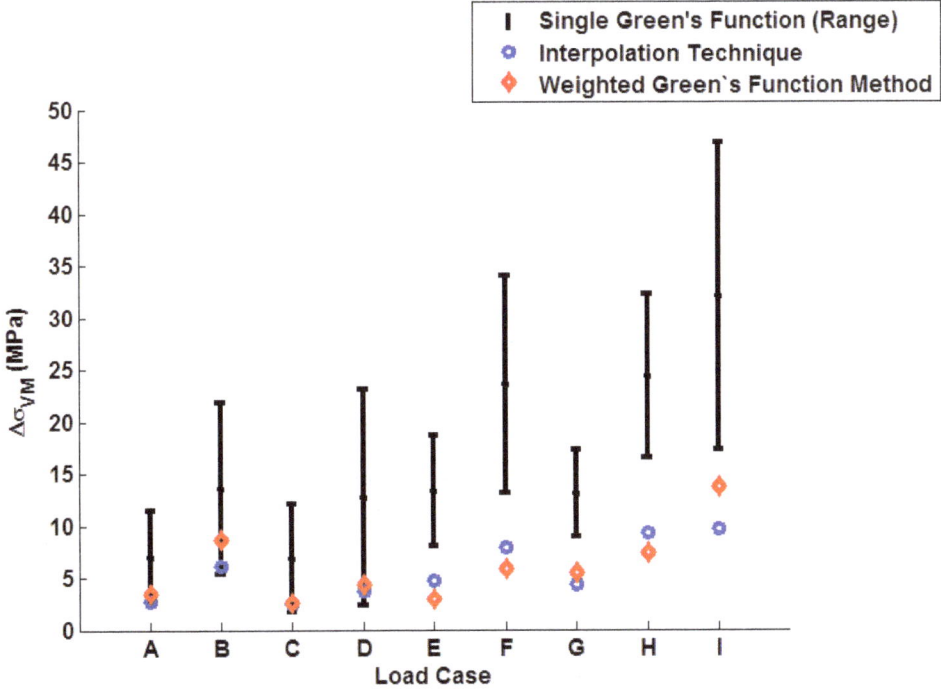

Figure 15. A plot to show the relative performance of the various implementation methods (using $\Delta\sigma_{VM}$).

Potential explanations for these behaviours can be found from analysis of the individual Green's function components themselves and, in particular, the relative effects of the seven exponential components. These components are plotted in Figure 16 using the Green's function coefficients given in Table 4. Of interest for the present work are the fifth and sixth order components (the C_6 and C_7 terms, respectively; most other components vary little over the given temperature range), which can be seen to (in part) control stress decay in the unit Green's functions. Furthermore, there is a shift at around 450 °C, with the sixth order component becoming more dominant in the decay characteristics and the fifth order component becoming positive (leading to a small contribution in stress development). Over large temperature ranges, multiple Green's functions may be used with a wide range of decay characteristics. The "ripple" features discussed previously are therefore due to the complex (time dependent) interaction of these decay functions. While varying decay characteristics can be accounted for in the proposed interpolation method, the time-independent weight function assumes a scaling dependent only on temperature (hence, the superior fit observed for the interpolation method in Cases A, B, G and H). Cases C–F used temperature ranges between 450 and 650 °C, where the relationship between components in Green's function is reasonably linear (with temperature) and can therefore be accounted for using the weighting function. As the weighting function coefficients are estimated by an optimisation procedure using all unit responses, this linear relationship can be approximated, and local errors in the (particularly in the 450 °C and 650 °C profiles) can be minimised.

The proposed interpolation method has been shown to be adept at predicting thermal stress histories in thick-walled components. Over large temperature ranges (>150 °C), where material properties may vary significantly, the interpolation procedure has out performed the weighting method. For more modest temperature ranges, the two methods are generally comparable. In addition to the increased generality that the interpolation procedure offers, it is worth highlighting important practical advantages. The order of the weighting function polynomial given in Equation (12) is ultimately dependent on the number of reference solutions generated. In the present work, nine unit temperature steps were applied to FEA models and used in the temperature dependency methods. Fewer reference solutions may be used in practice, however, due to practical limitations, potentially limiting the applicability of the weighting function. While a greater number of reference solutions is beneficial to the interpolation procedure, even a small number can be used to understand core variations in stress profiles with representative temperature. Some aspects of the second order effects may therefore also be captured. Future work will look to quantify the effect the number of available reference solutions has on the performance of the temperature dependency methods.

In conclusion, the present work has highlighted the importance of including temperature dependency in the Green's function method in order to better estimate transient thermal stresses for realistic bulk steam temperature increments in thick-walled components. The proposed interpolation technique provides a general procedure to incorporate temperature dependency in Green's function analysis. There is some suggestion from the results that the relative importance of each Green's function component varies with temperature and leads to complex interactions between thermal stress development and decay terms. Over larger temperature ranges, this has been seen to have an effect; however, for small temperature ranges, the weighting function method proposed by Koo *et al.* appears to be satisfactory [17]. Future work will therefore focus on increasing the number of load cases considered in order to verify the suggested phenomenon and in accounting for spatial variations in material properties and chosen analysis points.

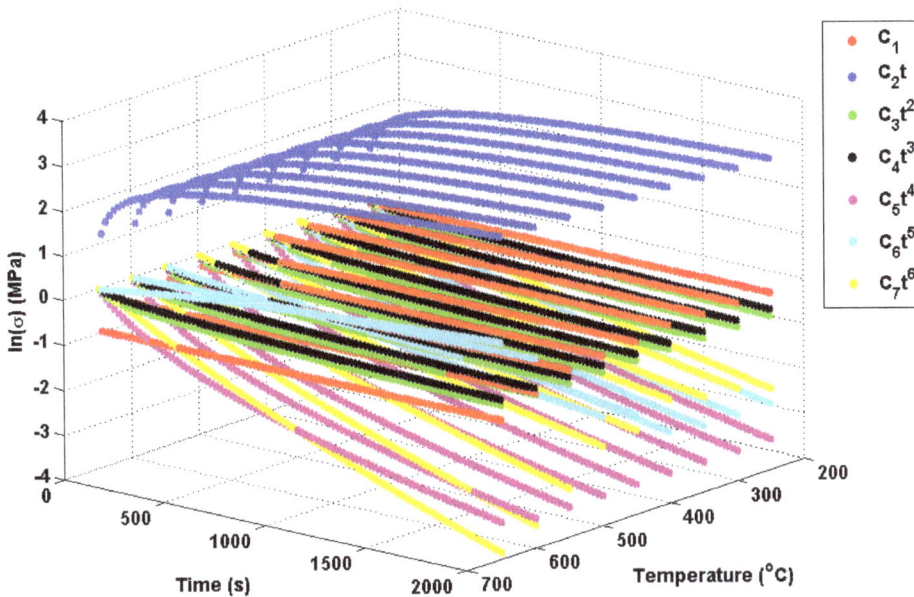

Figure 16. Decomposed Green's functions, showing the relative importance of the exponential components.

Acknowledgments: The authors would like to express their sincere gratitude to Andy Morris and EDF (Électricité de France) UK for their tremendous assistance and openness with information.

Author Contributions: The present work was written jointly between James Rouse and Christopher Hyde.

Conflicts of Interest: The authors declare no conflict of interest.

References

1. Beatt, R.J.I.; Birch, W.L.; Hinton, S.E.; Kelly, M.; Pully, M.J.; Wright, G.R.L.; Wright, W. Two-shift operation of 500 MW boiler/turbine generating units. *Proc. Inst. Mech. Eng. Part A J. Power Energy* **1983**, *197*, 247–255.

2. Shibli, A.; Ford, J. Damage to coal power plants due to cyclic operation. In *Coal Power Plant Materials and Life Assessment: Developments and Applications*; Shibli, A., Ed.; Woodhead Publishing: Cambridge, UK, 2014; pp. 333–357.

3. Samal, M.K.; Dutta, B.K.; Guin, S.; Kushwaha, H.S. A finite element program for on-line life assessment of critical plant components. *Eng. Fail. Anal.* **2009**, *16*, 85–111.

4. Mukhopadhyay, N.K.; Dutta, B.K.; Kushwaha, H.S. On-line fatigue-creep monitoring system for high-temperature components of power plants. *Int. J. Fatigue* **2001**, *23*, 549–560.

5. Daga, R.; Samal, M.K. Real-Time Monitoring of High Temperature Components. In Proceedings of the 6th International Conference on Creep, Fatigue and Creep-Fatigue Interaction, Kalpakkam, India, 22–25 January 2012; Volume 63, pp. 87–89.

6. Benato, A.; Bracco, S.; Stoppato, A.; Mirandola, A. LTE: A procedure to predict power plants dynamic behaviour and components lifetime reduction during transient operation. *Appl. Energy* **2016**, *162*, 880–891.

7. Wermelinger, T.; Bruckmüller, F.; Heinz, B. Fatigue Monitoring in the Context of Long-Term Operation of the Goesgen Nuclear Power Plant Using AREVA's FAMOSi. In Proceedings of the ASME 2015 Pressure Vessels and Piping Conference, Boston, MA, USA, 19–23 July 2015; Volume 7, doi:10.1115/PVP2015-45272.

8. Heinz, B.; Wu, D. AREVA's Modularized Fatigue Monitoring for Lifetime Extension and Flexible Plant Operation. In Proceedings of the ASME 2014 Pressure Vessels and Piping Conference, Anaheim, CA, USA, 20–24 July 2014; Volume 7, doi:10.1115/PVP2014-28726.

9. Farragher, T.P.; Scully, S.; O'Dowd, N.P.; Leen, S.B. Development of life assessment procedures for power plant headers operated under flexible loading scenarios. *Int. J. Fatigue* **2013**, *49*, 50–61.

10. Paterson, I.R.; Wilson, J.D. Use of damage monitoring systems for component life optimisation in power plant. *Int. J. Press. Vessels Pip.* **2002**, *79*, 541–547.

11. Ainsworth, R.A. R5 procedures for assessing structural integrity of components under creep and creep-fatigue conditions. *Int. Mater. Rev.* **2006**, *51*, 107–126.

12. Ainsworth, R.A.; Booth, S.E. Use of R5 in plant defect assessment. *Mater. High Temp.* **1998**, *15*, 299–302.

13. Ainsworth, R.A.; Hooton, D.G. R6 and R5 procedures: The way forward. *Int. J. Press. Vessels Pip.* **2008**, *85*, 175–182.

14. Sakai, K.; Hojo, K.; Kato, A.; Umehara, R. On-line fatigue-monitoring system for nuclear power plant. *Nuclear Eng. Des.* **1994**, *153*, 19–25.

15. Maekawa, O.; Kanazawa, Y.; Takahashi, Y.; Tani, M. Operating data monitoring and fatigue evaluation systems and findings for boiling water reactors in Japan. *Nuclear Eng. Des.* **1995**, *153*, 135–143.

16. Stevens, G.L.; Ranganath, S. Use of on-line fatigue monitoring of nuclear reactor components as a tool for plant life extension. *J. Press. Vessel Technol.* **1991**, *113*, 349–357.

17. Koo, G.H.; Kwoon, J.J.; Kim, W. Green's function method with consideration of temperature dependent material properties for fatigue monitoring of nuclear power plants. *Int. J. Press. Vessels Pip.* **2009**, *86*, 187–195.

18. Boley, B.A.; Weiner, J.H. *Theory of Thermal Stresses*; John Wiley and Sons, Inc.: Hoboken, NJ, USA, 1960.

19. *Abaqus*; Version 6.9; Dassault Systèmes: Paris, France, 2009.

20. Nakoneczny, G.J.; Schultz, C.C. Life Assessment of High Temperature Headers. In Proceedings of the 57th American Power Conference, Chicago, IL, USA, 18–20 April 1995; Volume 63, pp. 87–89.

21. Yaghi, A.; Hyde, T.H.; Becker, A.A.; Williams, J.A.; Sun, W. Residual stress simulation in P91 pipe welds. *J. Mater. Process. Technol.* **2005**, *167*, 480–487.

22. *Optimisation Toolbox User's Guide, Matlab*; R2014a; MathWorks: Natick, MA, USA, 2008.

23. Navidi, W. *Statistics for Engineers and Scientists*; McGraw-Hill: New York, NY, USA, 2008.

9

Facile Synthesis of $SrCO_3$-$Sr(OH)_2$/PPy Nanocomposite with Enhanced Photocatalytic Activity under Visible Light

Alfredo Márquez-Herrera [1,*], Victor Manuel Ovando-Medina [2], Blanca Estela Castillo-Reyes [2], Martin Zapata-Torres [3], Miguel Meléndez-Lira [4] and Jaquelina González-Castañeda [5]

Academic Editor: Klara Hernad

[1] Departamento de Ingeniería Agrícola, DICIVA, Campus Irapuato-Salamanca, Universidad de Guanajuato, Ex Hacienda el Copal, Carr. Irapuato-Silao km 9, Irapuato Gto 36500, Mexico

[2] Ingeniería Química, COARA, Universidad Autónoma de San Luis Potosí, Carr. a Cedral Km 5+600, San José de las Trojes, Matehuala, San Luis Potosí 78700, Mexico; victor.ovando@uaslp.mx (V.M.O.-M.); becr_iq@yahoo.com.mx (B.E.C.-R.)

[3] Centro de Investigación en Ciencia Aplicada y Tecnología Avanzada, Unidad Legaría IPN, Calzada Legaría 694, Col. Irrigación, México D.F. 11500, Mexico; mzapatat@ipn.mx

[4] Departamento de Física, CINVESTAV-IPN, Apartado Postal 14-740, México D.F. 07000, Mexico; mlira@fis.cinvestav.mx

[5] Departamento de Ingeniería Ambiental, DICIVA, Campus Irapuato-Salamanca, Universidad de Guanajuato, Ex Hacienda el Copal, Carr. Irapuato-Silao km 9, Irapuato Gto 36500, Mexico; jaquegc1@hotmail.com

* Correspondence: amarquez@ugto.mx

Abstract: Pyrrole monomer was chemically polymerized onto $SrCO_3$-$Sr(OH)_2$ powders to obtain $SrCO_3$-$Sr(OH)_2$/polypyrrole nanocomposite to be used as a candidate for photocatalytic degradation of methylene blue dye (MB). The material was characterized by Fourier transform infrared (FTIR) spectroscopy, UV/Vis spectroscopy, and X-ray diffraction (XRD). It was observed from transmission electronic microscopy (TEM) analysis that the reported synthesis route allows the production of $SrCO_3$-$Sr(OH)_2$ nanoparticles with particle size below 100 nm which were embedded within a semiconducting polypyrrole matrix (PPy). The $SrCO_3$-$Sr(OH)_2$ and $SrCO_3$-$Sr(OH)_2$/PPy nanocomposites were tested in the photodegradation of MB dye under visible light irradiation. Also, the effects of MB dye initial concentration and the catalyst load on photodegradation efficiency were studied and discussed. Under the same conditions, the efficiency of photodegradation of MB employing the $SrCO_3$-$Sr(OH)_2$/PPy nanocomposite increases as compared with that obtained employing the $SrCO_3$-$Sr(OH)_2$ nanocomposite.

Keywords: composite materials; inorganic compounds; nanostructures; chemical synthesis; X-ray photo-emission spectroscopy (XPS)

1. Introduction

During the past decade, photocatalytic degradation has proven to be a promising technology for the removal of various organic pollutants in waste water because of its many attractive advantages, including its environmental friendly feature, relatively low cost, and low energy consumption [1–10]. Photocatalytic processes are methods that utilize the solar radiation energy to perform catalytic processes such as water splitting, waste mineralization, recovery of precious metals, *etc.* [11,12]. Many photocatalytic materials have wide bad-gap values and require ultraviolet light (UV) to be photoactive. However, the need of UV light for activating the photocatalyst greatly limits practical applications because of the low content of UV light in the solar spectrum (about 4%) [13]. Therefore to take complete advantage of the sunlight one needs to make a visible light activated photocatalyst or increase its efficiency in the UV light region. In order to narrow the band gap of these materials, several researchers have focused on modifications by doping with appropriate ions [6,14–21]. Also, it has been reported that by using composite films or powders consisting of two semiconducting photocatalysts the absorption edge is shifted to the visible light region, e.g., TiO_2-$SrTiO_{3-\delta}$ [22], $BiVO_4$-$SrTiO_3$:Rh [23], Ag_3PO_4-Cr-$SrTiO_3$ [24], Fe_2O_3-$SrTiO_3$ [25], $SrCO_3$-$SrTiO_3$ [26], TiO_2-SO_4 [27], g-C_3N_4/Fe_3O_4/Ag_3VO_4 [28].

Conducting polymers (e.g., polyaniline, polypyrrole, and polythiophene) with delocalized conjugated structures have been widely studied due to their rapid photoinduced charge separation and relatively slow charge recombination [29,30]. In particular, polypyrrole (PPy) with extended p-conjugated electron systems has recently shown great promises to enhance photocatalytic activity owing to its unique electrical and optical properties, such as high absorption coefficients in the visible light, high mobility of charge carriers, and excellent stability [31,32]. Furthermore, PPy is also an efficient electron donor and good hole transporter upon visible light excitation. It was proposed that polypyrrole has the ability to channel the photoinduced holes from the surface of the semiconductor to the polymer/solution interface at a fast rate, which can then oxidize the pollutants [33,34]. The photocatalytic activity of semiconductors modified with PPy have shown that PPy can effectively enhance the photoactivity of TiO_2 [35–37], Ag-TiO_2 [38], Bi_2WO_6 [39], Fe_3O_4/ZnO [40], $Bi_2O_2CO_3$ [41], etc.

Taking into account some reports about $SrCO_3$-$Sr(OH)_2$ composite as background [42–47], and due the $SrCO_3$-$Sr(OH)_2$/PPy nanocomposite has not been studied as a photocatalyst candidate, this manuscript describes a modified strategy for the preparation of $SrCO_3$-$Sr(OH)_2$ nanocomposite, followed by coating with the semiconducting polypyrrole (PPy) to increase its photoactivity in the visible light range. Both $SrCO_3$-$Sr(OH)_2$ and $SrCO_3$-$Sr(OH)_2$/PPy nanocomposites were tested for photodegradation of MB dye under visible light irradiation. The effects on MB dye initial concentration and the catalyst load on photodegradation efficiency were studied and discussed.

2. Results and Discussion

The process here described to obtain $SrCO_3$-$Sr(OH)_2$/PPy nanocomposite consists of two straightforward steps. The first step implies the production of $Sr(OH)_2$ powders as a water insoluble white dust, which precipitates from the reaction medium according to the double-displacement chemical reaction in which the hydrated form of $Sr(OH)_2$ can be formed.

In the search to find a cheap and straight route to obtain the $SrCO_3$ phase, it was chose to dry $Sr(OH)_2$ at ambient atmosphere to take advantage of the reaction with CO_2 present in the air [42–47].

The second step implied the chemical polymerization of pyrrole monomer dispersing $SrCO_3$-$Sr(OH)_2$ nanocomposite using sodium dodecyl sulfate (SDS) surfactant. Since a practical point of view, the $SrCO_3$-$Sr(OH)_2$/PPy nanocomposite can be easily removed from MB aqueous solutions due to its water insolubility, facilitating its recovery. The most interesting characteristic of the $SrCO_3$-$Sr(OH)_2$/PPy composite is its high photoactivity under visible light as will be discussed later.

2.1. Characterization

2.1.1. Chemical Composition

Figure 1 shows the FTIR spectra of $SrCO_3$-$Sr(OH)_2$ and $SrCO_3$-$Sr(OH)_2$/PPy nanocomposites, also it can be observed the characteristic signals of PPy chains. The peak at 1480 cm^{-1} is ascribed to C–C ring stretching; the peak around 1560 cm^{-1} is due to C=C backbone stretching; and the peaks at 1300 and 1120 cm^{-1} are due to C–H in-plane and C–N stretching vibrations, respectively [48]. The peak located at 1560 cm^{-1} is considered as a reflection of the conducting polymer. Combined signals of $SrCO_3$-$Sr(OH)_2$/PPy were observed in Figure 1b indicating the interaction of $SrCO_3$, $Sr(OH)_2$ and PPy in the composite. The spectrum corresponding to $SrCO_3$-$Sr(OH)_2$ nanocomposite shows three main peaks, Figure 1a. The Peak at 3350 cm^{-1} is due to O–H physically adsorbed on the surface, the signal at 3500 cm^{-1} is ascribed to O–H bonds in $Sr(OH)_2$ phase. The peak at 1440 cm^{-1} is usually observed when C=O bonds are present [46]; in our case this signal can be due to the presence of the $SrCO_3$ phase. When the $SrCO_3$-$Sr(OH)_2$ nanocomposite is coated with PPy, the signals corresponding to $SrCO_3$-$Sr(OH)_2$ are masked by the semiconducting PPy, Figure 1b.

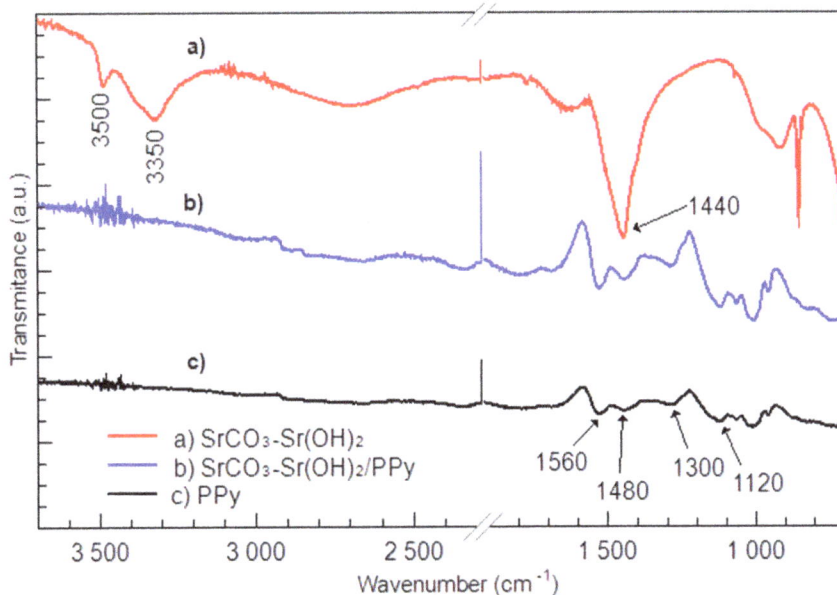

Figure 1. Fourier transform infrared (FTIR) spectra of (**a**) $SrCO_3$-$Sr(OH)_2$; (**b**) $SrCO_3$-$Sr(OH)_2$/PPy nanocomposites; and the characteristic signals of (**c**) polypyrrole matrix (PPy) chains.

In order to obtain insights into the chemical environment of the elements of the $SrCO_3$-$Sr(OH)_2$ powders a study using X-ray photoelectron spectroscopy (XPS) was performed. The general survey XPS spectrum, Figure 2, shown peaks related with Sr, O and C.

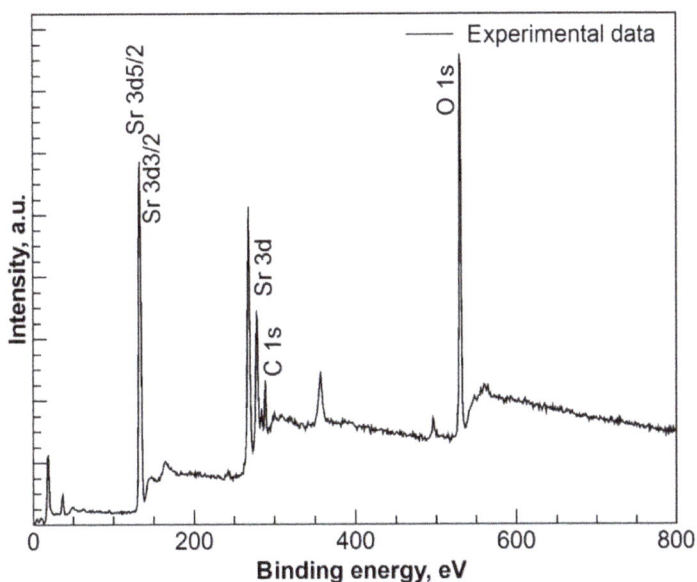

Figure 2. X-ray Photoelectron Spectroscopy (XPS) spectrum for $SrCO_3$-$Sr(HO)_2$ nanocomposite.

Figure 3 shows the high resolution XPS spectrum associated to the Sr binding energies for the $SrCO_3$-$Sr(OH)_2$ sample. It shows a peak fit analysis of the Sr $3d_{5/2}$ and Sr $3d_{3/2}$ signals with mixed Gaussian-Lorentzian profiles that reveals two underlying components of the binding energies at 133.2 eV and 134.8 eV. The inset in Figure 3 shows only the deconvolution of the Sr $3d_{5/2}$ signal that is attributed to Sr bonded to $Sr(HO)_2 \cdot 8H_2O$ and $SrCO_3$. The deconvolution for the Sr $3d_{3/2}$ signal was not carried out because there is no reported information about its strength in the compound $Sr(OH)_2$. The positions of the peaks were obtained from the X-ray Photoelectron Spectroscopy Database of NIST [49]. This result confirms that $Sr(HO)_2 \cdot 8H_2O$ and $SrCO_3$ phases are present in the composite.

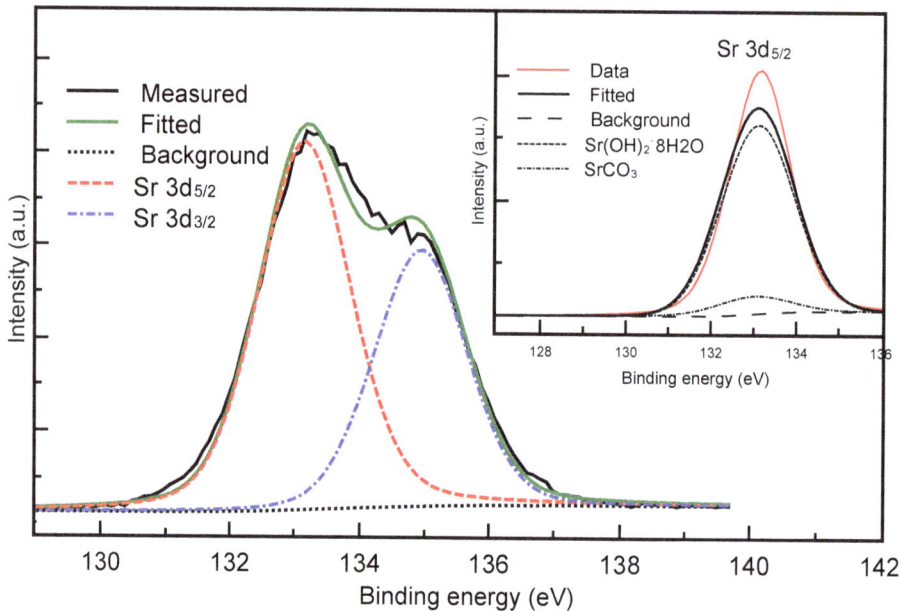

Figure 3. XPS spectrum of Sr $3d_{5/2}$ and Sr $3d_{3/2}$ for the $SrCO_3$-$Sr(HO)_2$ sample. The inset shows the deconvolution of the Sr $3d_{5/2}$ signal.

2.1.2. Crystallinity and Morphology

Figure 4 shows the transmission electron micrograph corresponding to $SrCO_3$-$Sr(OH)_2$ sample without polypyrrole. As can be seen, the powders consisted of clusters of $SrCO_3$-$Sr(OH)_2$ nanoparticles. Although the morphology of nanoparticles is not clearly defined, it looks like circular shapes. It can be observed that the particle size is below 100 nm. Due TEM technique burns the polypyrrole, it is not appropriate to verify the existence of PPy on the surface of $SrCO_3$-$Sr(OH)_2$ sample with TEM images. However it is worth to mentioning that the composite $SrCO_3$-$Sr(OH)_2$/PPy has a core-shell structure [50].

BET area for $SrCO_3$-$Sr(OH)_2$ particles was found to be 8.5 m^2/g, this high surface area is already evident from TEM image. This value is similar to the reported by Viriya-Empikul, *et al.* [51] for $SrCO_3$-$Sr(OH)_2 \cdot H_2O$ composite (5.2 m^2/g). This high surface area has a relevance because the surface of the photocatalytic material in contact with the contaminant plays an important role in determining the photocatalytic activity of the composite powders [52].

Figure 5 shows X-ray diffractogram of the $SrCO_3$-$Sr(OH)_2$ nanocomposite; the corresponding to the $SrCO_3$-$Sr(OH)_2$/PPy composite just incorporated a broad signal characteristic of amorphous polypyrrole. The positions of the diffraction peaks associated to the orthorhombic $Sr(OH)_2 \cdot 8H_2O$, $Sr(OH)_2 \cdot H_2O$ and $SrCO_3$ from the 271438, 281222 and 050418 cards of the Powder Diffraction File database (PDF card) are also shown. The close coincidence with the reported positions allows to establish that the peaks presented in the experimental diffractogram are due to the diffraction from the planes of the $Sr(OH)_2 \cdot H_2O$ and $Sr(OH)_2 \cdot 8H_2O$. A small signal from the planes of the $SrCO_3$ phase was found. The X-ray diffraction analysis corroborates that the powders are composed by $Sr(OH)_2 \cdot H_2O$, $Sr(OH)_2 \cdot 8H_2O$ and $SrCO_3$ (called $SrCO_3$-$Sr(OH)_2$) highly ordered crystals because there is a complete correspondence between the experimental diffraction peaks and the data base positions (for annealed sample see Figure S1).

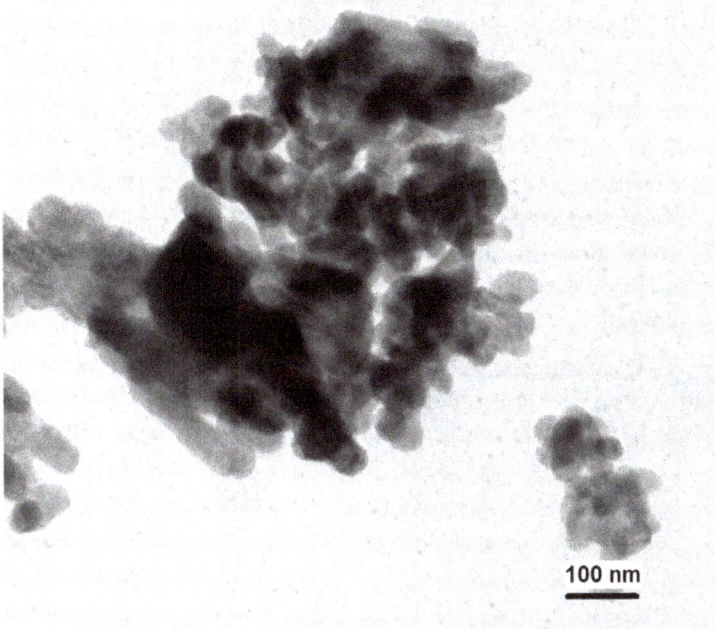

Figure 4. Transmission electron microscopy (TEM) image of the as-prepared $SrCO_3$-$Sr(OH)_2$ nanoparticles without polypyrrole.

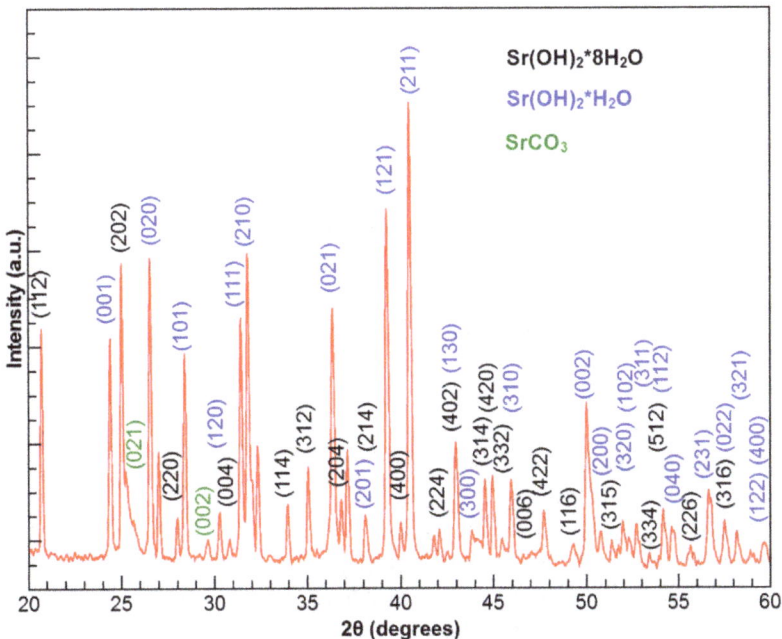

Figure 5. The X-ray diffraction (XRD) pattern obtained for the $SrCO_3$-$Sr(OH)_2$ nanocomposite.

From Rietveld refinement with an adjust factor R_{WP} better than 10%, the percentage of phases found in the composite were 59.3% $\pm$ 1.2%, 32.3% $\pm$ 0.7% and 8.4% $\pm$ 0.3% for $Sr(OH)_2 \cdot 8H_2O$, $Sr(OH)_2 \cdot H_2O$ and $SrCO_3$, respectively.

2.1.3. Photocatalytic Activity

The photocatalytic performances of the $SrCO_3$-$Sr(OH)_2$ and $Sr(OH)_2$/PPy nanocomposites were studied following the degradation process of aqueous solutions of MB dye under visible light irradiation. Figure 6 shows the UV/Vis spectra of MB aqueous solutions at different times. Solutions were prepared employing an initial MB concentration of 20 mg/L and 0.2 g of both (a) $SrCO_3$-$Sr(OH)_2$ and (b) $SrCO_3$-$Sr(OH)_2$/PPy nanocomposite load. It should be noted that there is a small difference in the $SrCO_3$-$Sr(OH)_2$ weight of the catalyst employed because the PPy. However, it highlights the positive effect of PPy on the photocatalytic activity of $SrCO_3$-$Sr(OH)_2$. It can be seen that the peak at $\lambda = 665$ nm decreases with the visible light irradiation time, reaching a minimum after 30 min for the $SrCO_3$-$Sr(OH)_2$/PPy nanocomposite. Insets in Figure 6 clearly shown a more discolored solution for the catalyst containing PPy. Based in the above results, photodegrading kinetics studies were made considering 30 min of reaction.

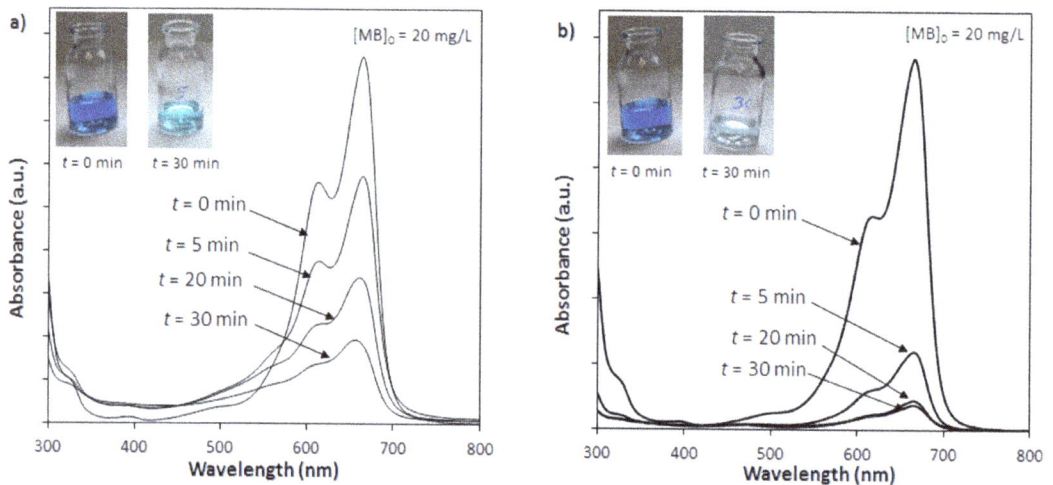

Figure 6. Ultraviolet-visible (UV/Vis) spectra of methylene blue dye (MB) aqueous solutions at different times for a 0.2 g of (a) $SrCO_3$-$Sr(OH)_2$; and (b) $SrCO_3$-$Sr(OH)_2$/PPy nanocomposites.

Figure 7 shows the ratio of residual to initial MB concentration (C/C_0) as a function of time using the $SrCO_3$-$Sr(OH)_2$ and the $SrCO_3$-$Sr(OH)_2$/PPy nanocomposites (0.2 g for both cases) at different MB initial concentrations. It can be observed that for the lower value of MB initial concentration, similar degradation efficiencies can be achieved after 30 min of visible light exposition for both materials (efficiency around 85%). However when the MB initial concentration was increased up to 50 mg/L, maintaining constant the catalyst load to 0.2 g, lower degradation efficiencies were obtained; 9.8% for bare $SrCO_3$-$Sr(OH)_2$ nanocomposite compared to 75.6% for $SrCO_3$-$Sr(OH)_2$/PPy nanocomposite. Furthermore, only 10 min were needed to achieve 71% of degradation for 10 mg/L of initial MB concentration using the $SrCO_3$-$Sr(OH)_2$/PPy nanocomposite.

By other hand, Figure 8 shows the effect of $SrCO_3$-$Sr(OH)_2$ and $SrCO_3$-$Sr(OH)_2$/PPy nanocomposite photocatalyst load on the photodegradation kinetics with a fixed initial MB concentration (20 mg/L). Employing 0.3 g of catalyst load, it can be observed for the $SrCO_3$-$Sr(OH)_2$ nanocomposite that 43.1% of degradation efficiency was achieved after 10 min of visible light irradiation and 83.6% after 30 min, whereas for the $SrCO_3$-$Sr(OH)_2$/PPy nanocomposite the corresponding efficiencies were 73.6% and 93.2%, respectively. Decreasing the amount of catalyst to

0.1 g and after 30 min of degradation, it resulted in a decrease in degradation efficiency from 93.2% to 75.1% for the $SrCO_3$-$Sr(OH)_2$/PPy nanocomposite; while for $SrCO_3$-$Sr(OH)_2$ nanocomposite dropped the degradation efficiency to 43.1%, thus, the photocatalyst amount strongly affects the efficiency of MB degradation, and show the enhanced performance of the $SrCO_3$-$Sr(OH)_2$/PPy nanocomposite under the studied conditions. These results show that $SrCO_3$-$Sr(OH)_2$ based nanocomposites are promising materials with excellent performance in photocatalytic applications, and the incorporation of PPy enhances noticeably their performance.

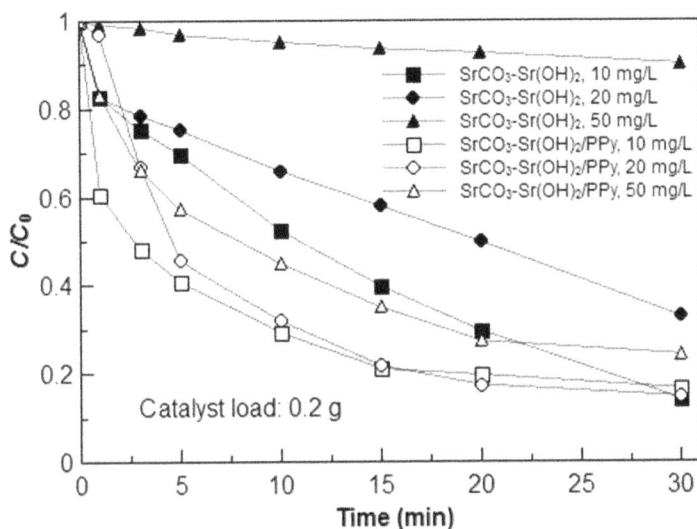

Figure 7. Kinetics of MB dye photodegradation under visible light irradiation using $SrCO_3$-$Sr(OH)_2$ and $SrCO_3$-$Sr(OH)_2$/PPy nanocomposites for the MB initial concentrations indicated in the label. The catalyst load was 0.2 g for each case.

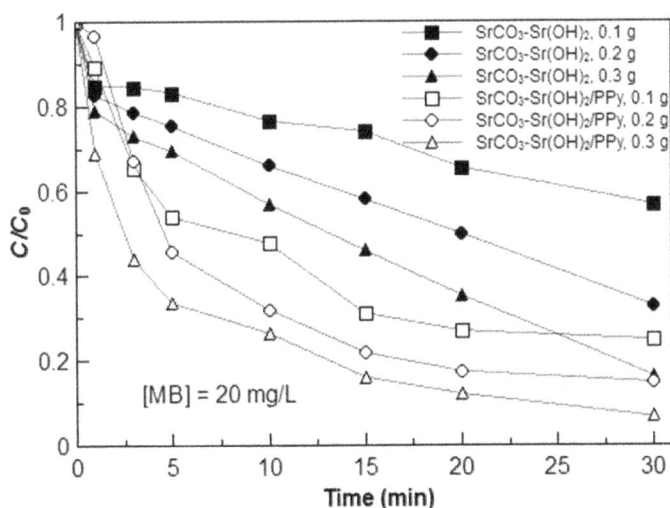

Figure 8. MB dye photodegradation kinetics under visible light irradiation using 20 mg/L of MB initial concentration and different $SrCO_3$-$Sr(OH)_2$ and $SrCO_3$-$Sr(OH)_2$/PPy nanocomposites loading.

Blank measurements were carried out employing both catalysts and MB solutions at low concentrations without observe any degradation at all confirming that these catalysts are activated by visible light (Figures S2–S4).

Because $Sr(OH)_2$ is an ionic compound, not a semiconductor, the photocatalytic activity of the composite containing $Sr(OH)_2$ cannot explained by changes in the band structure of it [52]. It is possible that the presence of the $SrCO_3$ phase is the responsible of the improvement in the photocatalytic activity of the composite [26]. The increase in carriers due to the absorption process in the semiconductor PPy coating injects more electrons to the $SrCO_3$ compound [48,53,54] increasing its photodegradation activity efficiency. However, the principle of the photocatalytic oxidation due to the $Sr(OH)_2$ phase still needs to be clarified.

In our particular case, the formation of semiconducting $SrCO_3$ (which has a reported band gap energy of 3.17 eV) [54,55] during both, drying of pure $Sr(OH)_2$ and dye photodegradation due to the CO_2 adsorption from air and water, respectively, permits the explanation of dye photodegradation mechanism as shown in the Figure 9. The performance of this composite is determined by the relative positions of the bands of nanoparticles and PPy. The values of each bandgap are reported in Figure 9, however, the exact determination of their position is beyond the scope of this work.

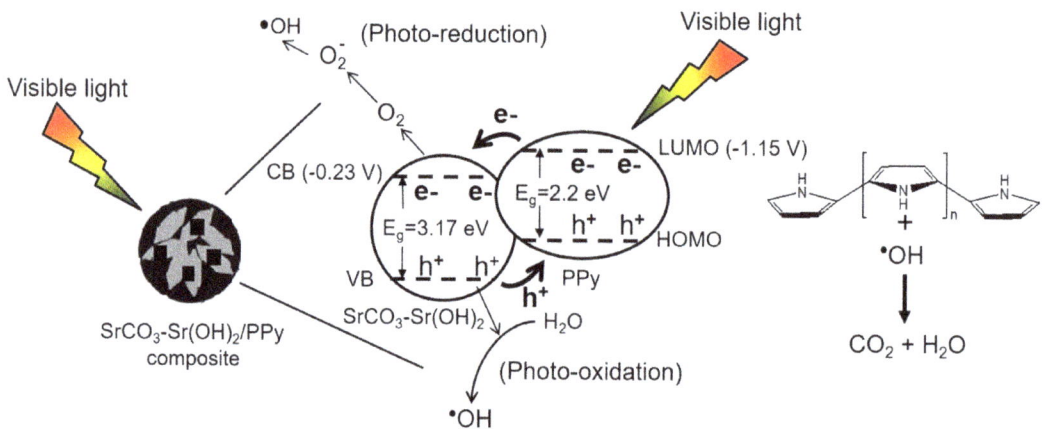

Figure 9. Possible MB dye photodegradation process.

When visible light impinges on the composite surface, electrons are promoted from HOMO to LUMO of PPy (which has a reported band gap energy of 2.2 eV) [56], generating holes in the PPy chain (h+), electrons in the LUMO (-1.15 V versus NHE) [56], travel through PPy chains to conduction band (CB), (-0.23 V versus NHE) [57], of inorganic material, which can react with oxygen solved in the aqueous phase initiating photo-reduction. On the other hand, electrons in the valence band (VB) of inorganic material travels to h+ in the HOMO of PPy, generating a hole in the VB in the inorganic material. These holes can react with water generating $\cdot$OH radicals, which attack organic molecules (photo-oxidation) until mineralization is done.

In the photocatalytic degradation of methylene blue, not only do O_2^- and $\cdot$OH play important roles, but the holes generated in the HOMO band of PPy also play a role, however they have a lower oxidative capability than those in the valence band of $SrCO_3$, as shown in Figure 9. It is energetically unfavorable to use pure PPy to oxidize methylene blue molecules to form $\cdot$OH radicals, because the methylene blue molecules need to be attacked by hydroxyl radicals to generate organic radicals or other intermediates.

3. Materials and Methods

3.1. Materials

In the present study, all chemicals used were analytical reagent grade. Strontium nitrate hexahydrate ($Sr(NO_3)_2 \cdot 6H_2O$) and sodium hydroxide (NaOH) were purchased from Onyx-Met, Inc. (Olsztyn, Poland). Methylene blue dye was purchased from Fluka (Toluca, Mexico). Pyrrole monomer and ammonium persulfate (APS) were purchased from Sigma-Aldrich (Toluca, Mexico). Sodium dodecyl sulfate (SDS) was acquired from Hycel (Guadalajara, Mexico). Deionized water was used in all the experiments.

3.2. Methods

3.2.1. Synthesis of $SrCO_3$-$Sr(OH)_2$/PPy Nanocomposite

The $SrCO_3$-$Sr(OH)_2$/PPy nanocomposite was prepared as described in Figure 10: first, NaOH (2.0000 g) and $Sr(NO_3)_2 \cdot 6H_2O$ (10.5814 g) were mixed together in distilled water (30 mL) (molar ratio of NaOH/$Sr(NO_3)_2$ of 2:1) under 450 rpm magnetic stirring by 2 h, resulting in a precipitate as a fine white powder of $Sr(OH)_2$ which was water insoluble. Afterward, the precipitates were filtered using Whatman 42 filter paper, washed several times with de-ionized water. Then, sample was dried at 90 °C in air for 2 h without annealing (Figures S5 and S6). Afterwards, dried sample of $SrCO_3$-$Sr(OH)_2$ (0.2500 g) was well dispersed in an aqueous solution of SDS (consisting in 30 mL of water and 0.8 g of SDS). This mixture was ultrasonicated (Cole-Parmer Instruments, CPX 130, Vernon Hill, IL, USA) by 10 min for homogenization; 0.4 g of pyrrole monomer was added and homogenized under magnetic stirring through 2 h. Then, APS was dissolved in 10 mL of water (0.6 M) and added to the reaction mixture to start pyrrole polymerization. The reaction proceeded under magnetic stirring for 1 h. The reaction mixture was poured into an excess of methanol to precipitate the $SrCO_3$-$Sr(OH)_2$/PPy composite (black dust). The sample was decanted and dried at 60 °C in an oven for 24 h.

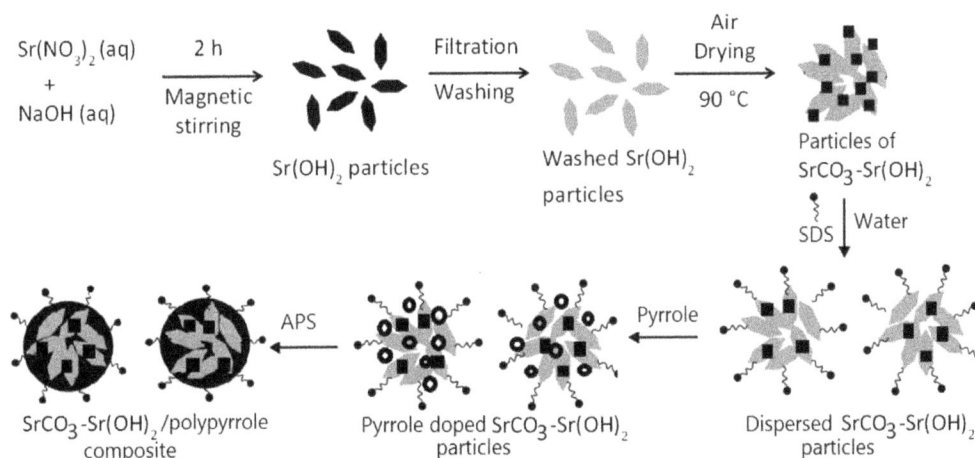

Figure 10. Experimental process to obtain SrCO3-Sr(OH)2/PPy nanocomposite.

3.2.2. Characterization

The chemical environment structures of strontium, carbon and oxygen were analyzed by X-ray Photoelectron Spectroscopy (XPS) (model K alpha, Thermo Scientific, Waltham, CT, USA). The general survey, as well as the high resolution spectra in the regions of the C 1s, O 1s and Sr 3d were obtained. The binding energy of the C 1s line at 284.5 eV was taken as the reference peak to calibrate the obtained spectra. The background subtraction was performed using the mathematical model derived by Shirley [58]. The Sr signal curve was fitted with an asymmetric Gaussian-Lorentzian function. The X-ray diffraction (XRD) measurement was performed with a Rigaku X'pert diffractometer (Rigaku, Tokio, Japan) using the $Cu_{K\alpha}$ line ($\lambda_{k\alpha 1}$ = 1.54056 Å and $\lambda_{k\alpha 2}$ = 1.54439 Å) and the correspondence between the experimental diffraction peaks and database position was made using the Match! 3 phase identification from powder diffraction software (Crystal Impact, Bonn, Germany). In order to determinate the percentage of each phase in the composite, the quantitative phase composition was analyzed according to the Rietveld refinement method [59] using the software Maud (University of Trento, Trento, Italy) [60]. The crystal data for each phase used in the quantitative phase analysis were obtained from Inorganic Crystal Structure Database (ICSD). By other hand, the particle size of representative $SrCO_3$-$Sr(OH)_2$ powders were observed via transmission electron microscopy (TEM) using a JEOL-2010 system (Jeol, Pleasanton, CA, USA) operated at 200 kV where the powders were dispersed in distilled water and deposited on carbon foil on copper grids. The particle size was calculated using the ImageJ 1.46c software (National Institute of Mental Health, Rockville, MD, USA) in TEM images. The average surface area (S_{BET}) of the $SrCO_3$-$Sr(OH)_2$ particles was obtained using a Brunauer-Emmett-Teller (BET) method [61]. For measuring nitrogen adsorption, 68.7 mg of sample, was used. It was dehydrated for four hours at 200 °C, then the adsorption of nitrogen was measured at liquid nitrogen temperature (-197.392 °C).

3.2.3. Photoactivity in the Visible Light of Synthesized Materials

The synthesized $SrCO_3$-$Sr(OH)_2$ and the $SrCO_3$-$Sr(OH)_2$/PPy nanocomposites were tested by photodegradation of aqueous solutions of MB dye under visible light irradiation. The reactor consisted of a glass vessel with two quartz bulbs, the first for water recirculation at constant temperature (20 °C) and the second to insert the visible light source. The effect of the catalyst load on MB degradation was studied using 0.1 g, 0.2 g and 0.3 g of $SrCO_3$-$Sr(OH)_2$ and $SrCO_3$-$Sr(OH)_2$/PPy nanocomposites; catalyst were mixed with 150 mL of aqueous solutions of MB at 20 mg/L of initial concentration (C_0). Afterwards, the mixture was charged to the reactor. In each case, the tested solutions were exposed to a visible light source from a halogen lamp with tungsten filament (Philips LongLife EcoVision H7, 12 V, and 55 W) and a cutoff filter ($\lambda > 400$ nm). Aliquots of 1.5 mL were obtained at different times, centrifuged and poured into a quartz cuvette to determine UV/Vis spectra (250 nm to 800 nm of wavelength) and absorbance (Genesys 10, Thermo-Spectronic) at a wavelength of 664 nm to calculate residual MB concentrations (C) from a calibration curve. In addition, initial MB concentrations were varied from 10 mg/L to 50 mg/L when working with $SrCO_3$-$Sr(OH)_2$ and $SrCO_3$-$Sr(OH)_2$/PPy composites at a fixed load of 2.0 g/L.

4. Conclusions

On the basis of FTIR spectroscopy, XPS, XRD and TEM results, the successful synthesis of $SrCO_3$-$Sr(OH)_2$/PPy nanocomposite was obtained using $Sr(NO_3)_2 \cdot 6H_2O$, NaOH, SDS and pyrrole monomer as precursors. The measurements indicate that the obtained material corresponds to $SrCO_3$-$Sr(OH)_2$ nanocomposite with particle size below 100 nm, which were immersed into a semiconducting polypyrrole matrix. The $SrCO_3$-$Sr(OH)_2$ particles showed only 9.7% of MB dye photodegradation after 30 min of visible light irradiation using a MB initial concentration of 50 mg/L and a catalyst load of 1.3 g/L of solution; and for the same conditions but with 20 mg/L of MB dye initial concentration, the efficiency was 67.0%. The corresponding efficiencies using the $SrCO_3$-$Sr(OH)_2$/PPy composite were 75.6% and 85.2%, respectively. It was also observed that using a catalyst load of 2.0 g/L of solution with 20 mg/L of MB dye initial concentration and after 30 min of photodegradation, 83.6% and 93.2% of efficiency were obtained for $SrCO_3$-$Sr(OH)_2$ and $SrCO_3$-$Sr(OH)_2$/PPy nanocomposites, respectively. In summary, the results obtained in the present study indicate that $SrCO_3$-$Sr(OH)_2$/PPy nanocomposite increased the catalytic efficiency of $SrCO_3$-$Sr(OH)_2$ nanocomposite and it may serve as a promising efficient photocatalyst for the degradation of organic contaminants as the methylene blue. Also, it is important to note that this nanocomposite meets at least four requirements: easy preparation/synthesis with availability of the raw materials, low cost, and highly effective.

Acknowledgments: The work was supported by the Universidad de Guanajuato through project 525/2015. The Technical support from Eng. Wilian Cahuich, José Bante, Marcela Guerrero, Angel Guillen-Cervantes, Rogelio Fragoso-Soriano, Laura Lopez, Daniel Perez-Escamilla and Luis Hernández-Hernandez is acknowledged.

Victor Manuel Ovando-Medina acknowledges the hospitality of Lorena. Farías-Cepeda at the sabbatical leave in FCQ-UA de Coahuila, Saltillo.

Author Contributions: Alfredo Márquez-Herrera conceived and designed the experiments; Alfredo Márquez-Herrera and Blanca Estela Castillo-Reyes performed the experiments; Alfredo Márquez-Herrera, Victor Manuel Ovando-Medina, Martin Zapata-Torres and Miguel Meléndez-Lira analyzed the data; Alfredo Márquez-Herrera, Martin Zapata-Torres, Miguel Meléndez-Lira and Jaquelina González-Castañeda contributed reagents/materials/analysis tools; Alfredo Márquez-Herrera wrote the paper.

Conflicts of Interest: The authors declare no conflict of interest.

References

1. Asahi, R.; Morikawa, T.; Ohwaki, T.; Aoki, K.; Taga, Y. Visible-light photocatalysis in nitrogen-doped titanium oxides. *Science* **2001**, *293*, 269–271. [CrossRef] [PubMed]

2. Tang, J.; Zou, Z.; Ye, J. Efficient photocatalytic decomposition of organic contaminants over $CaBi_2O_4$ under visible-light irradiation. *Angew. Chem. Int. Ed.* **2004**, *43*, 4463–4466. [CrossRef] [PubMed]

3. Kim, T.W.; Hwang, S.-J.; Jhung, S.H.; Chang, J.-S.; Park, H.; Choi, W.; Choy, J.H. Photolysis: Nickel oxide-containing porous nanocomposite. *Adv. Mater.* **2008**, *20*, 539–542. [CrossRef]

4. Hameed, A.; Montini, T.; Gombac, V.; Fornasiero, P. Surface phases and photocatalytic activity correlation of Bi_2O_3/Bi_2O_{4-x} nanocomposite. *J. Am. Chem. Soc.* **2008**, *130*, 9658–9659. [CrossRef] [PubMed]

5. Zeng, J.; Wang, H.; Zhang, Y.; Zhu, M.K.; Yan, H. Hydrothermal synthesis and photocatalytic properties of pyrochlore $La_2Sn_2O_7$ nanocubes. *J. Phys. Chem. C* **2007**, *111*, 11879–11887. [CrossRef]

6. Liu, J.W.; Chen, G.; Li, Z.H.; Zhang, Z.G. Electronic structure and visible light photocatalysis water splitting property of chromium-doped $SrTiO_3$. *J. Solid State Chem.* **2006**, *179*, 3704–3708. [CrossRef]

7. Fujishima, A.; Honda, K. Electrochemical photolysis of water at a semiconductor electrode. *Nature* **1972**, *238*, 37–38. [CrossRef] [PubMed]

8. Yang, J.; Li, D.; Wang, X.; Yang, X.J.; Lu, L.D. Rapid synthesis of nanocrystalline TiO_2/SnO_2 binary oxides and their photoinduced decomposition of methyl orange. *J. Solid State Chem.* **2002**, *165*, 193–198. [CrossRef]

9. Wang, C.; Zhao, J.C.; Wang, X.M.; Mai, B.X.; Sheng, G.Y.; Peng, P.A.; Fu, J.M. Preparation, characterization and photocatalytic activity of nano-sized ZnO/SnO_2 coupled photocatalysts. *Appl. Catal. B Environ.* **2002**, *39*, 269–279.

10. Yamashita, H.; Harada, M.; Misaka, J.; Takeuchi, M.; Ikeue, K.; Anpo, M. Degradation of propanol diluted in water under visible light irradiation using metal ion-implanted titanium dioxide photocatalysts. *J. Photochem. Photobiol. A Chem.* **2002**, *148*, 257–261. [CrossRef]

11. Herrmann, J.M. Heterogeneous photocatalysis: Fundamentals and applications to the removal of various types of aqueous pollutants. *Catal. Today* **1999**, *53*, 115–129. [CrossRef]

12. Matos, J.; Garcia, A.; Zhao, L.; Magdalena-Titirici, M. Solvothermal carbon-doped TiO_2 photocatalyst for the enhanced methylene blue degradation under visible light. *Appl. Catal. A Gen.* **2010**, *390*, 175–182. [CrossRef]

13. Kachina, A.; Puzenat, E.; Ould-Chikh, S.; Geantet, C.; Delichere, P.; Afanasiev, P. A new approach to the preparation of nitrogen-doped titania visible light photocatalyst. *Chem. Mater.* **2012**, *24*, 636–642. [CrossRef]

14. Wang, J.; Yin, S.; Komatsu, M.; Zhang, Q.; Saito, F.; Sato, T. Preparation and characterization of nitrogen doped $SrTiO_3$ photocatalyst. *J. Photochem. Photobiol. A Chem.* **2004**, *165*, 149–156. [CrossRef]

15. Chang, C.H.; Shen, Y.H. Synthesis and characterization of chromium doped $SrTiO_3$ photocatalyst. *Mater. Lett.* **2006**, *60*, 129–132. [CrossRef]

16. Xie, T.H.; Sun, X.; Lin, J. Enhanced photocatalytic degradation of RhB driven by visible light-induced MMCT of $Ti(IV)-O-Fe(II)$ formed in Fe-doped $SrTiO_3$. *J. Phys. Chem. C* **2008**, *112*, 9753–9759. [CrossRef]

17. Sulaeman, U.; Yin, S.; Suehiro, T.; Sato, T. Solvothermal synthesis of $SrTiO_3$-$LnTiO_2N$ solid solution and their visible light responsive photocatalytic properties. *IOP Conf. Ser. Mater. Sci. Eng.* **2009**, *1*. [CrossRef]

18. Ouyang, S.; Tong, H.; Umezawa, N.; Cao, J.; Li, P.; Bi, Y.; Zhang, Y.; Ye, J. Surface alkalinization induced enhancement of photocatalytic H_2 evolution over $SrTiO_3$ based photocatalysts. *J. Am. Chem. Soc.* **2012**, *134*, 1974–1977. [CrossRef] [PubMed]

19. Nishiro, R.; Tanaka, S.; Kudo, A. Hydrothermal-synthesized $SrTiO_3$ photocatalyst codoped with rhodium and antimony with visible-light response for sacrificial H_2 and O_2 evolution and application to overall water splitting. *Appl. Catal. B Environ.* **2014**, *150*, 187–196. [CrossRef]

20. Shen, S.; Jia, Y.; Fan, F.; Feng, Z.; Li, C. Time-resolved infrared spectroscopic investigation of roles of valence states of Cr in (La,Cr)-doped $SrTiO_3$ photocatalysts. *Chin. J. Catal.* **2013**, *34*, 2036–2040. [CrossRef]

21. Subramanian, V.; Roeder, R.K.; Wolf, E.E. Synthesis and UV-visible-light photoactivity of noble-metal-$SrTiO_3$ composites. *Ind. Eng. Chem. Res.* **2006**, *45*, 2187–2193. [CrossRef]

22. Ueda, M.; Otsuka-Yao-Matsuo, S. Preparation of tabular TiO_2–$SrTiO_{3-\delta}$ composite for photocatalytic electrode. *Sci. Technol. Adv. Mater.* **2004**, *5*, 187–193. [CrossRef]

23. Jia, Q.; Iwase, A.; Kudo, A. $BiVO_4$-Ru/$SrTiO_3$: Rh composite Z-scheme photocatalyst for solar water splitting. *Chem. Sci.* **2014**, *5*, 1513–1519. [CrossRef]

24. Guo, J.; Ouyang, S.; Li, P.; Zhang, Y.; Kako, T.; Ye, J. A new heterojunction Ag_3PO_4/Cr-$SrTiO_3$ photocatalyst towards efficient elimination of gaseous organic pollutants under visible light irradiation. *Appl. Catal. B Environ.* **2013**, *134*, 286–292. [CrossRef]

25. Zhang, H.; Wu, X.; Wang, Y.; Chen, X.; Li, Z.; Yu, T.; Ye, J.; Zou, Z. Preparation of Fe_2O_3/$SrTiO_3$ composite powders and their photocatalytic properties. *J. Phys. Chem. Solids* **2007**, *68*, 280–283. [CrossRef]

26. Márquez-Herrera, A.; Ovando-Medina, V.M.; Castillo-Reyes, B.E.; Meléndez-Lira, M.; Zapata-Torres, M.; Saldaña, N. A novel synthesis of $SrCO_3$–$SrTiO_3$ nanocomposites with high photocatalytic activity. *J. Nanopart Res.* **2014**, *16*, 1–10. [CrossRef]

27. Del-Ángel-Sánchez, M.T.; García-Alamilla, P.; Lagunes-Gálvez, L.M.; García-Alamilla, R.; Cabrera-Culebro, E.G. Aplicación de metodología de superficie de respuesta para la degradación de naranja de metilo con TiO_2 sol-gel sulfatado. *Rev. Int. Contam. Ambient.* **2015**, *31*, 99–106.

28. Mousavi, M.; Habibi-Yangjeh, A. Ternary g-C_3N_4/Fe_3O_4/Ag_3VO_4 nanocomposites: Novel magnetically separable visible-light-driven photocatalysts for efficientlydegradation of dye pollutants. *Mater. Chem. Phys.* **2015**, *163*, 421–430. [CrossRef]

29. Zhu, S.B.; Xu, T.G.; Fu, H.B.; Zhao, J.C.; Zhu, Y.F. Synergetic effect of Bi_2WO_6 photocatalyst with C60 and enhanced photoactivity under visible irradiation. *Environ. Sci. Technol.* **2007**, *41*, 6234–6239. [CrossRef] [PubMed]

30. Shang, M.; Wang, W.Z.; Sun, S.M.; Ren, J.; Zhou, L.; Zhang, L. Efficient visible light-induced photocatalytic degradation of contaminant by spindle-like PANI/$BiVO_4$. *J. Phys. Chem. C* **2009**, *113*, 20228–20233. [CrossRef]

31. Liang, H.C.; Li, X.Z. Visible-induced photocatalytic reactivity of polymer-sensitized titania nanotube films. *Appl. Catal. B* **2009**, *86*, 8–17. [CrossRef]

32. Luo, Q.Z.; Li, X.Y.; Wang, D.S.; Wang, Y.H.; An, J. Photocatalytic activity of polypyrrole/TiO_2 nanocomposites under visible and UV light. *J. Mater. Sci.* **2011**, *46*, 1646–1654. [CrossRef]

33. Kandiel, T.A.; Dillert, R.; Bahnemann, D.W. Enhanced photocatalytic production of molecular hydrogen on TiO_2 modified with Pt-polypyrrole nanocomposites. *Photochem. Photobiol. Sci.* **2009**, *8*, 683–690. [CrossRef] [PubMed]

34. Cooper, G.; Noufi, R.; Frank, A.J.; Nozik, A.J. Oxygen evolution on tantalum-polypyrrole-platinum anodes. *Nature* **1982**, *295*, 578–580. [CrossRef]

35. Chowdhury, D.; Paul, A.; Chattopadhyay, A. Photocatalytic polypyrrole-TiO_2-nanoparticles composite thin film generated at the air-water interface. *Langmuir* **2005**, *21*, 4123–4128. [CrossRef] [PubMed]

36. Zhang, C.R.; Li, Q.L.; Li, J.Q. Synthesis and characterization of polypyrrole/TiO_2 composite by *in situ* polymerization method. *Synth. Met.* **2010**, *160*, 1699–1703. [CrossRef]

37. Wang, D.S.; Wang, Y.H.; Li, X.Y.; Luo, Q.Z.; An, J.; Yue, J.X. Sunlight photocatalytic activity of polypyrrole-TiO_2 nanocomposites prepared by "*in situ*" method. *Catal. Comm.* **2008**, *9*, 1162–1166. [CrossRef]

38. Yang, Y.; Wen, J.; Wei, J.; Xiong, R.; Shi, J.; Pan, C. Polypyrrole-decorated Ag-TiO_2 nanofiber exhibiting enhanced photocatalytic activity under visible-light illumination. *Appl. Mater. Interfaces* **2013**, *5*, 6201–6207. [CrossRef] [PubMed]

39. Zhang, Z.; Wang, W.; Gao, E. Polypyrrole/Bi_2WO_6 composite with high charge separation efficiency and enhanced photocatalytic activity. *J. Mater. Sci.* **2014**, *49*, 7325–7332. [CrossRef]

40. An, L.; Wang, G.; Shi, X.; Su, M.; Gao, F.; Cheng, Y. Recyclable Fe_3O_4/ZnO/PPy composite photocatalyst: Fabrication and photocatalytic activity. *Russ. J. Phys. Chem. A* **2014**, *88*, 2419–2423. [CrossRef]

41. Wang, Q.; Zheng, L.; Chen, Y.; Fang, J.; Huang, H.; Su, B. Synthesis and characterization of novel PPy/$Bi_2O_2CO_3$ composite with improved photocatalytic activity for degradation od Rhodamine-B. *J. Alloy. Compd.* **2015**, *637*, 127–132. [CrossRef]

42. Song, L.; Zhang, S.; Chen, B. A novel visible-light-sensitive strontium carbonate photocatalyst with high photocatalytic activity. *Catal. Commun.* **2009**, *10*, 1565–1568. [CrossRef]

43. Alavi, M.A.; Morsali, A. Syntheses and characterization of $Sr(OH)_2$ and $SrCO_3$ nanostructures by ultrasonic method. *Ultrason. Sonochem.* **2010**, *17*, 132–138. [CrossRef] [PubMed]

44. Li, L.; Lin, R.; Tong, Z.; Feng, Q. Facile synthesis of $SrCO_3$ nanostructures in methanol/water solution without additives. *Nanoscale Res. Lett.* **2012**, *7*, 305. [CrossRef] [PubMed]

45. Mondal, M.K.; Lenka, M. Solubility of CO_2 in aqueous strontium hydroxide. *Fluid Ph. Equilib.* **2012**, *336*, 59–62. [CrossRef]

46. Momenian, H.R.; Gholamrezaei, S.; Salavati-Niasari, M.; Pedram, B.; Mozaffar, F.; Ghanbari, D. Sonochemical synthesis and photocatalytic properties of metal hydroxide and carbonate (M:Mg, Ca, Sr or Ba) Nanoparticles. *J. Clust. Sci.* **2013**, *24*, 1031–1042. [CrossRef]

47. Song, L.; Li, Y.; He, P.; Zhang, S.; Wu, X.; Fang, S.; Shan, J.; Sun, D. Synthesis and sonocatalytic property of rod-shape $Sr(OH)_2 \cdot 8H_2O$. *Ultrason. Sonochem.* **2014**, *21*, 1318–1324. [CrossRef] [PubMed]

48. Ovando-Medina, V.M.; Martínez-Gutiérrez, H.; Corona-Rivera, M.A.; Cervantes-González, E.; Flores-Mejía, J.; Farías-Cepeda, L. Silver/silver bromide/polypyrrole nanoparticles obtained by microemulsion photopolymerization in the presence of a cationic surfactant. *Colloid. Polym. Sci.* **2013**, *291*, 605–615. [CrossRef]

49. Naumkin, A.V.; Kraut-Vass, A.; Gaarenstroom, S.W.; Powell, C.J. NIST X-ray Photoelectron Spectroscopy Database, Version 4.1. National Institute of Standards and Technology: Gaithersburg, 2012. Available online: http://srdata.nist.gov/xps/ (accessed on 22 September 2015).

50. Castillo-Reyes, B.E.; Ovando-Medina, V.M.; Omar González-Ortega, O.; Alonso-Dávila, P.A.; Juárez-Ramírez, I.; Martínez-Gutiérrez, H.; Márquez-Herrera, A. TiO_2/polypyrrole nanocomposites photoactive under visible light synthesized by heterophase polymerization in the presence of different surfactants. *Res. Chem. Intermed.* **2015**, *41*, 8211–8231. [CrossRef]

51. Viriya-empikul, N.; Changsuwan, P.; Faungnawakij, K. Preparation of strontium-based fibers via electrospinning technique. *Ceram. Int.* **2012**, *38*, 2633–2636. [CrossRef]

52. Hoffmann, M.R.; Martin, S.T.; Choi, W.; Bahnemann, D.W. Environmental applications of semiconductor photocatalysis. *Chem. Rev.* **1995**, *95*, 69–96. [CrossRef]

53. Hassan, M.E.; Chen, J.; Liu, G.; Zhu, D.; Cai, J. Enhanced photocatalytic degradation of methyl orange dye under the daylight irradiation over CN-TiO_2 modified with OMS-2. *Materials* **2014**, *7*, 8024–8036. [CrossRef]

54. Ni, S.; Yang, X.; Li, T. Hydrothermal synthesis and photoluminescence properties of $SrCO_3$. *Mater. Lett.* **2010**, *65*, 766–768. [CrossRef]

55. Cho, M.H.; Lee, Y.S. Optical detection of Mn impurity in oxides: A case study of $Sr(NO_3)_2$ and $SrCO_3$. *New Phys. Sae Mulli* **2014**, *64*, 891–895. [CrossRef]

56. Li, S.; Chen, M.; He, L.; Xu, F.; Zhao, G. Preparation and characterization of polypyrrole/TiO_2 nanocomposite and its photocatalytic activity under visible light irradiation. *J. Mater. Res.* **2009**, *24*, 2547–2554. [CrossRef]

57. Scaife, D.E. Oxide semiconductors in photoelectrochemical conversion of solar energy. *Solar Energy* **1980**, *25*, 41–54. [CrossRef]

58. Shirley, D.A. High-resolution X-Ray photoemission spectrum of the valence bands of gold. *Phys. Rev. B* **1972**, *5*, 4709–4714. [CrossRef]

59. Rietveld, H.M. Line profiles of neutron powder-diffraction peaks for structure refinement. *Acta Cryst.* **1967**, *22*, 151–152. [CrossRef]

60. Lutterotti, L.; Bortolotti, M.; Ischia, G.; Lonardelli, I.; Wenk, H.-R. Rietveld texture analysis from diffraction images. *Z. Kristallogr. Suppl.* **2007**, *26*, 125–130. [CrossRef]

61. Leite, E.R.; Nobre, M.A.L.; Cerqueira, M.; Longo, E. Particle growth during calcination of polycation oxides synthesized by the polymeric precursors method. *J. Am. Ceram. Soc.* **1997**, *80*, 2649–2657. [CrossRef]

10

Improved Electrochemical Detection of Zinc Ions Using Electrode Modified with Electrochemically Reduced Graphene Oxide

Jiri Kudr [1,2], Lukas Richtera [1,2], Lukas Nejdl [1,2], Kledi Xhaxhiu [1,2], Petr Vitek [3], Branislav Rutkay-Nedecky [1,2], David Hynek [1,2], Pavel Kopel [1,2], Vojtech Adam [1,2] and Rene Kizek [4,*]

Academic Editor: Jung Ho Je

[1]   Department of Chemistry and Biochemistry, Mendel University in Brno, Zemedelska 1, Brno CZ-613 00, Czech Republic; george.kudr@centrum.cz (J.K.); oliver@centrum.cz (L.R.); lukasnejdl@gmail.com (L.N.); kledi.xhaxhiu@fshn.edu.al (K.X.); brano.ruttkay@seznam.cz (B.R.-N.); d.hynek@email.cz (D.H.); paulko@centrum.cz (P.K.); vojtech.adam@mendelu.cz (V.A.)
[2]   Central European Institute of Technology, Brno University of Technology, Technicka 3058/10, Brno CZ-616 00, Czech Republic
[3]   Global Change Research Institute, The Czech Academy of Sciences, v.v.i., Bělidla 4a, Brno CZ-603 00, Czech Republic; vitek.p@czechglobe.cz
[4]   Department of Biomedical and Environmental Analysis, Wroclaw Medical University, Borowska 211, Wrocław PL-50 556, Poland
*   Correspondence: kizek@sci.muni.cz

Abstract: Increasing urbanization and industrialization lead to the release of metals into the biosphere, which has become a serious issue for public health. In this paper, the direct electrochemical reduction of zinc ions is studied using electrochemically reduced graphene oxide (ERGO) modified glassy carbon electrode (GCE). The graphene oxide (GO) was fabricated using modified Hummers method and was electrochemically reduced on the surface of GCE by performing cyclic voltammograms from 0 to -1.5 V. The modification was optimized and properties of electrodes were determined using electrochemical impedance spectroscopy (EIS) and cyclic voltammetry (CV). The determination of Zn(II) was performed using differential pulse voltammetry technique, platinum wire as a counter electrode, and Ag/AgCl/3 M KCl reference electrode. Compared to the bare GCE the modified GCE/ERGO shows three times better electrocatalytic activity towards zinc ions, with an increase of reduction current along with a negative shift of reduction potential. Using GCE/ERGO detection limit 5 ng·mL^{-1} was obtained.

Keywords: carbon; cyclic voltammetry; electrochemical impedance spectroscopy; electrochemistry; graphene oxide; heavy metal detection; reduced graphene oxide

1. Introduction

Heavy metal pollution has become a major concern all over the world. Anthropogenic processes like urbanization and industrialization have led to their release from Earth's crust and their accumulation in the biosphere. The long-term monitoring of heavy metal pollution is the only way to meet the legislative demands and decrease pressure on the environment. However, most heavy metals like lead or cadmium are toxic even at low concentrations, others, which belong to a group of essential micronutrients, pose health risks in high supplementation only, but their monitoring is also needed [1–3]. Among essential micronutrients, zinc(II) plays one of the most important role. Zinc analysis is appealing not only from the environmental point of view but also from the biochemical one. Zinc(II) ions play an important role in cell replication and nucleic acid metabolism, and its deficiency is connected with some pathological processes like retarded growth and immunity dysfunction [4]. As was shown recently, the enhanced zinc intake by drinking water in the case of mice caused zinc deficiency in the hippocampus, associated with memory deficit and decreased expression levels of learning and memory related receptors [5]. Zinc has these important roles and effects mainly as a co-factor of numerous proteins, therefore it is not surprising that metallomics and proteomics of zinc-containing proteins are emerging fields of science [6–8]. From these, metallothioneins are highlighted as maintainers and transporters of these proteins and their importance in zinc metabolism belongs to the interest of numerous researchers [9–18].

Atomic absorption spectrometry (AAS) and inductively coupled plasma mass spectrometry (ICP-MS) represent a gold standard in detection of trace heavy metals concentrations. Nevertheless, they require expensive instrumentation, experienced operators, and the analyses are time-consuming. On the contrary, electrochemistry offers superior features like portability, easy use, low price, miniaturization, and high sensitivity [19–23]. The great advantage of electroanalysis is also the possibility of electrode surface modification [24].

Mercury electrodes have been widely used in trace heavy metal analysis for decades; however, they do not correspond with current trends including miniaturization. Whereas material sciences are a rapidly developing field of science, several micro to nanosized materials like liquid metals/metal oxides in order to improve electrode properties are attracting the attention of analysts [25–28]. Graphene, theoretically perfect two-dimensional (one-atom-thick) material, is the ideal choice for electrochemistry since it possesses unusual electronic conductivity and high surface area [29]. However, it is worth noting that a one-atom-thick, defect-less graphene monolayer is difficult to prepare and standard graphene materials are far from perfectly structured, and therefore more often reduced graphene oxide (rGO) is used. The procedure of GO reduction influences subsequent rGO properties. Electrodes modified with rGO obtained using constant potential chemical and thermal reduction was previously compared [30]. From electrochemical methods for GO reduction cyclic voltammetry was also used [31]. Electrodes modified with rGO are not only desirable for just electroanalytical chemistry, but also for the removal of organic pollutants from wastewaters [32]. Various methods have been used to prepare electrodes modified with GO [33–36]. Among others, electrodeposition of GO or rGO attracted interest due to its efficiency, ease of use, and rapid procedure [37]. Potentiostatic methods and cyclic voltammetry (CV) were shown to be suitable tool for electrodeposition of these materials on electrodes [38,39]. Recently, pulse potential method based on changing of anodic deposition and cathodic reduction periods was developed too [40]. Moreover, an electrode surface modification with biomolecules or graphene-like nanomaterials can significantly improve detection sensitivity and selectivity [41–43].

In this work the GO film on glassy carbon electrode (GCE) was fabricated by the potentiostatic deposition of GO. Electrochemically deposited GO was subsequently subjected to electrochemical reduction to produce electrochemically reduced graphene oxide (ERGO) using CV. The properties of this modified electrode were compared with standard bare GCE using CV and electrochemical impedance spectroscopy (EIS). $[Fe(CN)_6]^{3-}/[Fe(CN)_6]^{4-}$ was used as a redox probe for electrode characterization and the performance of GCE/ERGO on detection of Zn(II) using differential pulse voltammetry (DPV) was examined.

2. Results and Discussion

2.1. Preparation of GCE/ERGO

Since the discovery of graphene, it has been attracting great attention due to its high conductivity and surface to volume ratio [44]. However, from an electrochemical point of view graphene suffers from a limited number of hydrophilic moieties and electroactive sites [45]. GO with randomly distributed oxygen groups benefits from structural similarity with graphene, nevertheless structure breaks cause the decrease of conductivity [46]. Partially reduced GO represents an intermediate between ideal graphene structure and GO, whereas the amount of electroactive surface and reactive functionalities (epoxy, hydroxyl, carboxyl) are balanced.

The common method to fabricate rGO is exfoliation of graphite to produce GO followed by thermal or chemical reduction [47]. ERGO represents an alternative since no expensive equipment or use of toxic compounds is needed during its fabrication. Several procedures have been introduced to cover electrodes with a GO or GO/ERGO layer [34,38,48,49]. Direct electrodeposition from solution or drop-casting of GO or rGO on the surface of electrode can be used. If GO is used as a source material for electrode modification, deposition is followed by electrochemical reduction of GO to prepare ERGO. Previously, CV, potentiostatic or pulsed methods (several cycles of deposition in positive potential followed by GO reduction in negative potential) were used in order to cover the electrode with ERGO [50,51].

Here, GO was prepared according to the simplified Hummer's method (Figure 1A). It was revealed that our GO sample contains particles with hydrodynamic diameter of 848 ± 290 nm (Figure 1B). Negative charge of GO was confirmed using measurement of zetapotential ($\zeta = -43$ mV), which enables to deposit GO particles on the electrode using application of positive potential on it. This value also suggests that GO particles possess good stability in colloid phase.

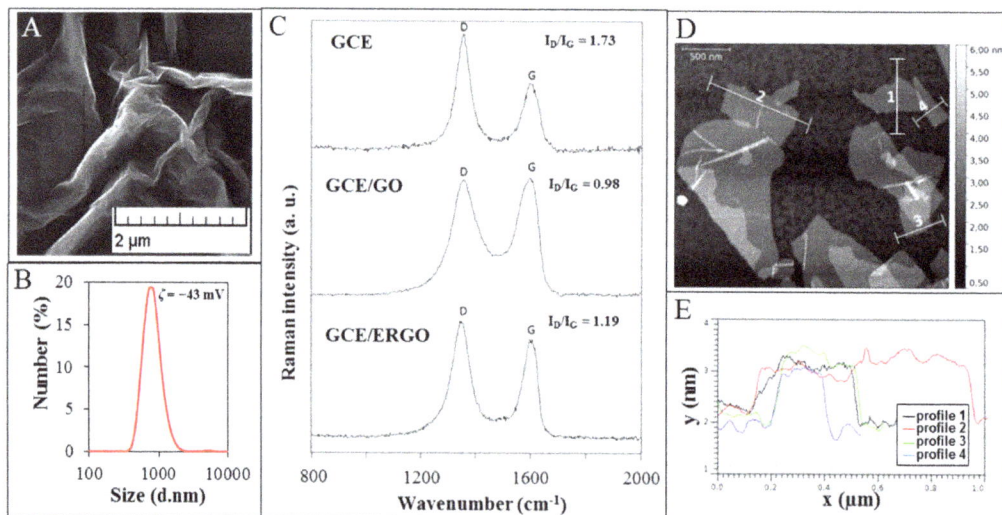

Figure 1. (**A**) Micrograph of GO used to modify GCE obtained by SEM; (**B**) GO size distribution including zetapotential; (**C**) Raman spectra of GCE, GCE modified with GO and GCE modified with ERGO; (**D**) AFM image of GO and (**E**) the height profiles along lines displayed in AFM image.

For the fabrication of GCE/ERGO, the constant potential +1.0 V *vs.* Ag/AgCl/3 M KCl reference electrode were applied to GCE in a previously sonicated water solution of GO (0.5 mg· mL^{-1}). Due to the presence of oxygen-containing functionalities negatively charged GO is electrostatically attracted to positively charge electrode. Subsequently, the working electrode was gently rinsed with water, transferred to acetate buffer and five CV cycles (from 0 to -1.50 V) were performed [52]. The irreversible reduction signals at -1.05 V and -0.85 V were observed in first cycle and completely disappeared in subsequent cycles (Figure 2A). It was shown previously that reduction of GO provides peak around -1.10 V [8,45,51]. Nevertheless different oxygen-containing moieties can be presented within GO, which can result in different reduction signals. The deposition (0–480 s) time of GO on electrode was optimized using detection of 20 µmol· L^{-1} Zn(II) signal and the deposition for 60 s was found as an optimal (Figure 2B). It was shown that although deposition of GO increased reduction signal of zinc slightly for 15 s, deposition of GO increased signal nearly three-fold for 30 s when compared with bare GCE cathodized for precise time in water. The deposition time 60 s was able to sufficiently modify the surface of GCE with GO, and the increase of deposition time did not result in the increase of detection signal.

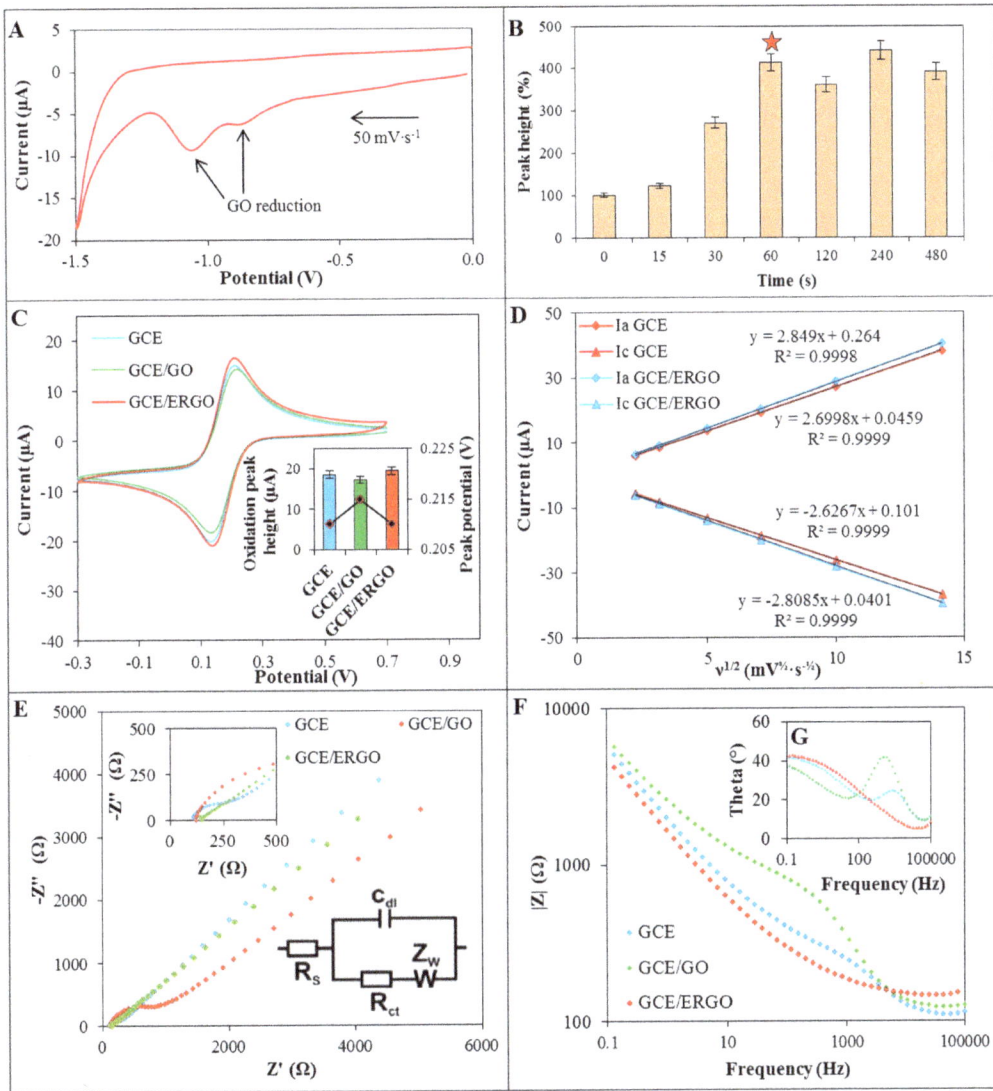

Figure 2. (**A**) The CV $(0 - (-1.5))$ V) of GCE/GO in acetate buffer; (**B**) Dependence of Zn(II) reduction signal obtained using GCE/ERGO on deposition time of GO (0.5 mg· mL^{-1}) on the electrode (deposition time selected as optimal is marked with star); (**C**) CV of 2 mM $[Fe(CN)_6]^{3-}/[Fe(CN)_6]^{4-}$ in 0.1 M KCl (50 mV·s^{-1}) recorded on bare GCE (blue line), GCE/GO (red line) and GCE/ERGO (green line) and corresponding peak current levels; (**D**) The dependence of $[Fe(CN)_6]^{3-}/[Fe(CN)_6]^{4-}$ anodic (Ia) and cathodic (Ic) peak heights on the square root of scan rate; (**E**) Nyquist plot, detail of nyquist plot high frequency region and equivalent circuit used for data evaluation in insets; (**F**) Bode modulus plot of bare GCE (blue line), GCE/GO (red line) and GCE/ERGO (green line); and (**G**) corresponding Bode phase diagram (same colours as previous figure).

Further, we analyzed the surface of the modified electrode by Raman spectroscopy. The D and G Raman bands were detected at 1355 cm^{-1} and 1595 cm^{-1} for GO and 1348 cm^{-1} and 1600 cm^{-1} for ERGO, both deposited on GCE. The Raman intensity ratio of the D and G bands (I_D/I_G) is increased in the case of GCE/ERGO (1.19) compared to GCE/GO (0.98) (Figure 1C), which is in accordance with literature [53,54]. It is attributed to the modification of the GO structure by reduction resulting in removal of functional groups and creation of defects between the sp^2 domains [55]. Change of full width at half maximum (FWHM) was observed from 115 cm^{-1} for GO towards 78 cm^{-1} in the case of ERGO for D band. The value for ERGO points at high disorder with low distances between defects [55]. The image of GO obtained using AFM suggests that GO is presented within sample in sheet-like shapes (Figure 1D). The thickness of GO, deduced from the height profile of AFM image, is about 1 nm, which is comparable to GO monolayer thickness published previously [56,57].

2.2. Characterization of GCE/ERGO

In order to characterize GCE/ERGO, cyclic voltammograms of equimolar 2 mmol·L^{-1} [Fe(CN)$_6$]$^{3-}$/[Fe(CN)$_6$]$^{4-}$ as a redox probe was measured using bare GCE and GCE/GO and compared with the record measured using GCE/ERGO. As it is shown in Figure 2C, deposition of GO on GCE reduces the peak current by 5%. On the contrary, GCE/ERGO exhibited better detection properties by 10% (inset in Figure 2C). Based on these data, the Randles-Sevcik equation (Equation (1)) was used to calculate the electroactive surface area of bare GCE and subsequently compare it with GCE/ERGO. The areas of 6.4 mm^2 and 7.0 mm^2 were acquired, respectively, which means increase for about 9.4% and confirm successful deposition and reduction of GO. The values of reduction and oxidation peaks of [Fe(CN)$_6$]$^{3-}$/[Fe(CN)$_6$]$^{4-}$ were plotted against the square root of scan rates (Figure 2D). The linear dependence revealed diffusion controlled processes for both GCE and GCE/ERGO and slightly improved sensitivity of detection in case of GCE/ERGO. These results were also confirmed by EIS (Figure 2E).

The Randles circuit was used as an equivalent circuit for fitting the EIS data. It consisted of solution resistance Rs, charge transfer resistance Rct, double layer capacitance Cdl and Warburg impedance ZW (inset in Figure 2E). Nyquist diagram showed in case of bare GCE depressed semicircle with charge transfer resistance 2.1 kΩ·cm^{-1}. After deposition of GO on GCE, charge transfer resistance increased four-fold to 8.5 kΩ·cm^{-1}. Very small depressed semicircle was observed in the case of GCE/ERGO, where Rct decreased to 0.6 kΩ·cm^{-1} (32% of GCE Rct). Significantly lower charge transfer resistance of GCE/ERGO in comparison with GCE/GO was previously reported [58]. In Bode diagram the frequency dependence on absolute magnitudes of impedance modulus |Z| was plotted (Figure 2F). The peaks of Bode phase diagram in case of GCE and GCE/GO (1–3 kHz) suggests that charge transfer resistance takes place in the electrode/electrolyte interface. Phase peak of Bode plot of GCE/ERGO disappeared at higher frequencies as a result of high electron transfer, where charge transfer resistance decreased (Figure 2G).

2.3. Detection of Zn(II)

The detection of Zn(II) was performed using DPV. Firstly, deposition potentials ((−1.45) − (−0.65) V) of Zn(II) on the surface of GCE/ERGO was optimized. As it can be seen in Figure 3A, the obtained Zn(II) signal increased from potential −0.65 to −1.25 V. At potential −1.25 V the Zn(II) signal reached its higher value and was choose as an optimal. As the next step, deposition time (0–90 s) of Zn(II) was optimized (Figure 3B). It was revealed that the signal increased by 73% using deposition time 60 s and deposition potential −1.25 V in comparison with deposition time 0 s. After this optimization, different concentrations of Zn(II) were measured using GCE/ERGO and calibration curve was determined (Figure 3C). It exhibited linear section between 1.0 μmol·L^{-1} and 62.5 μmol·L^{-1} and other analytical parameters of detection are displayed in Table 1. The modification of GCE with ERGO improved the detection of zinc ions (35 μmol·L^{-1}) four-fold in comparison with bare GCE and slightly shifted peak potential from −1.18 V to −1.2 V (Figure 3D). Using GCE/ERGO, we obtained limit of detection (LOD) 0.1 μmol·L^{-1} Zn(II) (~5 ng·mL^{-1}).

Table 1. Analytical parameters of electrochemical detection of Zn(II).

Substance	Working Electrode	Regression Equation	Linear Dynamic Range (μmol·L^{-1})	$R^{2\ a}$	LOD [b] (μmol·L^{-1})	LOQ [c] (μmol·L^{-1})	RSD (%)
Zn(II)	GCE/ERGO	$y = 0.1608x - 0.1231$	62.5 – 1.0	0.9999	0.1	0.4	4.8
Zn(II)	GCE	$y = 0.0539x - 0.0916$	500.0 – 2.0	0.9992	0.5	2.0	5.2

[a] Regression coefficient; [b] LOD ($S/N = 3$); [c] LOQ ($S/N = 10$).

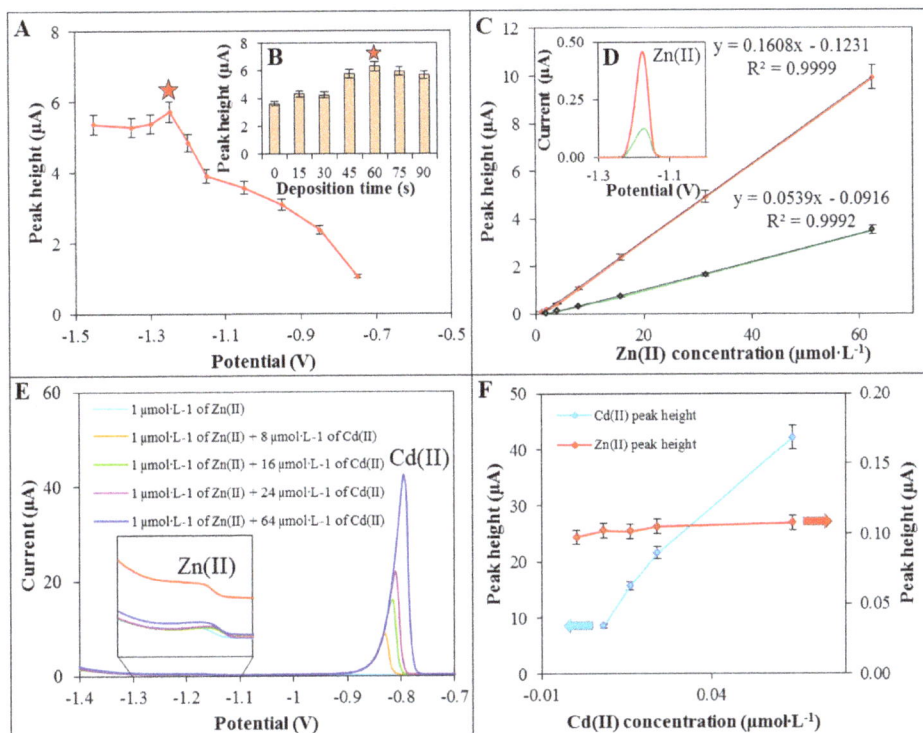

Figure 3. Dependence of Zn(II) reduction signal on deposition potential (**A**) and deposition time (**B**) of Zn(II) (35 μmol·L^{-1}) on GCE/ERGO (parameters marked with star was selected as an optimal); (**C**) Dependence of electrochemical signal on Zn(II) concentration (1.0–62.5 μmol·L^{-1}) and comparison of DPV reduction signals of Zn(II) (4 μmol·L^{-1}) (**D**) obtained using GCE/ERGO (red line) and bare GCE (blue line); (**E**) DPV voltammograms $((-1.40) - (-0.70))$ V) and comparison of Zn(II) and Cd(II) peak heights (**F**) of Zn(II) solution (1 μmol·L^{-1}) with different concentrations of Cd(II) (0–64 μmol·L^{-1}). Comparison of 10 μmol·L^{-1} Zn(II) electrochemical signal in acetate buffer with added 50 μmol·L^{-1} K(I), Ca(II) and Mg(II) in inset.

As the final step, the effect of interference with Zn(II) detections was examined. We chose Cd(II) since it is quite often presented in environmental samples and may affect Zn(II) detection [41,59]. Zn(II) solutions (1 μmol·L^{-1}) with different concentrations of Cd(II) (0–64 μmol·L^{-1}) were measured and the peak heights of Zn(II) (potential -1.15 V) were compared (Figure 3E). As it can be seen, peak of Cd(II) (about potential -0.80 V) is well separated from Zn(II) peak and did not significantly affect Zn(II) peak heights even at a 64-times higher concentration (Figure 3F). In addition, other monovalent and bivalent ions were tested as a possible interference in real sample. To Zn(II) solution five-times higher concentrations of K(I), Ca(II), and Mg(II) ions were added and their effects on Zn(II) were evaluated. All Zn(II) analysis presented here were performed in the acetate buffer where Na(I) ions were present in high concentration. Other tested ions showed no apparent interference in Zn(II) detection (Figure 3F inset).

3. Experimental Section

3.1. Chemicals and Material

ACS purity (*i.e.*, chemicals meet the specifications of the American Chemical Society) sodium acetate trihydrate, acetic acid, zinc nitrate, potassium hexacyanoferrate(III), potassium hexacyanoferrate(II) trihydrate, potassium chloride, water, and other chemicals were purchased from Sigma-Aldrich (St. Louis, MO, USA) unless noted otherwise.

3.2. Preparation of GO

GO was synthesized using chemical oxidation of graphite flakes (5.0 g, Sigma-Aldrich, and 100 mesh, $\geqslant$75% min) in a mixture of concentrated H_2SO_4 (670 mL, ACS reagent 95.0%–98.0%) and 30.0 g $KMnO_4$ (>99%) according to the simplified Hummer's method [60]. The reaction mixture was stirred vigorously. After four days, the oxidation of graphite was terminated by slow adding of H_2O_2 solution (250 mL, 30 wt % in H_2O) and the colour of the mixture turned to bright yellow, indicating high oxidation level of graphite. Formed graphite oxide was washed three times with 1 M of HCl and washed with water several times (total volume used 12 L) until constant pH value (4–5) was achieved using a simple decantation. Then, it was possible to centrifuge this solution. During the washing process with deionized water, exfoliation of graphite oxide led to the thickening of solution and formation of a stable colloid of GO.

3.3. Glassy Carbon Electrode Modification with Graphene

GCE was mechanically polished by the 1.0 μm and 0.3 μm alumina suspension (CH Instruments, Austin, TX, USA) on polishing cloth to produce mirror-like surface. Then, the electrode was sonicated for 3 min in distilled water (25 °C) and acetone successively in the Sonorex digital 10 P ultrasonic bath (Bandelin, Berlin, Germany). As prepared, the electrode was rinsed with water solution of GO (0.5 mg·mL^{-1}) and potential +1.0 V was applied on working electrode *vs.* Ag/AgCl/3 M KCl. The deposited film of GO was reduced by performing CV from 0.0 V to −1.5 V in acetate buffer (0.2 M, pH = 5) to produce ERGO.

3.4. Instrumentation

Determination of Zn(II) and $[Fe(CN)_6]^{3-}$/$[Fe(CN)_6]^{4-}$ by DPV and CV respectively was performed using PGSTAT302N (Metrohm, Herisau, Switzerland) using a three electrode system. A 3 mm diameter GCE (CH Instruments, Austin, TX, USA) was employed as the working electrode. An Ag/AgCl/3 M KCl electrode was used as the reference and platinum wire served as auxiliary. For data processing NOVA 1.8 (Metrohm, Herisau, Switzerland) was employed. Acetate buffer (0.2 mol·L^{-1} CH_3COONa and CH_3COOH, pH = 5) and 0.2 mol·L^{-1} KCl were used as a supporting electrolyte in cases of Zn(II) and $[Fe(CN)_6]^{3-}$/$[Fe(CN)_6]^{4-}$ determination, respectively.

The parameters of the measurement by DPV were as it follows: initial potential −1.3 V, end potential −1.0 V, deposition time 60 s, time interval 0.03 s, step potential 5 mV, scan rate 50 mV·s^{-1}. Parameters of the measurement by CV were as it follows: initial potential of −0.3 V, upper vertex potential 0.7 V, lower vertex potential −0.3 V, step potential 2.4 mV, scan rate 50 mV·s^{-1}. All measurements were carried out at 25 ± 1 °C.

The value of formal potential of $[Fe(CN)_6]^{3-}$ in 0.1 mol·L^{-1} KCl was 0.25 V and we also adopted it at impedance measurements. Impedance spectra were measured from 0.1 Hz to 10^5 Hz with alternating current (AC) amplitude of 10 mV. PGSTAT302N (Metrohm, Herisau, Switzerland) was used for impedance measurements with the same three electrode system as mentioned previously. Individual elements of equivalent circuit were calculated using NOVA 1.8 (Metrohm, Herisau, Switzerland).

3.5. The Electroactive Surface Determination

In order to determine electroactive area of GCE and to compare it with GCE/ERGO, cyclic voltammograms of 2 mM $[Fe(CN)_6]^{3-}/[Fe(CN)_6]^{4-}$ in 0.1 M KCl were recorded using the aforementioned electrodes. Electroactive surface was calculated according to Randles-Sevcik equation:

$$I_p = 2.69 \cdot 10^5 A \cdot D^{\frac{1}{2}} n^{\frac{3}{2}} v^{\frac{1}{2}} C \qquad (1)$$

where I_p is anodic current peak (A), A is the electroactive area (cm^2), D is the diffusion coefficient of $[Fe(CN)_6]^{4-}$ in solution ($6.1 \times 10^{-6} \, cm^2 \cdot s^{-1}$ was taken according to Prathish et al. [61]), n is the number of electrons transferred in half-reaction (1 in case of $[Fe(CN)_6]^{4-}$), v is scan rate (0.05 $V \cdot s^{-1}$ was chosen) and C is $[Fe(CN)_6]^{4-}$ concentration ($mol \cdot L^{-1}$).

3.6. Scanning Electron Microscopy (SEM)

Structure of carbon materials were characterized by SEM. For documentation of the structure, a MIRA3 LMU (Tescan, Brno, Czech Republic) was used. The SEM was fitted with In-Beam SE detector. For automated acquisition of selected areas a TESCAN proprietary software tool called Image Snapper was used. The software enabled automatic acquisition of selected areas with defined resolution. An accelerating voltage of 15 kV and beam currents about 1 nA gave satisfactory results.

3.7. Dynamic Light Scattering (DLS)

Average particle size, size distribution, and particle zetapotential were determined by dynamic light scattering method by Zetasizer Nano-ZS (Malvern Instruments Ltd., Worchestershire, UK) with a scattering angle θ = 173°. Samples were measured in water solution.

3.8. Raman Spectroscopy

All carbonaceous materials, bare GCE, GO, and ERGO deposited on GCE were characterized by Raman spectroscopy. Measurements were performed on a Renishaw InVia Reflex Raman microspectrometer equipped by the 514.5 nm line of an argon laser for excitation. A Leica microscope equipped with a standard 50× objective were used. The laser power was set to 1–2 mW at source to obtain an optimal Raman signal and simultaneously avoid any thermal alteration of the sample. Scans of 5–8 s were accumulated 10 times. Resulting spectra were baseline-corrected in GRAMS/AI 9.1.

3.9. Interference Measurement

$Zn(NO_3)_2$ was mixed with KCl, $CaCl_2$, and $MgCl_2$ to obtain a final concentration of 10 $\mu mol \cdot L^{-1}$ Zn(II) and 50 $\mu mol \cdot L^{-1}$ K(I), Ca(II) and Mg(II), respectively, in acetate buffer. Obtained Zn(II) reduction signals were compared with a signal of 10 $\mu mol \cdot L^{-1}$ Zn(II) in acetate buffer.

3.10. Atomic Force Microscopy Measurement

3.10.1. GO Immobilization

GO was immobilized on freshly cleaved mica surfaces grade V-1 (Structure Probe/SPI Supplies, West Chester, PA, USA). The mica surface was first modified by silanization in vapours of N-aminopropyldimethylethoxysilane (APDMES) with catalysis of N,N-diisopropylethylamine (DiPEA, both from Sigma Aldrich). Fifty microliters of GO stock solution 10-times diluted in double distilled water was subsequently transferred onto the modified mica surface and left to incubate for 15 min in a wet chamber under laboratory temperature. Then the surface was carefully washed with double distilled water and left to dry in desiccators for another 30 min (10 Pa vacuum).

3.10.2. Visualization of GO

The AFM images of GO fixed on mica sheets were taken by Bruker Dimension FastScan atomic force microscope (Bruker Nano Surface, Santa Barbara, CA, USA) operated in tapping mode. Basic parameters of the visualization process were as follows: set point value 3.5 nm, iGain 0.8, PGain 5.5, piezo Z scale range 500 nm. All images were collected under ambient conditions at 38% relative humidity and 22.5 °C with a scanning raster rate of 2.0 Hz. Silicon nitride triangular cantilevers "FastScan A" (Bruker Nano Surface) characterized by spring constant of 17 N/m and resonant frequency of 1397 kHz equipped with tetrahedral silicon tip with nominal tip radius 5 nm were used for imaging.

Gwyddion software [62] version 2.43 was used for AFM data post processing and for graphical output.

4. Conclusions

Modifications of electrode surface, where redox processes in electrochemical measurements take place, are promising techniques to improve detection sensitivity. Nanomaterials and different carbon materials among others are nowadays frequently used to meet this goal. As it was evident from our measurements, graphene modification of working electrodes is an easy way to enhance electrode properties. GO was electrodeposited from solution on GCE at constant positive potential and subsequently electrochemically reduced using cyclic voltammetry measurement at negative potentials. Proved by Raman and electrochemical impedance spectroscopy, successful modification of the electrode resulted in an increase of electroactive surface area by 9.4% compared with bare GCE. We found that GCE/ERGO possesses three-fold higher sensitivity for zinc ions in comparison with bare GCE. Acceptable selectivity towards interfering ions, such as K(I), Cd(II), Mg(II), and Ca(II) was achieved.

Acknowledgments: Financial support from IGA MENDELU TP-01-15 is greatly acknowledged. Petr Vitek gives thanks for the NPU I grant of Ministry of education, youth and sport (grant number LO1415). Authors would like to thank to Jan Pribyl from CEITEC Masaryk University for the AFM measurements.

Author Contributions: Jiri Kudr prepared the GCE/ERGO, performed the EIS measurements, and drafted the manuscript; Lukas Richtera prepared the GO for electrode modification; Lukas Nejdl performed DPV and CV measurements; Kledi Xhaxhiu carried out the dynamic light scattering measurements; Petr Vitek characterized the material by Raman spectroscopy; Branislav Rutkay-Nedecky was involved in DPV and CV measurements and manuscript correction; David Hynek provided SEM imaging; Pavel Kopel was involved in the design of the study; Vojtech Adam and Rene Kizek coordinated experiments and participated in design of the study.

Conflicts of Interest: The authors declare no conflict of interest.

References

1.   Haider, S.; Anis, L.; Batool, Z.; Sajid, I.; Naqvi, F.; Khaliq, S.; Ahmed, S. Short term cadmium administration dose dependently elicits immediate biochemical, neurochemical and neurobehavioral dysfunction in male rats. *Metab. Brain Dis.* **2015**, *30*, 83–92. [CrossRef] [PubMed]

2.   Mazzei, V.; Longo, G.; Brundo, M.V.; Sinatra, F.; Copat, C.; Conti, G.O.; Ferrante, M. Bioaccumulation of cadmium and lead and its effects on hepatopancreas morphology in three terrestrial isopod crustacean species. *Ecotoxical. Environ. Saf.* **2014**, *110*, 269–279. [CrossRef] [PubMed]

3.   Blazovics, A.; Szentmihalyi, K.; Vinkler, P.; Kovacs, A. Zn overdose may cause disturbance in iron metabolism in inactive inflammatory bowel diseases. *Trace Elem. Electrolytes* **2004**, *21*, 240–247. [CrossRef]

4.   Murakami, M.; Hirano, T. Intracellular zinc homeostasis and zinc signaling. *Cancer Sci.* **2008**, *99*, 1515–1522. [CrossRef] [PubMed]

5.   Yang, Y.; Jing, X.P.; Zhang, S.P.; Gu, R.X.; Tang, F.X.; Wang, X.L.; Xiong, Y.; Qiu, M.; Sun, X.Y.; Ke, D.; *et al.* High dose zinc supplementation induces hippocampal zinc deficiency and memory impairment with inhibition of BDNF signaling. *PLoS ONE* **2013**, *8*, e55384. [CrossRef] [PubMed]

6. Krizkova, S.; Ryvolova, M.; Hynek, D.; Eckschlager, T.; Hodek, P.; Masarik, M.; Adam, V.; Kizek, R. Immunoextraction of zinc proteins from human plasma using chicken yolk antibodies immobilized onto paramagnetic particles and their electrophoretic analysis. *Electrophoresis* **2012**, *33*, 1824–1832. [CrossRef] [PubMed]

7. Ryvolova, M.; Hynek, D.; Skutkova, H.; Adam, V.; Provaznik, I.; Kizek, R. Structural changes in metallothionein isoforms revealed by capillary electrophoresis and Brdicka reaction. *Electrophoresis* **2012**, *33*, 270–279. [CrossRef] [PubMed]

8. Frederickson, C.J.; Koh, J.Y.; Bush, A.I. The neurobiology of zinc in health and disease. *Nat. Rev. Neurosci.* **2005**, *6*, 449–462. [CrossRef] [PubMed]

9. Masarik, M.; Gumulec, J.; Sztalmachova, M.; Hlavna, M.; Babula, P.; Krizkova, S.; Ryvolova, M.; Jurajda, M.; Sochor, J.; Adam, V.; *et al.* Isolation of metallothionein from cells derived from aggressive form of high-grade prostate carcinoma using paramagnetic antibody-modified microbeads off-line coupled with electrochemical and electrophoretic analysis. *Electrophoresis* **2011**, *32*, 3576–3588. [CrossRef] [PubMed]

10. Krizkova, S.; Ryvolova, M.; Gumulec, J.; Masarik, M.; Adam, V.; Majzlik, P.; Hubalek, J.; Provaznik, I.; Kizek, R. Electrophoretic fingerprint metallothionein analysis as a potential prostate cancer biomarker. *Electrophoresis* **2011**, *32*, 1952–1961. [CrossRef] [PubMed]

11. Krizkova, S.; Ryvolova, M.; Hrabeta, J.; Adam, V.; Stiborova, M.; Eckschlager, T.; Kizek, R. Metallothioneins and zinc in cancer diagnosis and therapy. *Drug Metab. Rev.* **2012**, *44*, 287–301. [CrossRef] [PubMed]

12. Gumulec, J.; Masarik, M.; Krizkova, S.; Adam, V.; Hubalek, J.; Hrabeta, J.; Eckschlager, T.; Stiborova, M.; Kizek, R. Insight to physiology and pathology of zinc(II) ions and their actions in breast and prostate carcinoma. *Curr. Med. Chem.* **2011**, *18*, 5041–5051. [CrossRef] [PubMed]

13. Adam, V.; Petrlova, J.; Wang, J.; Eckschlager, T.; Trnkova, L.; Kizek, R. Zeptomole electrochemical detection of metallothioneins. *PLoS ONE* **2010**, *5*, e11441. [CrossRef] [PubMed]

14. Babula, P.; Masarik, M.; Adam, V.; Eckschlager, T.; Stiborova, M.; Trnkova, L.; Skutkova, H.; Provaznik, I.; Hubalek, J.; Kizek, R. Mammalians' metallothioneins and their properties and functions. *Metallomics* **2012**, *4*, 739–750. [CrossRef] [PubMed]

15. Sobrova, P.; Vyslouzilova, L.; Stepankova, O.; Ryvolova, M.; Anyz, J.; Trnkova, L.; Adam, V.; Hubalek, J.; Kizek, R. Tissue specific electrochemical fingerprinting. *PLoS ONE* **2012**, *7*, e49654. [CrossRef] [PubMed]

16. Bonaventura, P.; Benedetti, G.; Albarede, F.; Miossec, P. Zinc and its role in immunity and inflammation. *Autoimmun. Rev.* **2015**, *14*, 277–285. [CrossRef] [PubMed]

17. Haase, H.; Rink, L. Zinc signals and immune function. *Biofactors* **2014**, *40*, 27–40. [CrossRef] [PubMed]

18. Zalewska, M.; Trefon, J.; Milnerowicz, H. The role of metallothionein interactions with other proteins. *Proteomics* **2014**, *14*, 1343–1356. [CrossRef] [PubMed]

19. Nejdl, L.; Kudr, J.; Cihalova, K.; Chudobova, D.; Zurek, M.; Zalud, L.; Kopecny, L.; Burian, F.; Ruttkay-Nedecky, B.; Krizkova, S.; *et al.* Remote-controlled robotic platform ORPHEUS as a new tool for detection of bacteria in the environment. *Electrophoresis* **2014**, *35*, 2333–2345. [CrossRef] [PubMed]

20. Prasek, J.; Adamek, M.; Hubalek, J.; Adam, V.; Trnkova, L.; Kizek, R. New hydrodynamic electrochemical arrangement for cadmium ions detection using thick-film chemical sensor electrodes. *Sensors* **2006**, *6*, 1498–1512. [CrossRef]

21. Krystofova, O.; Trnkova, L.; Adam, V.; Zehnalek, J.; Hubalek, J.; Babula, P.; Kizek, R. Electrochemical microsensors for the detection of cadmium(II) and lead(II) ions in plants. *Sensors* **2010**, *10*, 5308–5328. [CrossRef] [PubMed]

22. Krizkova, S.; Krystofova, O.; Trnkova, L.; Hubalek, J.; Adam, V.; Beklova, M.; Horna, A.; Havel, L.; Kizek, R. Silver(I) ions ultrasensitive detection at carbon electrodes—Analysis of waters, tobacco cells and fish tissues. *Sensors* **2009**, *9*, 6934–6950. [CrossRef] [PubMed]

23. Nejdl, L.; Ruttkay-Nedecky, B.; Kudr, J.; Kremplova, M.; Cernei, N.; Prasek, J.; Konecna, M.; Hubalek, J.; Zitka, O.; Kynicky, J.; *et al.* Behaviour of zinc complexes and zinc sulphide nanoparticles revealed by using screen printed electrodes and spectrometry. *Sensors* **2013**, *13*, 14417–14437. [CrossRef] [PubMed]

24. Adam, V.; Baloun, J.; Fabrik, I.; Trnkova, L.; Kizek, R. An electrochemical detection of metallothioneins at the zeptomole level in nanolitre volumes. *Sensors* **2008**, *8*, 2293–2305. [CrossRef]

25. Zhang, W.; Ou, J.Z.; Tang, S.Y.; Sivan, V.; Yao, D.D.; Latham, K.; Khoshmanesh, K.; Mitchell, A.; O'Mullane, A.P.; Kalantar-zadeh, K. Liquid metal/metal oxide frameworks. *Adv. Funct. Mater.* **2014**, *24*, 3799–3807. [CrossRef]

26. Cincotto, F.H.; Martinez-Garcia, G.; Yanez-Sedeno, P.; Canevari, T.C.; Machado, S.A.S.; Pingarron, J.M. Electrochemical immunosensor for ethinylestradiol using diazonium salt grafting onto silver nanoparticles-silica-graphene oxide hybrids. *Talanta* **2016**, *147*, 328–334. [CrossRef] [PubMed]

27. Hui, N.; Wang, S.Y.; Xie, H.B.; Xu, S.H.; Niu, S.Y.; Luo, X.L. Nickel nanoparticles modified conducting polymer composite of reduced graphene oxide doped poly(3,4-ethylenedioxythiophene) for enhanced nonenzymatic glucose sensing. *Sens. Actuators B Chem.* **2015**, *221*, 606–613. [CrossRef]

28. Campbell, J.L.; Breedon, M.; Latham, K.; Kalantar-Zadeh, K. Electrowetting of superhydrophobic ZnO nanorods. *Langmuir* **2008**, *24*, 5091–5098. [CrossRef] [PubMed]

29. Li, D.; Kaner, R.B. Materials science—Graphene-based materials. *Science* **2008**, *320*, 1170–1171. [CrossRef] [PubMed]

30. Le, T.X.H.; Bechelany, M.; Lacour, S.; Oturan, N.; Oturan, M.A.; Cretin, M. High removal efficiency of dye pollutants by electron-Fenton process using a graphene based cathode. *Carbon* **2015**, *94*, 1003–1011. [CrossRef]

31. Li, B.; Pan, G.H.; Avent, N.D.; Lowry, R.B.; Madgett, T.E.; Waines, P.L. Graphene electrode modified with electrochemically reduced graphene oxide for label-free DNA detection. *Biosens. Bioelectron.* **2015**, *72*, 313–319. [CrossRef] [PubMed]

32. Le, T.X.H.; Bechelany, M.; Champavert, J.; Cretin, M. A highly active based graphene cathode for the electro-Fenton reaction. *RSC Adv.* **2015**, *5*, 42536–42539. [CrossRef]

33. Kim, Y.R.; Bong, S.; Kang, Y.J.; Yang, Y.; Mahajan, R.K.; Kim, J.S.; Kim, H. Electrochemical detection of dopamine in the presence of ascorbic acid using graphene modified electrodes. *Biosens. Bioelectron.* **2010**, *25*, 2366–2369. [CrossRef] [PubMed]

34. Chen, L.Y.; Tang, Y.H.; Wang, K.; Liu, C.B.; Luo, S.L. Direct electrodeposition of reduced graphene oxide on glassy carbon electrode and its electrochemical application. *Electrochem. Commun.* **2011**, *13*, 133–137. [CrossRef]

35. Cheemalapati, S.; Palanisamy, S.; Chen, S.M. Electrochemical determination of isoniazid at electrochemically reduced graphene oxide modified electrode. *Int. J. Electrochem. Sci.* **2013**, *8*, 3953–3962.

36. Sehat, A.A.; Khodadadi, A.A.; Shemirani, F.; Mortazavi, Y. Fast immobilization of glucose oxidase on graphene oxide for highly sensitive glucose biosensor fabrication. *Int. J. Electrochem. Sci.* **2015**, *10*, 272–286.

37. Ping, J.F.; Wang, Y.X.; Fan, K.; Wu, J.; Ying, Y.B. Direct electrochemical reduction of graphene oxide on ionic liquid doped screen-printed electrode and its electrochemical biosensing application. *Biosens. Bioelectron.* **2011**, *28*, 204–209. [CrossRef] [PubMed]

38. Cui, F.; Zhang, X.L. A method based on electrodeposition of reduced graphene oxide on glassy carbon electrode for sensitive detection of theophylline. *J. Solid State Electrochem.* **2013**, *17*, 167–173. [CrossRef]

39. Li, G.N.; Li, T.T.; Deng, Y.; Cheng, Y.; Shi, F.; Sun, W.; Sun, Z.F. Electrodeposited nanogold decorated graphene modified carbon ionic liquid electrode for the electrochemical myoglobin biosensor. *J. Solid State Electrochem.* **2013**, *17*, 2333–2340. [CrossRef]

40. Wang, F.; Wu, Y.J.; Lu, K.; Ye, B.X. A sensitive voltammetric sensor for taxifolin based on graphene nanosheets with certain orientation modified glassy carbon electrode. *Sens. Actuator B Chem.* **2015**, *208*, 188–194. [CrossRef]

41. Adam, V.; Petrlova, J.; Potesil, D.; Zehnalek, J.; Sures, B.; Trnkova, L.; Jelen, F.; Kizek, R. Study of metallothionein modified electrode surface behavior in the presence of heavy metal ions-biosensor. *Electroanalysis* **2005**, *17*, 1649–1657. [CrossRef]

42. Kudr, J.; Nguyen, V.H.; Gumulec, J.; Nejdl, L.; Blazkova, I.; Ruttkay-Nedecky, B.; Hynek, D.; Kynicky, J.; Adam, V.; Kizek, R. Simultaneous automatic electrochemical detection of zinc, cadmium, copper and lead ions in environmental samples using a thin-film mercury electrode and an artificial neural network. *Sensors* **2015**, *15*, 592–610. [CrossRef] [PubMed]

43. Chao, M.Y.; Ma, X.Y.; Li, X. Graphene-modified electrode for the selective determination of uric acid under coexistence of dopamine and ascorbic acid. *Int. J. Electrochem. Sci.* **2012**, *7*, 2201–2213.

44. Novoselov, K.S.; Fal'ko, V.I.; Colombo, L.; Gellert, P.R.; Schwab, M.G.; Kim, K. A roadmap for graphene. *Nature* **2012**, *490*, 192–200. [CrossRef] [PubMed]

45. Davies, T.J.; Hyde, M.E.; Compton, R.G. Nanotrench arrays reveal insight into graphite electrochemistry. *Angew. Chem. Int. Ed.* **2005**, *44*, 5121–5126. [CrossRef] [PubMed]

46. Zhao, J.P.; Pei, S.F.; Ren, W.C.; Gao, L.B.; Cheng, H.M. Efficient preparation of large-area graphene oxide sheets for transparent conductive films. *ACS Nano* **2010**, *4*, 5245–5252. [CrossRef] [PubMed]

47. Chen, D.; Feng, H.B.; Li, J.H. Graphene oxide: Preparation, functionalization, and electrochemical applications. *Chem. Rev.* **2012**, *112*, 6027–6053. [CrossRef] [PubMed]

48. Castro, S.S.L.; de Oliveira, M.F.; Stradiotto, N.R. Study of the electrochemical behavior of histamine using a Nafion (R)-Copper(II) hexacyanoferrate film-modified electrode. *Int. J. Electrochem. Sci.* **2010**, *5*, 1447–1456.

49. Gilje, S.; Han, S.; Wang, M.; Wang, K.L.; Kaner, R.B. A chemical route to graphene for device applications. *Nano Lett.* **2007**, *7*, 3394–3398. [CrossRef] [PubMed]

50. Ye, W.C.; Zhang, X.J.; Chen, Y.; Du, Y.L.; Zhou, F.; Wang, C.M. Pulsed electrodeposition of reduced graphene oxide on glass carbon electrode as an effective support of electrodeposited Pt microspherical particles: Nucleation studies and the application for methanol electro-oxidation. *Int. J. Electrochem. Sci.* **2013**, *8*, 2122–2139.

51. Zhang, Z.P.; Yan, J.; Jin, H.Z.; Yin, J.G. Tuning the reduction extent of electrochemically reduced graphene oxide electrode film to enhance its detection limit for voltammetric analysis. *Electrochim. Acta* **2014**, *139*, 232–237. [CrossRef]

52. Guo, H.L.; Wang, X.F.; Qian, Q.Y.; Wang, F.B.; Xia, X.H. A green approach to the synthesis of graphene nanosheets. *ACS Nano* **2009**, *3*, 2653–2659. [CrossRef] [PubMed]

53. Stankovich, S.; Dikin, D.A.; Piner, R.D.; Kohlhaas, K.A.; Kleinhammes, A.; Jia, Y.; Wu, Y.; Nguyen, S.T.; Ruoff, R.S. Synthesis of graphene-based nanosheets via chemical reduction of exfoliated graphite oxide. *Carbon* **2007**, *45*, 1558–1565. [CrossRef]

54. Krishnamoorthy, K.; Veerapandian, M.; Mohan, R.; Kim, S.J. Investigation of Raman and photoluminescence studies of reduced graphene oxide sheets. *Appl. Phys. A Mater. Sci. Process.* **2012**, *106*, 501–506. [CrossRef]

55. Eigler, S.; Dotzer, C.; Hirsch, A. Visualization of defect densities in reduced graphene oxide. *Carbon* **2012**, *50*, 3666–3673. [CrossRef]

56. Bi, S.; Zhao, T.T.; Jia, X.Q.; He, P. Magnetic graphene oxide-supported hemin as peroxidase probe for sensitive detection of thiols in extracts of cancer cells. *Biosens. Bioelectron.* **2014**, *57*, 110–116. [CrossRef] [PubMed]

57. Huang, P.; Xu, C.; Lin, J.; Wang, C.; Wang, X.S.; Zhang, C.L.; Zhou, X.J.; Guo, S.W.; Cui, D.X. Folic acid-conjugated graphene oxide loaded with photosensitizers for targeting photodynamic therapy. *Theranostics* **2011**, *1*, 240–250. [CrossRef] [PubMed]

58. Casero, E.; Parra-Alfambra, A.M.; Petit-Dominguez, M.D.; Pariente, F.; Lorenzo, E.; Alonso, C. Differentiation between graphene oxide and reduced graphene by electrochemical impedance spectroscopy (EIS). *Electrochem. Commun.* **2012**, *20*, 63–66. [CrossRef]

59. Nejdl, L.; Kudr, J.; Ruttkay-Nedecky, B.; Heger, Z.; Zima, L.; Zalud, L.; Krizkova, S.; Adam, V.; Vaculovicova, M.; Kizek, R. Remote-controlled robotic platform for electrochemical determination of water contaminated by heavy metal ions. *Int. J. Electrochem. Sci.* **2015**, *10*, 3635–3643.

60. Hummers, W.S.; Offeman, R.E. Preparation of graphitic oxide. *J. Am. Chem. Soc.* **1958**, *80*, 1339–1339. [CrossRef]

61. Prathish, K.P.; Barsan, M.M.; Geng, D.S.; Sun, X.L.; Brett, C.M.A. Chemically modified graphene and nitrogen-doped graphene: Electrochemical characterisation and sensing applications. *Electrochim. Acta* **2013**, *114*, 533–542. [CrossRef]

62. Necas, D.; Klapetek, P. Gwyddion: An open-source software for SPM data analysis. *Cent. Eur. J. Phys.* **2012**, *10*, 181–188.

11

Heteroatom Doped-Carbon Nanospheres as Anodes in Lithium Ion Batteries

George S. Pappas [1,†], Stefania Ferrari [1,*,†], Xiaobin Huang [2], Rohit Bhagat [1], David M. Haddleton [3] and Chaoying Wan [1,*]

Academic Editor: Federico Bella

[1]   Warwick Manufacturing Group, University of Warwick, Coventry CV4 7AL, UK; g.pappas@warwick.ac.uk (G.S.P.); r.bhagat@warwick.ac.uk (R.B.)
[2]   School of Aeronautics and Astronautics, Shanghai Jiao Tong University, Shanghai 200240, China; xbhuang@sjtu.edu.cn
[3]   Department of Chemistry, University of Warwick, Coventry CV4 7AL, UK; d.m.haddleton@warwick.ac.uk
*   Correspondence: S.Ferrari@warwick.ac.uk (S.F.); chaoying.wan@warwick.ac.uk (C.W.);

†   These authors contributed equally to this work.

Abstract: Long cycle performance is a crucial requirement in energy storage devices. New formulations and/or improvement of "conventional" materials have been investigated in order to achieve this target. Here we explore the performance of a novel type of carbon nanospheres (CNSs) with three heteroatom co-doped (nitrogen, phosphorous and sulfur) and high specific surface area as anode materials for lithium ion batteries. The CNSs were obtained from carbonization of highly-crosslinked organo (phosphazene) nanospheres (OPZs) of 300 nm diameter. The OPZs were synthesized via a single and facile step of polycondensation reaction between hexachlorocyclotriphosphazene (HCCP) and 4,4′-sulphonyldiphenol (BPS). The X-ray Photoelectron Spectroscopy (XPS) analysis showed a high heteroatom-doping content in the structure of CNSs while the textural evaluation from the N_2 sorption isotherms revealed the presence of micro- and mesopores and a high specific surface area of 875 m^2/g. The CNSs anode showed remarkable stability and coulombic efficiency in a long charge–discharge cycling up to 1000 cycles at 1C rate, delivering about 130 mA· h· g^{-1}. This study represents a step toward smart engineering of inexpensive materials with practical applications for energy devices.

Keywords: carbon nanospheres; lithium batteries; organo phosphazene

1. Introduction

Scientific and industrial research in rechargeable batteries has been extended beyond the "conventional" lithium-ion batteries (LIBs) technology, driven by the ongoing development of high power demanding applications and tools [1]. Li-Air and Li-Sulfur (Li-S), having theoretical energy densities a few orders of magnitude larger than Li-ion commercial batteries, are the two major technologies predicted to move forward towards the realization of advanced batteries [1,2]. In the meantime, graphite-based LIBs still hold a prominent position since graphite is the only material to combine low cost, performance and easy processability—three factors which are crucial for industrial production and realization in true applications. The development and commercialization of advanced anode materials (silicon, transition metal oxides, various alloys, *etc.*) for Li-ion batteries is at the core of today's efforts [3]. These materials have larger theoretical capacity than graphite (372 mA· h· g^{-1}) but suffer mainly from low electrical conductivity and large volume expansion which lead to poor cycling performance.

Carbon (nano) spheres (C(N)Ss) have also been studied as active materials in various energy storage and conversion applications such as supercapacitors [4,5], hydrogen storage cells [6], catalysis [7,8] and lithium ion batteries [9,10]. In LIBs, Li-alloy-core-carbon-shell structures have been increasingly reported as composite anode materials, since the carbon layer can limit the issues of the alloy anodes mentioned above. Nevertheless, bare CSs (with or without heteroatom doping) have also attracted some interest in this field as an alternative anode to graphite. Wang *et al.*, prepared dense nitrogen-doped CNSs by carbonization of polypyrrole nanospheres. The CNSs, with diameters between 60 and 70 nm and specific area 59 $m^2 \cdot g^{-1}$, delivered a reversible capacity of 380 $mA \cdot h \cdot g^{-1}$ at a current density of 60 $mA \cdot g^{-1}$ after 60 cycles and showed good high-rate performance delivering 200 $mA \cdot h \cdot g^{-1}$ at 3 $A \cdot g^{-1}$ [11]. Xiao *et al.*, synthesized hydrogenated CNSs by a low temperature solvothermal method using $CHCl_3$ as carbon source. Compared to Wang's work, the obtained nanospheres had similar textural characteristics (particle size and specific area) and semi-graphitized structure but higher electrochemical performance (978 $mA \cdot h \cdot g^{-1}$@50 $mA \cdot g^{-1}$ after 50 cycles) which was attributed to the high hydrogen-doping favoring the Li binding [12].

In order to increase the storage capacity and shorten the pathway length of Li ion diffusion, many groups have explored more sophisticated structures such as hierarchically porous [13,14] and hollow [15,16] carbon spheres. Recently, porous CNSs with single or double hollow architecture were prepared via a hard template method followed by carbonization and subsequent etching of the inorganic template. Compared to the single hollow nanospheres, the double shelled nanospheres showed improved cycling and rate performance ascribed to their unique hollow-in-hollow structure [17]. An unfavorable consequence of these engineered CSs (besides the multistep synthesis) is the high irreversible capacity loss at the first cycle. The increased reactivity due to the high surface area usually causes the fast decomposition of the electrolyte and a thick solid-electrolyte interface (SEI) layer formation [18]. Zhang *et al.*, prepared double shelled CNSs doped with nitrogen and showed their better electrochemical performance than the un-doped nanospheres when tested both in lithium ion and sodium ion cells [19]. It is obvious that not only the control over the textural structure but also the substitution of carbon by heteroatoms has a significantly positive effect in the performance of carbon materials in LIBs.

The synthetic methods for the formation of spherical particles, organic, inorganic or hybrid, relies on the chemical nature of the precursors and the desirable structure and properties of the final material which subsequently define the reaction conditions [20]. The one-step synthesis of highly cross-linked hybrid organo(cyclotriphosphazenes) (OPZs) is an excellent and highly efficient method to prepare various nanostructured materials such as nanospheres, nanotubes, core-shell and hollow particles. This method was first reported by Tang and Huang for the synthesis of OPZ nanotubes and microspheres and since then it has been extensively studied and applied to various materials applications [21,22]. Briefly, the OPZs formation proceeds through a polycondensation reaction between the hexachlorocyclotriphosphazene monomer and a co-monomer (cross-linker) with two -OH or -NH_2 groups, in the presence of triethylamine (TEA). Control over the size and morphology can easily be achieved by varying the concentration and ratio of the monomers, the chemical structure of the co-monomer, the type of the solvent and sonication power [23,24]. The as-prepared OPZs have a hybrid organic–inorganic structure which, after carbonization under inert atmosphere, can provide heteroatom doped-carbons [25]. The ambient conditions, the fast rate of reaction and the low cost are the main advantages of this synthetic method.

Herein, we report the preparation of OPZs with diameters of approximately 300 nm and narrow dispersity by optimizing the reaction conditions. The equivalent heteroatom co-doped (nitrogen, phosphorous and sulfur) CNSs were obtained by carbonization of the as-prepared OPZs and their performance as anodes in a lithium half-cell is evaluated.

2. Results and Discussion

OPZs were successfully prepared by a single step polycondensation reaction of HCCP and BPS in acetonitrile under ultrasonication at ambient conditions. TEA was added to initiate the reaction which is characterized by a fast rate with completion within seconds. The HCCP/co-monomer ratio and HCCP concentration are two important factors controlling the size of the OPZ particles [26]. Here, a 3.5:1 BPS to HCCP molar ratio was selected while the HCCP concentration was 2.25 mg/mL. The resulting nanospheres have relatively narrow size dispersity with diameters ~330 nm and a relatively smooth surface (Figure 1a). The presence of some smaller nanospheres <170 nm is unavoidable but can be minimized by carefully positioning the flask in the sonicator in order that the sonication power is evenly distributed in the solution. The EDS (Electron Dispersive Spectroscopy) spectrum showed the presence of P, N, S, C, O and some remaining Cl originating from unreacted P-Cl sites of the HCCP ring (Figure 1b).

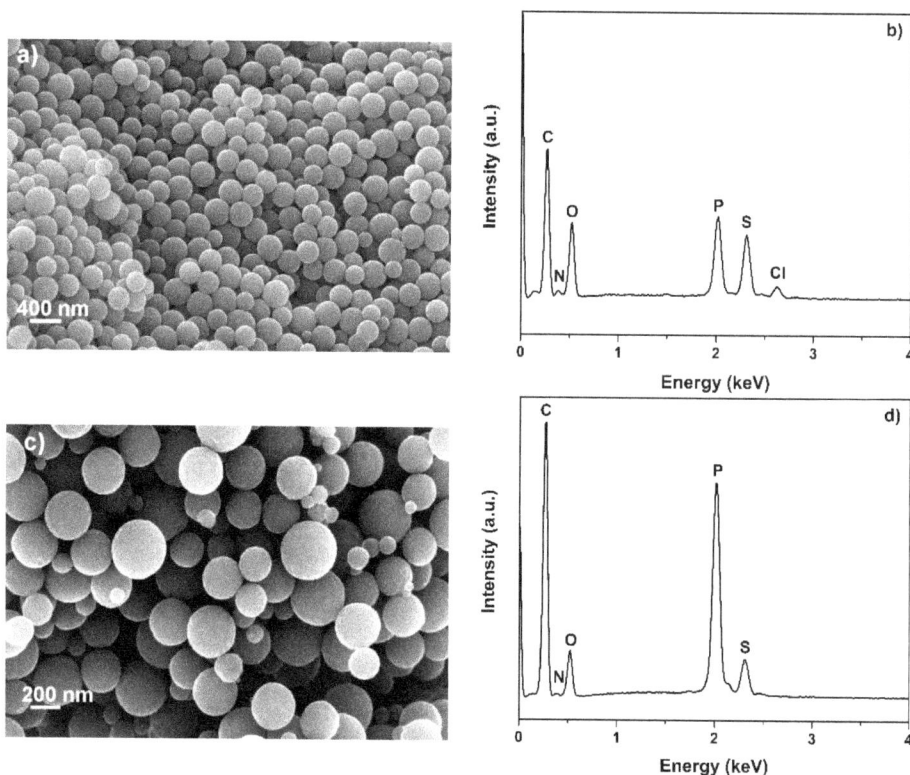

Figure 1. (a) SEM image of the organo (phosphazene) (OPZ) nanospheres ($\times$50 k magnification); (b) electron dispersive spectroscopy analysis (EDS) spectra of the OPZ nanospheres; (c) SEM image of the carbon nanospheres (CNSs) after carbonization at 850 °C ($\times$100 k magnification); (d) EDS spectra of the CNSs.

CNSs were obtained by carbonization of the as-prepared OPZs at 850 °C. The structure of the particles remained intact while an overall decrease in diameter is observed due to the weight loss and shrinkage during carbonization at high temperature (Figure 1c). From the EDS analysis (Figure 1d) it is evident that the derived carbon nanospheres are co-doped with N, P and S, showing the advantage of this method to produce carbons doped with multiple heteroatoms in one-pot synthesis. It is worth mentioning that the doping of carbon materials with multiple heteroatoms is desirable for energy storage applications since the presence of heteroatoms not only affects the textural characteristics but also alters the electronic properties of the carbon matrix [27]. The weight loss percentage between 120 and 850 °C calculated from the TGA curve is 56 wt % and two decomposition steps at T_{onset} = 478 °C and T_{onset} = 802 °C were observed (Figure S1). The first weight loss is attributed to the decomposition of the cross-linked HCCP/BPS structure and its initial conversion to an amorphous/low graphitized carbon structure. At higher temperature the transformation of amorphous to graphitic structure continued simultaneously with some heteroatom removal from the structure.

The chemical structure of the cross-linked OPZs was characterized by ATR-FTIR (Attenuated Total Reflectance Fourier Transform Infrared Spectroscopy) (Figure S2). The arrows indicate the major vibrational peaks attributed to the substituted phosphazene ring and the 4,4'-sulfonyl diphenol while, the peak at 935 cm^{-1} is assigned to the P-O-C$_{(aromatic)}$ asymmetric stretch vibration indicating the successful reaction of the phenolic hydroxyls with the reactive P-Cl from the HCCP ring. Complementary to infrared spectra the Raman spectra of the OPZ nanospheres before and after carbonization was recorded (Figure 2a). The near infrared laser (785 nm) was selected to record the spectra of OPZs since the visible laser (532 nm) produced significant fluorescence and the vibration peaks were not visible. There are three main peaks at 730, 1154 and 1588 cm^{-1} attributed to C-S, O=S=O, and aromatic C-C stretching, respectively. After carbonization, the peaks arising from HCCP and BPS are absent and the two broad peaks are observed centered at 1597 cm^{-1} (G band) and 1347 cm^{-1} (D band), which belong to carbon sp^2 and sp^3 electronic configurations, respectively. The ratio of I_D/I_G is 0.75 which is representative for pyrolized amorphous or partially graphitized carbons (hard carbons) and to some extent of defected turbostratic graphitic structure [28]. The results from Raman spectroscopy are further supported by the XRD pattern of the carbon nanospheres (Figure 2b). There are two broad peaks centered at 27 and 50° 2θ assigned to the (002) and (101) diffraction planes of hexagonal carbon layers (JCPDS, No. 75-1621). Similar to the Raman results, the broadening of the peaks is due to the disordered and highly defected structure of the material.

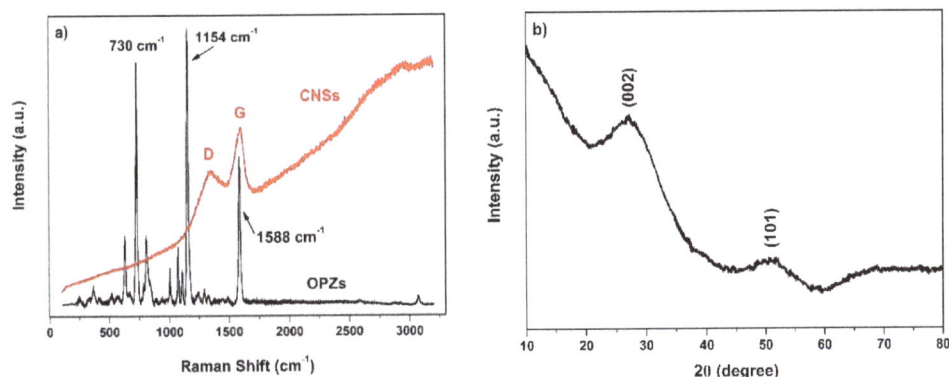

Figure 2. (a) Raman spectra of the OPZ and the CNSs; and (b) XRD pattern of the CNSs.

The N$_2$ adsorption isotherm of the OPZ nanospheres shows a very slow increase in N$_2$ adsorption up to 0.95 of the relative pressure (P/P$_0$) where a steep increase of the adsorbed volume is observed and capillary condensation takes place (Figure 3a). This behavior is common in non-porous and/or macroporous materials (<50 nm diameter). Since no pores of this scale are observed in SEM analysis, the existence of macropores is originated from the textural void space between the nanospheres. This result is also evident from the Barrett-Joyner-Halenda (BJH) pore size distribution which is very broad up to relatively unlimited pore sizes (Figure 3b, black line). The specific surface area, calculated by the Brunauer-Emmett-Teller (BET) equation in the range 0.05–0.2 P/P$_0$, is 19 m$^2 \cdot$g^{-1}. After carbonization, the CNSs showed a mixed type of isotherm curves. At low P/P$_0$ < 0.05 a type-I isotherm with a high initial N$_2$ uptake is observed which is characteristic of microporous materials. The adsorbed volume remains constant up to 0.9 P/P$_0$ where the filling of large pores takes place (type-IV). The desorption-branch forms a H1-type hysteresis loop which closes just below 0.8 P/P$_0$ indicating the presence of a wide range of mesopores. The specific surface area of CNSs is 875 $m^2 \cdot g^{-1}$ and was calculated by the Langmuir equation since the BET method resulted in a negative C constant value in the range 0.05–0.35 P/P_0 which significantly demeans the true result. According to V-t plot calculations between 0.1 and 0.2 P/P_0, the micropore area is 808 $m^2 \cdot g^{-1}$ and the major contributor of the total specific surface area. The pore size distribution of the CNSs shows a wide range of mesopores and an incomplete distribution in the micropore region (limitation of the BJH method). The high micropore surface area is a result of "defected" carbon structure due to the presence of heteroatoms. Additionally, the presence of the small quantity of Cl in the OPZs, could also affect the textural structure during carbonization. The large mesopores and macropores could be attributed to some aggregation between the CNSs during the carbonization. The total pore volume of CNSs is 0.43 $cm^3 \cdot g^{-1}$ and the micropore volume is 0.27 $cm^3 \cdot g^{-1}$.

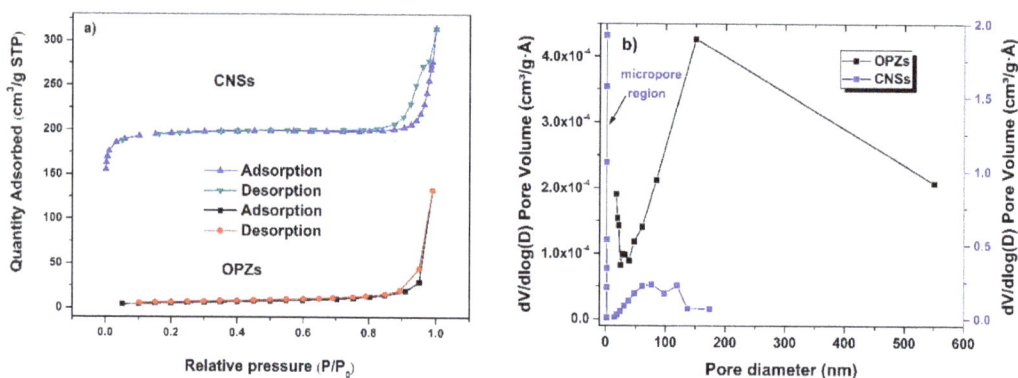

Figure 3. (a) N_2 adsorption-desorption isotherms; and (b) pore size distribution of the OPZs and CNSs.

XPS analysis was performed on the CNSs in order to investigate the chemical composition and the relative concentration, and give information regarding chemical bonding configurations in the doped carbon nanospheres. The survey spectrum of the sample (Figure 4a) evidenced peaks of the expected doping elements, N, S, P together with C and O. This result is a clear indication of the successful incorporation of N, S and P atoms within the carbon spheres (in agreement with EDS analyses). The atomic percentage of the elements is reported in Table 1 together with the peaks position and width.

Table 1. Atomic percentage, peak position and width obtained by fitting the high resolution XPS spectra.

Element	at %	Peak Position (eV)	Width (eV)
C 1s	67.0	(I) 284.3 (II) 289.8	1.2 2.8
O 1s	19.6	(I) 532.4 (II) 530.5 (III) 535.7 (IV) 537.9	2.0 1.4 1.9 1.9
P 2p	8.0	(3/2) 132.8 (1/2) 133.6	1.8 1.8
N 1s	4.7	(I) 397.6 (II) 399.6 (III) 401.2	1.6 1.6 1.6
S 2p	0.7	(3/2) 163.4 (1/2) 164.6	0.9 0.9

It can be observed that—apart from O—among the other heteroatoms, P has the highest atomic percentage, while S is present in less than 1 at %. The high at % of P is related to: (i) the higher bonding degree of P in the initial OPZs' structure; (ii) the lower "diffusion" ability through the carbon structure due to the larger atomic radius of P compared to N or S; and (iii) partial oxidation of P during carbonization. Peak fitting was performed using multiple components and the P $2p$ and S $2p$ curves were fitted taking into account the spin-orbit splitting and ratio $2p_{1/2}:2p_{3/2}$ components of 0.5.

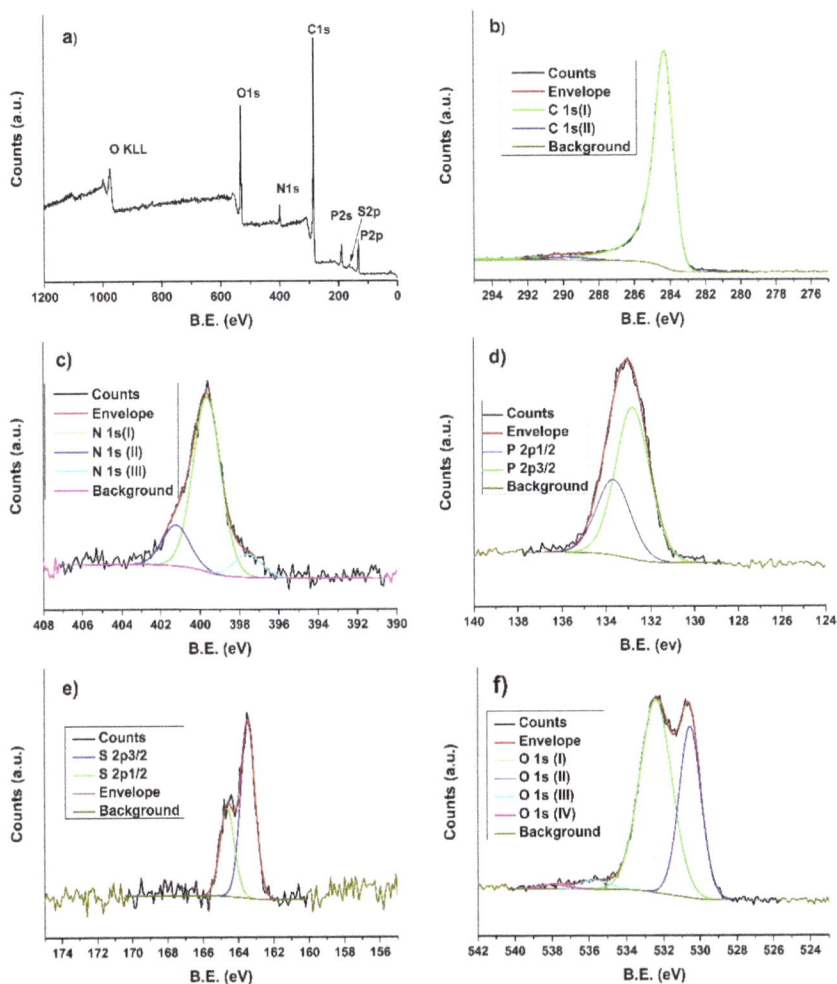

Figure 4. XPS of the CNSs sample: (**a**) survey spectrum and fitted high resolution spectra of; (**b**) C $1s$; (**c**) N $1s$; (**d**) P $2p$; (**e**) S $2p$; and (**f**) O $1s$.

The high resolution spectra of C $1s$ (Figure 4b) shows the typical graphitic carbon asymmetric peak-shape centered at 284.3 eV due to sp^2 bonding, with an associated shake-up feature at 289.8 eV originating from the pi to pi* transition [29]. Examination of the N $1s$ spectrum shown in Figure 4c revealed three different components that following previous interpretations [30] can be assigned to pyridinic (397.6 eV), pyrrolic (399.6 eV) and substitutional (401.2 eV) N. Phosphorous doped carbons have been reported in a very small number of studies [27,31] and for this reason clear evidence for the bonding of P even in graphitic carbon nanomaterials is missing [32]. The P $2p$ spectrum reported in Figure 4d shows the existence of a single P bonding environment, with the $2p_{3/2}$ component centered at 132.8 eV. According to some authors, this can be attributed to P-O bonds, while usually a P $2p$ signature between 131 and 129 eV is assigned to C-P species (substitutional P) [33]. Sulphur was also successfully incorporated in the CNSs as confirmed by XPS analysis; the S $2p$ spectrum (Figure 4e) shows the S $2p_{3/2}$ and S $2p_{1/2}$ peaks at 163.4 and 164.6 eV respectively, which are consistent with C-S-C bonds such as in thiophene-S [7,34]. SO_x groups, which usually show peaks at higher binding energy (about 168 eV), were not detected here [7,34]. The O $1s$ spectral envelope (Figure 4f) contains several different components pertaining to different oxygen-containing species. While the components at higher binding energies are ascribed to atmospheric contamination (H_2O, C-OH, COOH, *etc.*), the component at 530.5 eV is likely due to NO/CO groups as determined previously for carbon nanoparticles [35].

Electrochemical Behavior

The lithium storage properties of the heteroatom-doped CNSs were evaluated in Swagelok and coin lithium half-cells by cyclic voltammetry and galvanostatic cycling for their potential application as anode material for lithium batteries. In Figure 5 the cyclic voltammetry (CV) test recorded at 0.1 $mV \cdot s^{-1}$ is reported. The first cycle shows a first cathodic peak at about 0.8 V that can be related to the formation of the SEI layer [16]. Also a second peak from about 0.2 V to 0.005 V is clearly observed, which can be assigned to the insertion of the Li ions in the pores of the nanospheres. A very broad anodic peak is also detected at about 1.0 V which could indicate that a reversible oxidation of some SEI components was taking place. For the subsequent cycles no other relevant reduction/oxidation peaks were observed. The voltammogram of this sample is in full agreement with previous reports in which SEI formation and Li insertion peaks were reported in the initial cycles [16,19].

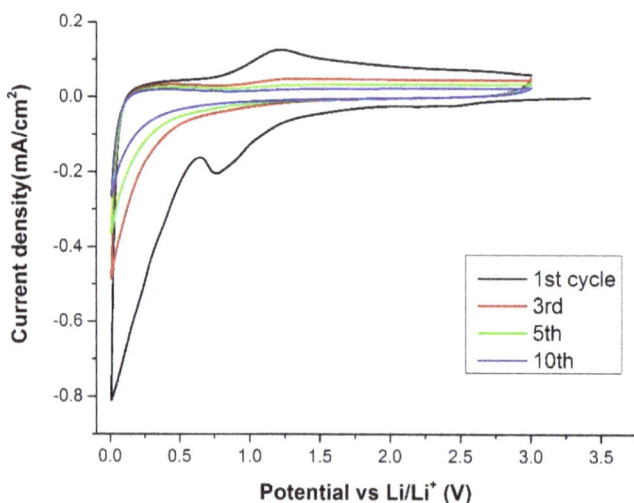

Figure 5. Cyclic voltammetry (CV) test of the CNSs. The graph shows some selected cycles between 5 mV and 3 V at a scan rate of 0.1 $mV \cdot s^{-1}$.

The galvanostatic discharge/charge results are shown in Figure 6. The cells were cycled at different current rates up to 1C (Figure 6a) after three initial formation cycles at a current of C/20 (the theoretical capacity was considered 372 $mA \cdot h \cdot g^{-1}$ as for a graphite electrode). A first discharge capacity of more than 1126 $mA \cdot h \cdot g^{-1}$ (Figure 6b) was found with a large irreversible capacity of about 440 $mA \cdot h \cdot g^{-1}$, as expected due to the large part of the initial discharge capacity related to the electrolyte decomposition and SEI formation on the anode. The high surface area of the CNSs can be considered responsible for the increased reactivity toward the electrolyte, which led to the observed irreversible capacity. In a previous work of Dahn's group [36] the important role of the S and O atoms on the irreversible capacity of disordered carbons was evidenced. The authors found that the irreversible capacity loss increased by increasing the chalcogen content. In this work the S content is low and should have a minor effect on the capacity loss; the O content instead is significant and together with the high surface area is associated with the observed capacity loss after the first cycle. The charge and discharge voltage profile of the second and subsequent cycles are very similar with a steep voltage increase/decrease at higher current rates, which mainly suggests a pseudo-capacitive behaviour. This cell was then cycled for a further 100 cycles at the medium-low C/5 current rate (Figure 6c). After an initial recovery of the capacity that reached 300 mA· h· g^{-1} a continuous decline was observed with an overall 7% loss in discharge capacity for these cycles. However, a remarkable coulombic efficiency approaching 100% was obtained.

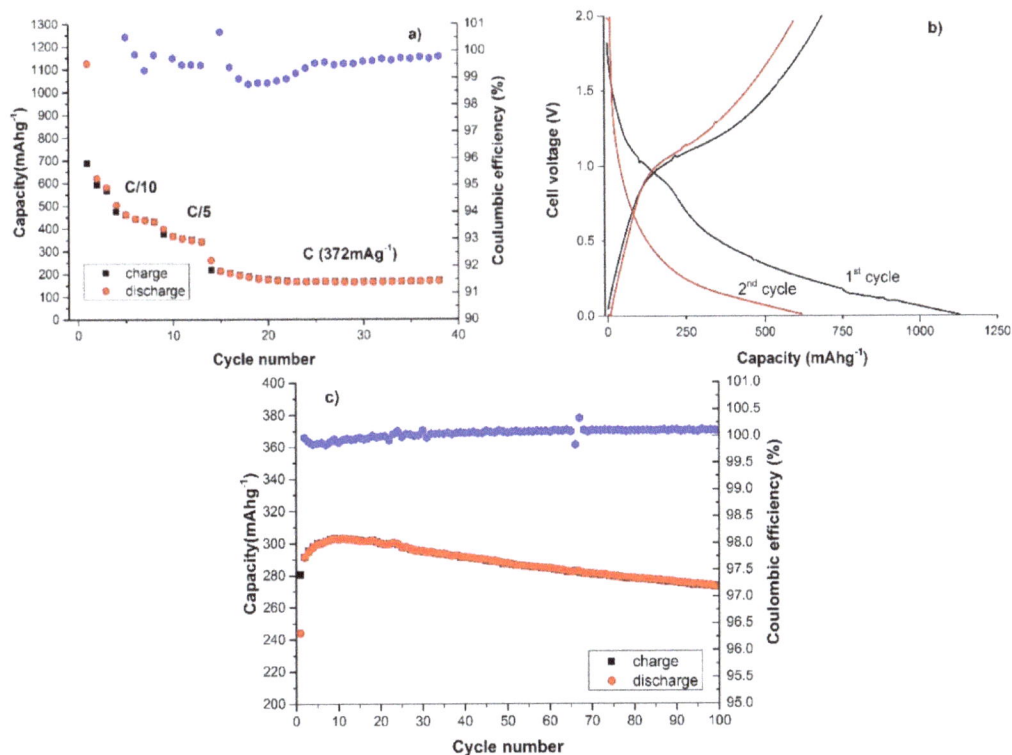

Figure 6. (a) Capacity vs. cycle number at C/20 (3 cycles), C/10 and C/5 (5 cycles each) and 1C current rate (25 cycles); (b) charge and discharge voltage profile of the first and second cycles at C/20; and (c) capacity vs. cycle number at C/5 current for 100 cycles.

The CNSs anode was then tested at 1C and higher C rates to check the rate capability. The cell was cycled at 1C for more than 1000 cycles (Figure 7a) and is well beyond the generally reported cycling for this type of materials for which no more than few hundreds of cycles are usually reported. The importance of this test has to be underlined since a novel anode material should be able to sustain a prolonged cycling for being considered competitive with the graphite anodes currently in use in commercial devices. The first few formation cycles confirmed the behaviour previously observed of an initial large irreversible capacity. A stable capacity value of about 180 mA· h· g^{-1} was reached and maintained for about 300 cycles, then a constant decay led to the final discharge value of 125 mA· h· g^{-1} for the last cycle.

The coulombic efficiency, although showing some fluctuation due to the experimental conditions, was definitely high with an average value of 99.992% (calculated over the whole cycling) that is the ideal efficiency required for a real application. In general, for all our tests, the efficiency was affected during the initial cycles by the formation of the SEI film and this is common for amorphous carbon also. The heteroatom-doped CNSs showed an efficiency among the highest reported in the literature so far indicating their excellent cycle stability. This good cycle performance could be achieved due to the unique porous structure of the CNSs which favoured strain relaxation during Li$^+$ insertion/extraction and also to an appropriate surface area. At a higher C rate (5C) the discharge capacity was still around 90 mA·h·g^{-1} (Figure 7b). Although similar carbon systems have shown higher rate capabilities [16,19,37], the CNSs characterized herein for Li-ion cells, doped with N, P and S showed an interesting electrochemical performance and could represent a promising material concept. A synergetic effect between structure and heteroatom doping might be responsible for the observed high cycle stability. Heteroatoms are known to affect the electronic properties of carbon, the textural structure (porosity, disorder degree, crystal size) and also increase the active sites of the anode material [27]. All these effects are concomitant thus making it complicated to interpret the single contribution of all those factors to the electrochemical behavior of the CNSs. Although the nitrogen-doped carbons have been well studied in energy storage applications (see references herein), there is limited knowledge for other heteroatoms' effect and even less for dual or ternary heteroatom doped carbons [27]. In this concept, further work is needed in order to clarify the doping effect on the electrochemical response optimizing the materials for real device applications.

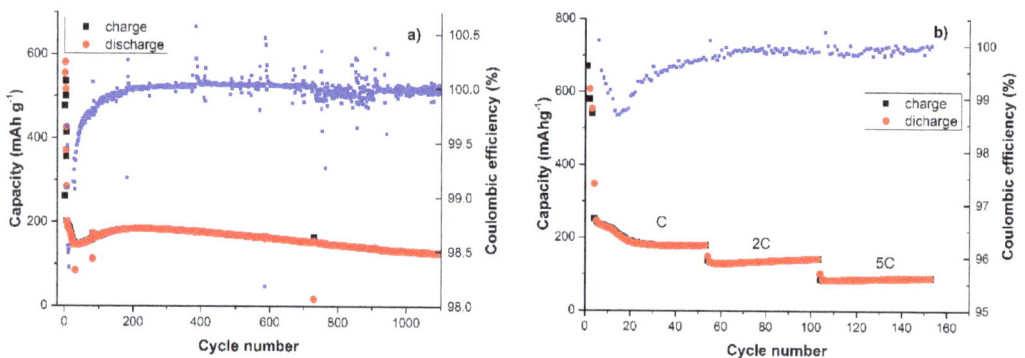

Figure 7. Capacity *vs.* cycle number at: (**a**) 1C rate for 1100 cycles; and (**b**) at 1C, 2C and 5C current rate for 50 cycles each.

3. Materials and Methods

3.1. Materials

Hexachlorocyclotriphosphazene (98%, Acros Organics, Pittsburgh, PA, USA), 4′-sulphonildiphenol (Fisher Scientific, Pittsburgh, PA, USA), triethylamine (Fisher Scientific) and acetonitrile (analytical grade, Fisher Scientific) were used as received. Acetone and deionized water were used for the washing steps.

3.2. Synthesis of Organophosphazene Nanospheres

The synthesis of the OPZs was performed according to literature [22,38] with a slightly modification in order to prepare particles of smaller diameter. Typically, a 250 mL round bottom flask containing 150 mL of acetonitrile, HCCP (2.25 mg/mL) and BPS (5.7 mg/mL), was placed in a sonicator bath (80 W, 37 kHz) and the precursors left to dissolve before 4 mL of TEA were added. The particle formation was observed within 10 s and the mixture left under sonication for 10 min. The solids were collected by centrifugation, washed several times with acetone and water, and finally dried in vacuum oven at 50 °C for 24 h (yield 70%).

3.3. Carbonization

The conversion of the as-prepared OPZ nanospheres to carbon nanospheres took place in a tube furnace from 25 °C to 850 °C under N$_2$ gas flow ~100 mL·min^{-1} and a heating rate of 2.5 °C·min^{-1}.

An isothermal step at 850 °C was maintained for 2 h and then the furnace left to cool down to room temperature.

3.4. Characterization

Field emission scanning electron microscopy (FE-SEM) and X-ray electron dispersive spectroscopy analysis (EDS) were performed on a Zeiss SIGMA SEM (Carl Zeiss AG, Oberkochen, Germany). Gold-Palladium (AuPd) sputtering was applied to the samples before observation. Carbonized samples did not need sputtering. FT-IR spectra were recorded on a Bruker Tensor 27 spectrometer equipped with an ATR cell (Bruker, Billerica, MA, USA). Raman spectra were recorded on a Renishaw in Via Confocal Raman Microscope equipped with 532 nm and 785 laser-lines, and a 0.75 NA (×50) lens was used (Renishaw, Wotton-under-Edge, UK). Thermogravimetric analysis curves were obtained with a Mettler Toledo instrument (Mettler-Toledo, Columbus, OH, USA) in the temperature range 25–850 °C at a heating rate of 10 °C· min^{-1} under gas N_2 flow. Specific surface area and pore size distribution were calculated from the nitrogen adsorption-desorption isotherm curves obtained with a Micromeritics ASAP 2020 Physisorption Analyzer (Micromeritics, Norcross, GA, USA). CNSs were degassed at 200 °C and OPZ at 60 °C for 16 h before measurements. X-ray Photoelctron spectroscopy (XPS) characterization was carried out with a Kratos Axis Ultra DLD spectrometer (Kratos Analytical Ltd, Manchester, UK) using monochromatic Al Kα source (hv = 1486.6 eV). Survey spectra were collected with a pass energy of 160 eV over a binding energy (BE) range of 1200–0 eV. High-resolution spectra were obtained using a 20 eV pass energy (resolution approximately 0.4 eV) and an analysis area of ~300 × 700 μm. The spectrometer was calibrated using the Fermi edge position of a polycrystalline Ag sample immediately prior to the experiments reported below. Peak fitting was performed by using CasaXPS software, using mixed Gaussian-Lorentzian (Voigt) line shapes and Shirley backgrounds.

3.5. Electrode Preparation and Electrochemical Characterization

To prepare the electrode, a slurry was made by mixing the carbon nanospheres with carbon black (Super P 65, TIMCAL, Cleveland, OH, USA) and poly(vinylidene fluoride) (PVDF, Solvay, Brussels, Belgium) in N-methyl-2-pyrrolidone (NMP, Sigma-Aldrich, St. Louis, MO, USA) with a weight ratio of 80:20:10. The obtained suspension was sonicated for one hour, then mixed on a magnetic stirrer for four hours and spread on a copper current collector by using a draw-down coater and a stainless steel applicator (bird applicator) with a 70 μm gap. The solvent was let to completely evaporate in a vacuum oven overnight and the foils were then transferred to a dry-room (humidity less than 1%, −45 °C dew point, Munters, Fort Myers, FL, USA). All cell components were dried in a vacuum oven (Binder Vacuum Drying Ovens with integrated vacuum pump system) at 50 °C overnight before assembly. The galvanostatic cycling were performed using 2032 coin cells from MTI (Richmond, CA 94804, USA); lithium metal was used as the counter electrode, and Celgard 2325 (Celgard, Charlotte, NC, USA) was used as the separator. To assemble the 2032 coin cells, the electrode foils were cut into disks of 1.2 cm diameter with a loading of about 3.5 mg/cm^2 of active material. The electrolyte was 1 M LiPF6 in Ethylene Carbonate/ Ethyl methyl Carbonate (EC/EMC) 3:7 v/v and 1 wt % VC (PuriEl, Soulbrain MI, Northville, MI, USA). The cell were cycled at different current rate (from C/20 to 5C) in the 0.005–2 V range by using a Maccor Series 4000 battery cycler (Maccor, Tulsa, OK, USA). Swagelok cells with Li as reference electrode were also assembled for the cyclic voltammetry (CV) test. The CV was performed at a scan rate of 0.1 mV/s in the potential range 0.005–3.0 V using a Biologic VMP3 (Bio-Logic, Grenoble, France). The cells were tested at ambient temperature.

4. Conclusions

Heteroatom co-doped carbon nanospheres were successfully prepared by a carbonization process of highly cross-linked poly(cyclotriphosphazene-co-4,4′ sulphonyl diphenol) nanospheres. The co-doped carbon nanospheres have a mixed microporous/mesoporous structure and a high surface area of 875 m^2/g. The XPS analysis showed a complex of heteroatoms in the carbon matrix and revealed the respective atomic concentration of S, N and P. The as-prepared CNSs showed remarkable cycling stability for more than 1000 cycles, delivering a capacity of about 130 mA· h· g^{-1} at a current rate of 1C. In addition, a remarkable coulombic efficiency as high as 99.99% was maintained during the long cycling thereby showing great promise for application in lithium batteries. The high specific surface area, the porous structure and multi-heteroatom doping all contributed to the electrochemical performance of the CNSs. Heteroatom doped CNSs are potential materials for various important applications in catalysis, supercapacitors and hydrogen storage and the synergistic effects of dopant atoms could be key to advanced technology.

Acknowledgments: The authors would like to thank Marc Walker (Warwick Photoemission Facility) for assistance during the XPS data collection and for helpful contribution for XPS data analysis.

Author Contributions: Chaoying Wan and George S. Pappas conceived and designed the experiments; George S. Pappas and Stefania Ferrari performed the experiments; George S. Pappas and Stefania Ferrari analyzed the data; David M. Haddleton, Rohit Bhagat, Xiaobin Huang and Chaoying Wan contributed reagents/materials/analysis tools; Xiaobin Huang contributed on the discussion; George S. Pappas and Stefania Ferrari wrote the paper; all the authors revised the paper.

Conflicts of Interest: The authors declare no conflict of interest.

References

1. Bini, M.; Capsoni, D.; Ferrari, S.; Quartarone, E.; Mustarelli, P. Rechargeable lithium batteries: Key scientific and technological challenges. In *Rechargeable Lithium Batteries: From Fundamentals to Applications*; Franco, A.A., Ed.; Woodhead Publishing Limited: Cambridge, UK, 2015; pp. 1–16.

2. Pappas, G.S.; Ferrari, S.; Wan, C. Recent advances in graphene-based materials for lithium batteries. *Curr. Org. Chem.* **2015**, *19*, 1838–1849. [CrossRef]

3. Obrovac, M.N.; Chevrier, V.L. Alloy negative electrodes for Li-ion batteries. *Chem. Rev.* **2014**, *114*, 11444–11502. [CrossRef] [PubMed]

4. Lee, W.; Moon, J.H. Monodispersed N-doped carbon nanospheres for supercapacitor application. *ACS Appl. Mater. Interfaces* **2014**, *6*, 13968–13976. [CrossRef] [PubMed]

5. Dai, Y.; Jiang, H.; Hu, Y.; Fu, Y.; Li, C. Controlled synthesis of ultrathin hollow mesoporous carbon nanospheres for supercapacitor Applications. *Ind. Eng. Chem. Res.* **2014**, *53*, 3125–3130. [CrossRef]

6. Jiang, J.; Gao, Q.; Zheng, Z.; Xia, K.; Hu, J. Enhanced room temperature hydrogen storage capacity of hollow nitrogen-containing carbon spheres. *Int. J. Hydrog. Energy* **2010**, *35*, 210–216. [CrossRef]

7. You, C.; Liao, S.; Li, H.; Hou, S.; Peng, H.; Zeng, X.; Liu, F.; Zheng, R.; Fu, Z.; Li, Y. Uniform nitrogen and sulfur co-doped carbon nanospheres as catalysts for the oxygen reduction reaction. *Carbon* **2014**, *69*, 294–301. [CrossRef]

8. Centi, G.; Barbera, K.; Perathoner, S.; Gupta, N.K.; Ember, E.E.; Lercher, J.A. Onion-like graphene carbon nanospheres as stable catalysts for carbon monoxide and methane chlorination. *ChemCatChem* **2015**, *7*, 3036–3046. [CrossRef]

9. Wang, H.; Abe, T.; Maruyama, S.; Iriyama, Y.; Ogumi, Z.; Yoshikawa, K. Graphitized carbon nanobeads with an onion texture as a Lithium-ion battery negative electrode for high-rate use. *Adv. Mater.* **2005**, *17*, 2857–2860. [CrossRef]

10. Jin, Y.Z.; Kim, Y.J.; Gao, C.; Zhu, Y.Q.; Huczko, A.; Endo, M.; Kroto, H.W. High temperature annealing effects on carbon spheres and their applications as anode materials in Li-ion secondary battery. *Carbon* **2006**, *44*, 724–729. [CrossRef]

11. Wang, Y.; Su, F.; Wood, C.D.; Lee, J.Y.; Zhao, X.S. Preparation and characterization of carbon nanospheres as anode materials in lithium-ion secondary batteries. *Ind. Eng. Chem. Res.* **2008**, *47*, 2294–2300. [CrossRef]

12. Xiao, J.; Yao, M.; Zhu, K.; Zhang, D.; Zhao, S.; Lu, S.; Liu, B.; Cui, W.; Liu, B. Facile synthesis of hydrogenated carbon nanospheres with a graphite-like ordered carbon structure. *Nanoscale* **2013**, *5*, 11306–11312. [CrossRef] [PubMed]

13. Wang, S.-X.; Chen, S.; Wei, Q.; Zhang, X.; Wong, S.Y.; Sun, S.; Li, X. Bioinspired synthesis of hierarchical porous graphitic carbon spheres with outstanding high-rate performance in lithium-ion batteries. *Chem. Mater.* **2015**, *27*, 336–342. [CrossRef]

14. Wang, F.; Song, R.; Song, H.; Chen, X.; Zhou, J.; Ma, Z.; Li, M.; Lei, Q. Simple synthesis of novel hierarchical porous carbon microspheres and their application to rechargeable lithium-ion batteries. *Carbon* **2015**, *81*, 314–321. [CrossRef]

15. Chen, X.; Kierzek, K.; Jianh, Z.; Chen, H.; Tang, T.; Wojtoniszak, M.; Kalenczuk, R.J.; Chu, P.C.; Borowiak-Palen, E. Synthesis, growth mechanism, and electrochemical properties of hollow mesoporous carbon spheres with controlled diameter. *J. Phys. Chem. C* **2011**, *115*, 17717–17724. [CrossRef]

16. Tang, K.; White, R.J.; Mu, X.; Titirici, M.-M.; van Aken, P.A.; Maier, J. Hollow carbon nanospheres with a high rate capability for lithium-based batteries. *ChemSusChem* **2012**, *5*, 400–403. [CrossRef] [PubMed]

17. Zang, J.; Ye, J.; Fang, X.; Zhang, X.; Zheng, M.; Dong, Q. Hollow-in-hollow carbon spheres for lithium-ion batteries with superior capacity and cyclic performance. *Electrochimica Acta* **2015**, *186*, 436–441. [CrossRef]

18. Vu, A.; Qian, Y.; Stein, A. Porous electrode materials for lithium-ion batteries—How to Prepare them and what makes them special. *Adv. Energy Mater.* **2012**, *2*, 1056–1085. [CrossRef]

19. Zhang, K.; Li, X.; Liang, J.; Zhu, Y.; Hu, L.; Cheng, Q.; Guo, C.; Lin, N.; Qian, Y. Nitrogen-doped porous interconnected double-shelled hollow carbon spheres with high capacity for lithium ion batteries and sodium ion batteries. *Electrochimica Acta* **2015**, *155*, 174–182. [CrossRef]

20. Nieto-Márquez, A.; Romero, R.; Romero, A.; Valverde, J.L. Carbon nanospheres: Synthesis, physicochemical properties and applications. *J. Mater. Chem.* **2011**, *21*, 1664–1672. [CrossRef]

21. Zhu, L.; Xu, Y.; Yuang, W.; Xi, J.; Huang, X.; Tang, X.; Zheng, S. One-pot synthesis of poly (cyclotriphosphazene-co-4,4'-sulfonyldiphenol) nanotubes via an *in situ* template approach. *Adv. Mater.* **2006**, *18*, 2997–3000. [CrossRef]

22. Zhu, Y.; Huang, X.; Li, W.; Fu, J.; Tang, X. Preparation of novel hybrid inorganic—Organic microspheres with active hydroxyl groups using ultrasonic irradiation via one-step precipitation polymerization. *Mater. Lett.* **2008**, *62*, 1389–1392. [CrossRef]

23. Zhu, Y.; Huang, X.; Fu, J.; Wang, G.; Tang, X. Morphology control between microspheres and nanofibers by solvent-induced approach based on crosslinked phosphazene-containing materials. *Mater. Sci. Eng. B* **2008**, *153*, 62–65. [CrossRef]

24. Wang, Y.; Shi, L.; Zhang, W.; Jiang, Z.; Mu, J. A comparative structure-property study of polyphosphazene micro-nano spheres. *Polym. Bull.* **2014**, *71*, 275–285. [CrossRef]

25. Fu, J.; Wang, M.; Zhang, C.; Zhang, P.; Xu, Q. High hydrogen storage capacity of heteroatom-containing porous carbon nanospheres produced from cross-linked polyphosphazene nanospheres. *Mater. Lett.* **2012**, *81*, 215–218. [CrossRef]

26. Zhang, P.; Huang, X.; Fu, J.; Huang, Y.; Zhu, Y.; Tang, X. A one-pot approach to novel cross-linked polyphosphazene microspheres with active amino groups. *Macromol. Chem. Phys.* **2009**, *210*, 792–798. [CrossRef]

27. Paraknowitsch, J.P.; Thomas, A. Doping carbons beyond nitrogen: An overview of advanced heteroatom doped carbons with boron, sulphur and phosphorus for energy applications. *Energy Environ. Sci.* **2013**, *6*, 2839–2855. [CrossRef]

28. Schwan, J.; Ulrich, S.; Batori, V.; Ehrhardt, H. Raman spectroscopy on amorphous carbon films. *J. Appl. Phys.* **1996**, *80*, 440–447. [CrossRef]

29. Xu, S.; Yan, X.B.; Wang, X.L.; Yang, S.R.; Xue, Q.J. Synthesis of carbon nanospheres from carbon-based network polymers. *J. Mater. Sci.* **2010**, *45*, 2619–2624. [CrossRef]

30. Zhu, D.; Wang, Y.; Gan, L.; Liu, M.; Cheng, K.; Zhao, Y.; Deng, X.; Sun, D. Nitrogen-containing carbon microspheres for supercapacitor electrodes. *Electrochim. Acta* **2015**, *158*, 166–174. [CrossRef]

31. Hasegawa, G.; Deguchi, T.; Kanamori, K.; Kobayashi, Y.; Kageyama, H.; Abe, T.; Nakanishi, K. High-level doping of nitrogen, phosphorus, and sulfur into activated carbon monoliths and their electrochemical capacitances. *Chem. Mater* **2015**, *27*, 4703–4712. [CrossRef]

32. Susi, T.; Pichler, T.; Ayala, P. X-ray photoelectron spectroscopy of graphitic carbon nanomaterials doped with heteroatoms. *Beilstein J. Nanotechnol.* **2015**, *6*, 177–192. [CrossRef] [PubMed]

33. Gorham, J.; Torres, J.; Wolfe, G.; d'Agostino, A.; Fairbrother, D.H. Surface reaction of molecular and atomic oxygen with carbon phosphide film. *J. Phys. Chem. B* **2005**, *109*, 20379–20386. [CrossRef] [PubMed]

34. Chen, X.; Chen, X.; Xu, X.; Yang, Z.; Liu, Z.; Zhang, L.; Xu, X.; Chen, Y.; Huang, S. Sulfur-doped porous reduced graphene oxide hollow nanosphere frameworks as metal-free electrocatalysts for oxygen reduction reaction and as supercapacitor electrode materials. *Nanoscale* **2014**, *6*, 13740–13747. [CrossRef] [PubMed]

35. Ray, S.C.; Saha, A.; Jana, N.R.; Sarkar, R. Fluorescent carbon nanoparticles: Synthesis, characterization and bioimaging application. *J. Phys. Chem. C* **2009**, *113*, 18546–18551. [CrossRef]

36. Larcher, D.; Mudalige, C.; Gharghouri, M.; Dahn, J.R. Electrochemical insertion of Li and irreversibility in disordered carbons prepared from oxygen and sulfur-containing pitches. *Electrochimica Acta* **1999**, *44*, 4069–4072. [CrossRef]

37. Li, D.; Ding, L.-X.; Chen, H.; Wang, S.; Li, Z.; Zhu, M.; Wang, H. Novel nitrogen-rich porous carbon spheres as a high-performance anode material for lithium-ion batteries. *J. Mater. Chem. A* **2014**, *2*, 16617–16622. [CrossRef]

38. Wei, W.; Huang, X.; Tao, Y.; Chen, K.; Tang, X. Enhancement of the electrocapacitive performance of manganese dioxide by introducing a microporous carbon spheres network. *Phys. Chem. Chem. Phys.* **2012**, *14*, 5966–5972. [CrossRef] [PubMed]

12

Preliminary Investigation of the Process Capabilities of Hydroforging

Bandar Alzahrani and Gracious Ngaile *

Academic Editor: Nooman Ben Khalifa

Department of Mechanical and Aerospace Engineering, North Carolina State University, 911 Oval Drive-3160 EB3, Raleigh, NC 27695-7910, USA; baalzahr@ncsu.edu
* Correspondence: gngaile@ncsu.edu

Abstract: Hydroforging is a hybrid forming operation whereby a thick tube is formed to a desired geometry by combining forging and hydroforming principles. Through this process hollow structures with high strength-to-weight ratio can be produced for applications in power transmission systems and other structural components that demands high strength-to-weight ratio. In this process, a thick tube is deformed by pressurized fluid contained within the tube using a multi-purpose punch assembly, which is also used to feed tube material into the die cavity. Fluid pressure inside the thick tube is developed by volume change governed by the movement of the punch assembly. In contrast to the conventional tube hydroforming (THF), the hydroforging process presented in this study does not require external supply of pressurized fluid to the deforming tube. To investigate the capability of hydroforging process, an experimental setup was developed and used to hydroforge various geometries. These geometries included hollow flanged vessels, hexagonal flanged parts, and hollow bevel and spur gears.

Keywords: hybrid process; hydroforging; tube hydroforming process; lightweight

1. Introduction

There is a growing demand for lightweight structures in the automotive, aerospace, and maritime industries. One of the ways to achieve significant weight reduction is exploring new and innovative manufacturing techniques. In metal forming, the combination of two or more processes is called hybrid forming process. Combining two or more forming processes have been used to develop metal forming processes thereby increasing the process productivity, enhancing part quality, increasing metal formability, reducing the overall process cost, and producing parts with features that would have not been feasible to produce. Attempts to hybridize metal forming processes have been done by several researchers; for example, Debin *et al.* [1] combined an isothermal closed forging and piercing process into a micro-scale hybrid forging process to produce a micro-double gear. A hybrid laser-assisted incremental sheet forming combined with stretch forming lead to reduction in cycle time and increase in formability [2]. A hybrid process combining warm and electromagnetic forming of magnesium alloy sheet has been reported [3]. Penda and Ngaile [4] introduced a new drawing process that incorporate attributes of hydroforming. Hybrid processes which combines deep drawing and cold forging, hot extrusion and integrated equal channel angular pressing, and a combination of tube spinning and a tube bending have been presented [5].

The tube hydroforming (THF) and forging processes are used widely in the industry. The THF process utilizes internal fluid pressure and restrictive die shape to form tubular shapes. In contrast with other many forming processes, THF uses a soft tool (fluid) to deform material, making it useful in producing hollow products with a high strength-to-weight ratio. However, the tube hydroforming process is limited to thin tubular components due to the load and power required to deform the tube [6,7]. Forging is a bulk metal forming process used to shape metal and increase its strength by hammering or pressing. Forging processes play a major role in the automotive industry where highly complex and dimensionally accurate parts are produced with enhanced material characteristics [8]. A large percentage of forged products for the automotive industry are used in power transmission units. Since forged products originate from solid billets, they are usually bulky. There is a great interest in reducing the weight of power transmission units by utilizing hollow structures. Combining unique features of THF and forging processes has the potential to open avenues for producing hollows structures that would have been difficult to produce if individual processes were to be used.

Hydroforging is a hybrid metal forming process which combines forging and hydroforming operations. The unique characteristics emanating from forging and hydroforming processes, make this hybrid process ideal for manufacturing of thick walled tubular components. In the last decade, extensive research work on a similar variant of hydroforging process was carried out at the University of Strathclyde in the UK [9–16]. The process was referred to as injection forging. In this process, a pressurized polymeric material is inserted inside a thick tube while a movable punch is used to feed the tube and pressurize the polymeric material. Different types of polymeric materials were studied and design guidelines for the process were outlined. While the use of polymeric materials as pressure medium was easier to implement, establishment of optimal pressure loading for a specific part is not feasible. Furthermore, the process require secondary operation to remove the injected material. Experimental investigations of hydroforging where fluid is used as a pressure medium were carried out to produce simple bulge shapes by several researchers [17,18]. The aim of this paper is to present experimental investigation on the process capabilities of hydroforging. The investigation was focused on hollow flanged vessels, hexagonal flanged parts, and hollow bevel and spur gears.

2. Hydroforging Process Window and Design Aspects

In order to provide a clear distinction between hydroforging and the conventional tube hydroforming, we first highlight the main characteristics of the two processes. Figure 1a shows a schematic diagram of a conventional tube hydroforming (THF). The major components of a THF process are the press, of which the function is to close the die set, and the pressure intensifier that is used to supply high pressure fluid to the tube. One of the limitations of THF is that it is difficult and cost prohibitive to hydroform thick tubes as the THF system would require very large pressure intensifier. A schematic diagram of the hydroforging process is given in Figure 1b. In this process, a thick tube is deformed by pressurized fluid contained within the tube during the upsetting process. A punch assembly is used to feed tube material into the die cavity. The pressure is generated by compressing the fluid volume contained within the tube. During deformation, no fluid is supplied to the deforming specimen. Principally, the pressure created by pushing the fluid should be greater than the required pressure to form the part without defects. To achieve the desired load path, a pressure relief valve is used to adjust the pressure by releasing fluid. The control architecture for this process is such that at time, t, the fluid pressure generated in the chamber is communicated to the data acquisition center. We envision that the architecture of a hydroforging press to be used in an industrial setting will be similar to conventional forging presses with few additional auxiliary units, namely punch assembly and fluid pressure control units.

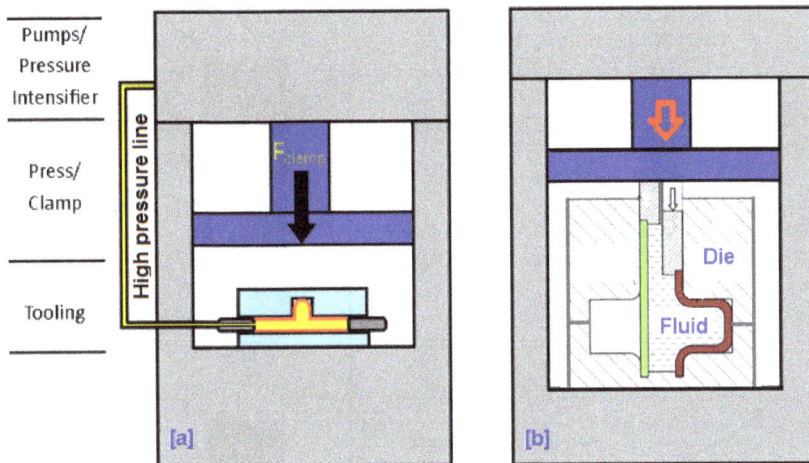

Figure 1. (a) Conventional THF; (b) Tube hydroforging process.

There is a potential to produce different types of products with high strength-to-weight ratio using hydroforging process. Figure 2 shows families of potential candidates for this process, which includes step shafts, polygonal shaped flanges, hollow gears, hollow branched components and hollow vessels.

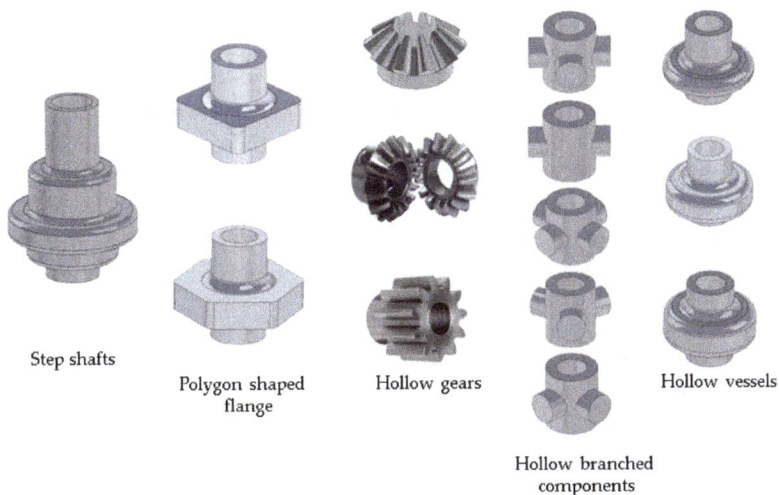

Figure 2. Families of potential candidate parts for hydroforging.

Since in hydroforging no external fluid is supplied to the tube, the process window for hydroforging is thus largely dependent on the initial fluid volume in the tube. A part would be feasible to form if the initial fluid volume is enough to generate the required pressure throughout the process. To investigate the potential candidates for this process, volume calculations for both the tube and the fluid at the initial and final forming stages were performed for several geometries. Flow diagram for volume calculation is given in Figure 3. The main output of the volume calculations was the tube length suitable to produce a product by the hydroforging process. For a specific part geometry, a process window can be built by plotting the flange to tube outer diameter ratio D_f/D_o *versus* the initial tube length to thickness ratio L_0/t_0 at several tube thicknesses. The generated curves provide process limit, *i.e.*, the regions below the curves represent non-feasible region while above the curves signify feasible regions.

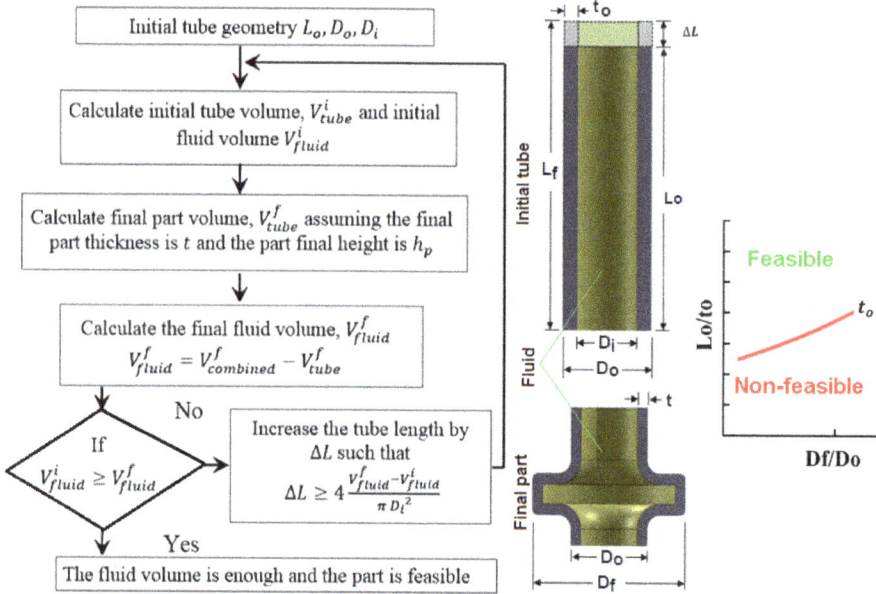

Figure 3. Volume calculation scheme.

2.1. Process Window for Hollow Vessels

Process windows were constructed for hollow vessels using three flange heights: $h_f = 12.7$ mm, $h_f = 19.05$ mm, and $h_f = 25.4$ mm. Figure 4 shows a process window for hollow vessels with a flange height of $h_f = 12.7$ mm at six different tube thicknesses (2.54 mm, 3.81 mm, 5.08 mm, 6.35 mm, 7.62 mm, and 8.89 mm). For a selected tube thickness, the part is considered to be feasible above the line and infeasible under the line. For instance, a hollow vessel with a flange to tube outer diameter ratio D_f/D_o of 2 is feasible to form using a 3.81 mm thick tube if the tube length $L_o \geqslant 105.5$ mm as shown in Figure 4. A shorter tube will not provide enough fluid to generate the required pressure to form the part.

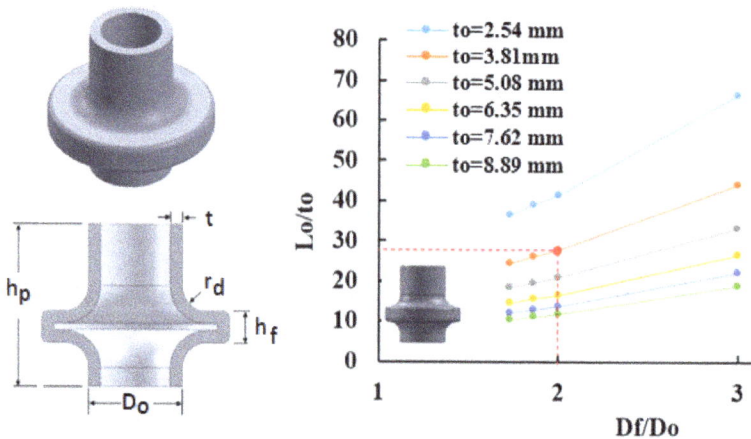

Figure 4. Process window for hollow vessel with flange height $h_f = 12.7$ mm.

2.2. Process Window for Hollow Hexagon Flanges

Process windows were constructed for hollow hexagon shaped flanges using three flange heights, $h_f = 12.7$ mm, $h_f = 19.05$ mm, and $h_f = 25.4$ mm. Figure 5 shows a constructed process window for a flange height of h_f = 25.4 mm at six different tube thicknesses (6.35 mm, 7.62 mm, 8.89 mm, 10.16 mm, 11.43 mm and 12.7 mm). For example, a hollow hexagon shaped flange with a flange to tube outer diameter ratio D_f/D_o of 3 is feasible to form using a 12.7 mm thick tube if the tube length $L_o \geqslant$ 266 mm. The flange diameter is defined as the distance between two parallel flange surfaces.

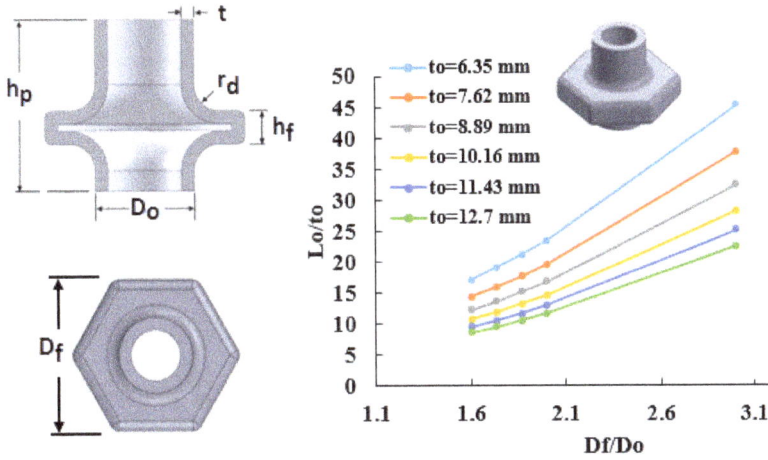

Figure 5. Process window for hollow hexagonal with flange height h_f = 25.4 mm.

2.3. Process Window for Hollow Gears

For a ten teeth hollow bevel gear, a process window for a flange height of h_f = 21.59 mm was constructed using seven tube thicknesses (2.54 mm, 3.81 mm, 5.08 mm, 6.35 mm, 7.62 mm, 8.89 mm, and 10.16 mm) as shown in Figure 6. For instance, a hollow bevel gear with a flange to tube outer diameter ratio D_f/D_o of 3 is feasible to form using a 3.81 mm thick tube if the tube length $L_o \geqslant$ 183 mm. Similarly, a process window was constructed for hollow spur gear with a flange height of h_f = 19.05 mm at six different tube thicknesses (2.54 mm, 3.81 mm, 5.08 mm, 6.35 mm, 7.62 mm, 8.89 mm, and 10.16 mm) as shown in Figure 7. With this example, a hollow spur gear with a flange to tube outer diameter ratio D_f/D_o of 3 is feasible to form using a 5.08 mm thick tube if the tube length $L_o \geqslant$ 190.5 mm.

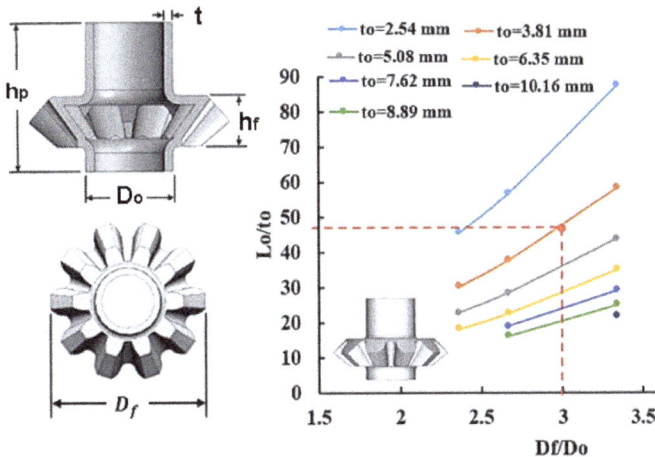

Figure 6. Process window for a ten teeth hollow bevel gear.

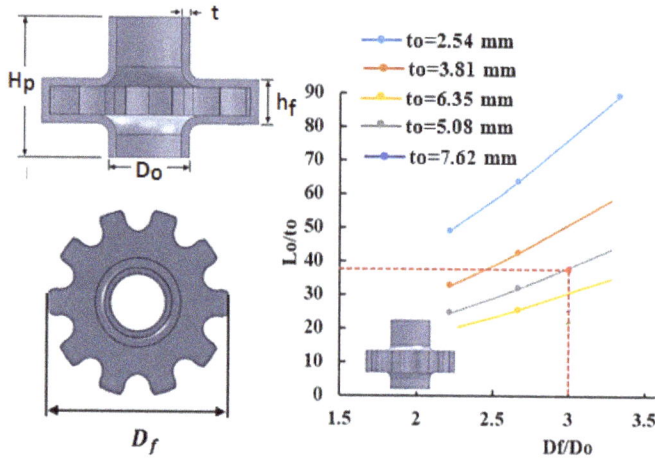

Figure 7. Process window for a ten teeth hollow spur gear.

It should be noted that the fluids used in hydroforging will be compressible, thus, slightly longer tube will be needed to account for the compressibility factor. Furthermore, control of the optimal pressure loading path may require release of fluid from the tube. Figure 8 shows conceptual two pressure profiles for a hydroforging process one representing an optimal pressure required to successfully form the part and the other profile represent induced fluid pressure due to upsetting of the tube. The control should be such that at time, t, some fluid is released to match the desired pressure. This implies that longer tubes beyond that shown in the process windows will be needed. These constraints indicate that hydroforging process will be practical and cost effective only for certain family of parts. One of the major benefits, however, is that very high pressure sufficient to hydroforge thick tubing can be generated during upsetting of the tube.

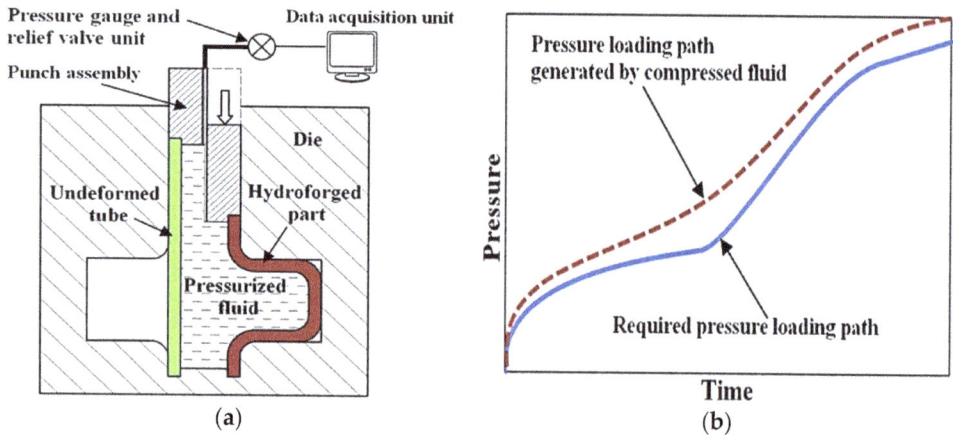

Figure 8. Hydroforging (a) and pressure loading schematic (b).

3. Experimental Setup Development

A schematic of the hydroforging setup is shown in Figure 9. The setup consists of a 150 ton hydraulic press that houses the tooling assembly. The tooling assembly is connected to the pressure lines, relief valve, pressure transducer, and data acquisition system. The hydroforging tooling assembly consists of die inserts, die housing, multi-purpose punch assembly, tube blank, and a sealing insert.

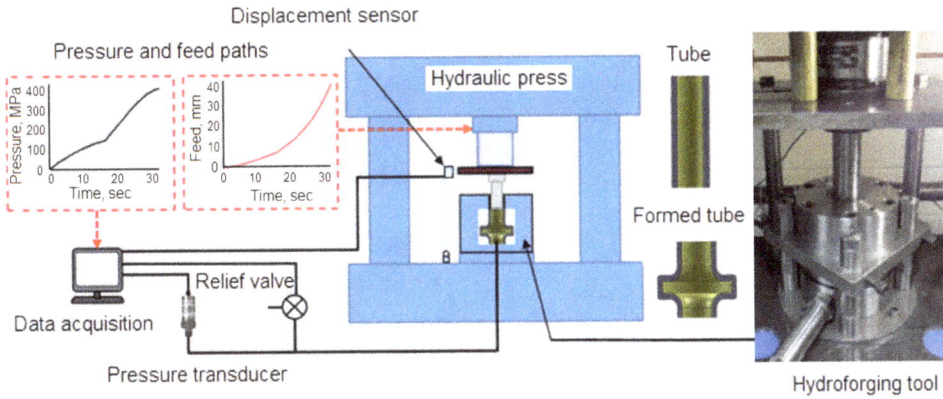

Figure 9. Schematic of hydroforging experimental setup.

The tube hydroforging tooling was designed with interchangeable die inserts so that a variety of tubular materials and geometries could be tested. An exploded view of the hydroforging tooling is shown in Figure 10. The figure shows lower die housing, upper die housing, bottom sealing insert, die insert, supporting ring, guiding zone inserts, punch assembly fastening bolts, tube sample, and a pressure transducer. The punch assembly consists of the punch body and the punch nose. The upper die housing, the lower die housing, and supporting ring are designed to be universal parts. These parts were fabricated to meet the requirement of several tube sizes up to 50.8 mm OD, flange die cavity up to 88.9 mm OD, and flange height of up to 25.4 mm. The punch nose, the guiding zone inserts, the bottom sealing inserts, and the die inserts were designed to meet the required tube sizes and thicknesses. The punch nose and guiding zone inserts were made out of A2 tool steel and hardened to 59 HRC to be able to handle high pressure loads. Punch bodies were fabricated from alloy steel for each tube size. Figure 11 shows a set of fabricated parts made to customize the tube sizes, tube thicknesses, and flange heights.

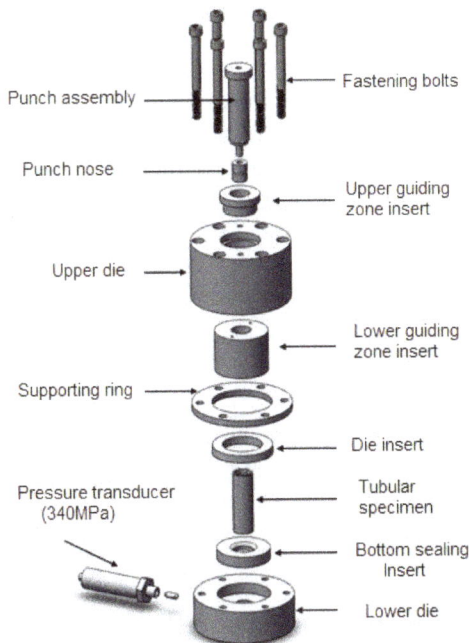

Figure 10. Exploded model for tube hydroforging tooling.

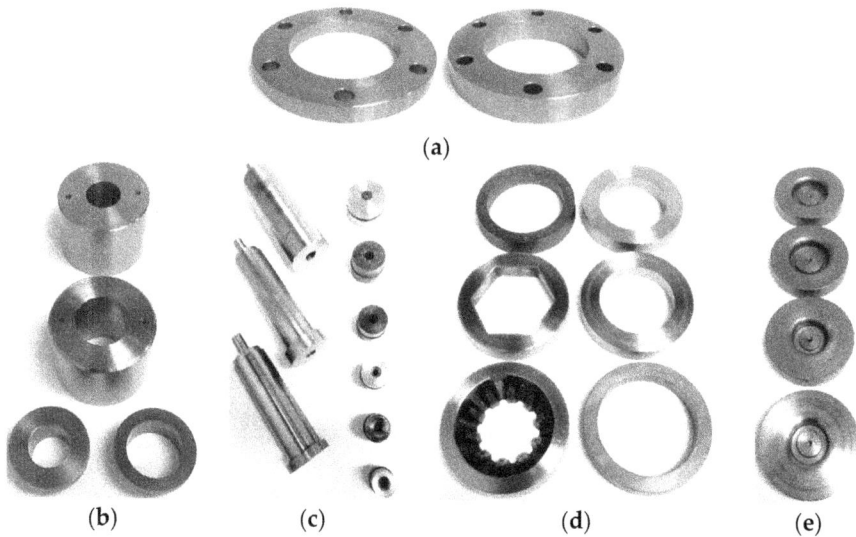

Figure 11. Experimental setup parts: (a) Support rings; (b) guiding zone inserts; (c) punch assemblies; (d) die inserts; and (e) bottom sealing inserts.

Experimental Matrix and Test Procedures

Aluminum (AL6061) and stainless steel (SS 304) tube samples were used to perform the experiments. The properties for these materials were as follows; for AL6061 the strength coefficient K = 118 MPa and strain hardening exponent n = 0.22. The strength coefficient and strain hardening exponent for SS304 were 1426 MPa and 0.5, respectively. The experimental matrix is given in Table 1. In the table, the tube dimensions are described by the tube outer diameter D_o, the tube thickness t_o, and the tube length, L_o. The die dimensions are defined by the flange height h_f, and the flange diameter D_f. Annealed AL6061 and annealed SS304 tubes were used to perform the experiments. Before the experiments all the tubular specimens were wrapped with Teflon sheet that acted as a lubricant. The tube was filled with oil and the punch is pre-assembled to establish contact with the tube end. ISO 32 light hydraulic oil with a viscosity of 32 centistoke was used. The punch is then pushed downward by a 150 ton press. As the tube is pushed towards the cavity, fluid pressure is generated inside the tube due to volume change, forcing the material to flow toward to die cavity wall. During the process, fluid pressure and punch stroke are recorded as a function of time. The pressure control system is installed with 350 MPa pressure transducer. Figure 12 shows a variety of parts that were hydroforged which included hollow vessels, hollow hexagonal shaped flanges, and hollow spur and bevel gears.

Table 1. Experimental matrix.

Tube Dimensions and Material				Die Dimensions		Hydroforged Parts
D_o, mm	t_o, mm	L_o, mm	Material	h_f, mm	D_f, mm	
38.1	6.35	127	AL6061 annealed	18.5	76.2	Hollow vessel-6.35 mm thick
38.1	6.35	127	AL6061 annealed	21.6	76.2	Hollow vessel-21.6 mm flange
38.1	8	127	AL6061 annealed	18.5	76.2	Hollow vessel-8 mm thick
38.1	6.35	127	AL6061 annealed	17.8	76.2	Hollow hexagon flange
50.8	6.35	114	AL6061 annealed	18.5	76.2	Hollow vessel-6.35 mm thick
50.8	9.5	120	AL6061 annealed	18.5	88.9	Hollow vessel-9.5 mm thick
50.8	6.35	101	AL6061 annealed	20.3	76.2	Hollow spur gear-AL6061
50.8	9.5	108	AL6061 annealed	17.8	76.2	Hollow bevel gear-Al6061
50.8	1.65	114	SS304 annealed	17.8	76.2	Hollow bevel gear-SS304

Figure 12. Hydroforged parts.

4. Experimental Results and Discussion

Figure 13a shows hydroforged hollow vessel with flange sizes of 18 mm and 21 mm. These parts were hydroforged from 38 mm OD × 127 mm long tube with a wall thickness of 6.35 mm. The loading paths that resulted from the hydroforging operations are given in Figure 13b. The experiments were carried out in 35 s and attained a maximum punch force of 325 kN and maximum fluid pressure of 180 MPa. To form this part a feed of 60 mm was used. Figure 14 shows hydroforged hexagonal shaped hollow parts with a flange thickness of 18 mm and the corresponding loading paths. These parts were hydroforged using 38 mm OD × 127 mm long tubular samples with a wall thickness of 6.35 mm. The maximum punch load and fluid pressure of 300 kN and 160 MPa, respectively, were attained at maximum punch stroke of 68 mm.

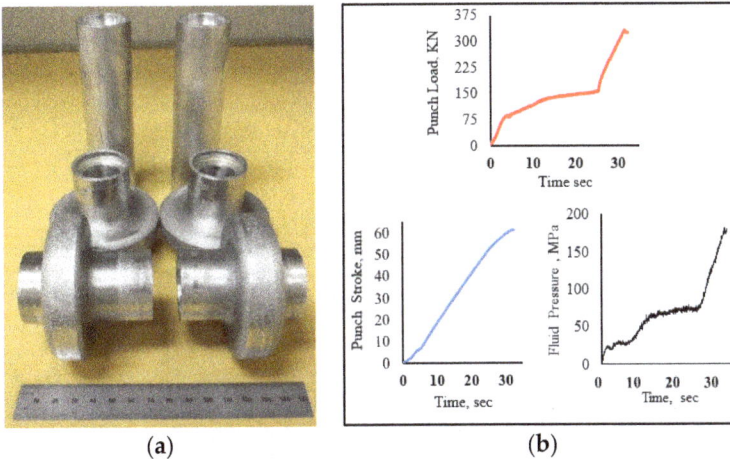

Figure 13. (a) Hollow vessel, flange = 18 mm and 21 mm; (b) loading for 18 mm flange.

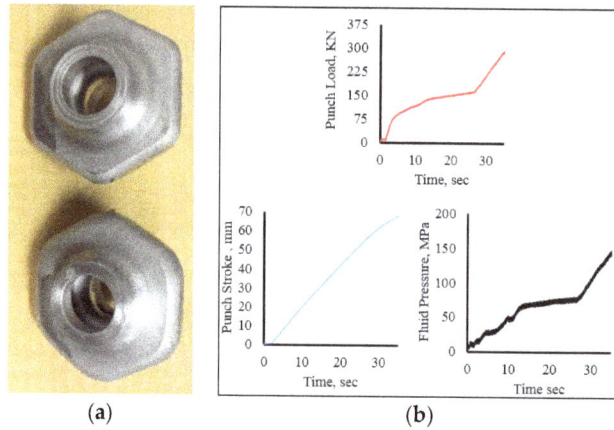

Figure 14. (a) Hexagonal hollow parts; (b) loading paths.

A ten teeth spur gear was hydroforged using a 50.8 mm OD × 6.35 mm aluminum tubes as shown in Figure 15. The corresponding loading paths are given in Figure 15b. The maximum press load reached 474 kN at the end of the process. During teeth forming, the press load increased from 130 kN to 474 kN and a maximum fluid pressure of 175 MPa was reached at punch stroke of 30 mm. In this test, the spur gear was not fully formed. The tube material at the tooth root was subjected to excessing shearing causing rupture. Hydroforging of bevel gears from AL6061 and SS304 was also conducted, as shown in Figure 16.

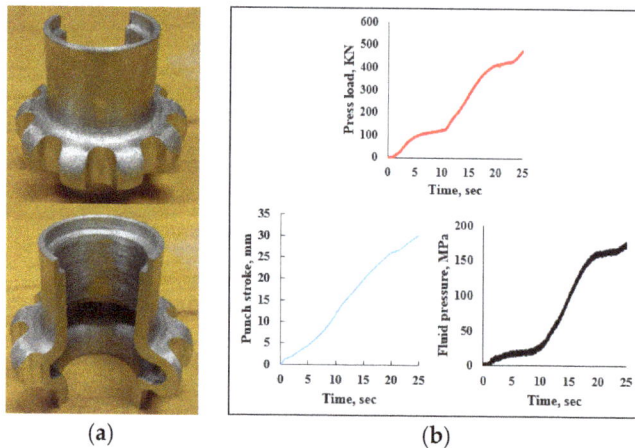

Figure 15. (a) Spur gear; (b) loading paths.

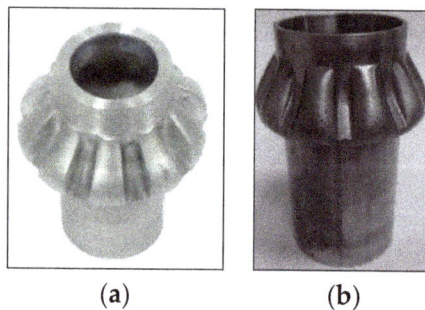

Figure 16. Bevel gear: AL6061(a) and SS304 (b).

Part failures were observed in several parts formed by the tube hydroforging process. The failure modes in these parts can be classified into three categories: tube rupture, surface cracks and excessive thinning. Figure 17 shows the experimental evidence of tube rupture in a bevel gear formed using a 6.35 mm thick aluminum tube. In this experiment, the tubular material was pushed by the generated pressure against the die teeth causing excessive shear as the tooth was being formed. The failure progression is illustrated in Figure 17. The tube material was exposed to shearing at the sharp die teeth edges leading to rupture. To reduce the shear effect, an extra forming stage may be needed with larger die radius. When a thicker aluminum tube (t_o = 9.5 mm) was used to form the bevel gear, no rupture was observed.

Few samples exhibited internal surface cracks at the bottom side of a hexagon shaped flange shown in Figure 18. The crack initiated during the expansion stage where the tube undergo tensile loading caused by excessive thinning at this location. Tube thickening can be observed at the top portion of the flange. This failure would be avoided if more material was fed inside the die cavity in advance of pressure. It should be noted that the pressure loading paths used in this study were not optimized. However, they were sufficient to demonstrate the feasibility of the hydroforging process. The above mentioned failures could be avoided by controlling the generated pressure in which more material is fed into the die cavity and subsequently delaying or completely eliminating crack initiation and tube thickening. Future work will include determination and control of optimal pressure loading that lead to uniform wall thickness distribution.

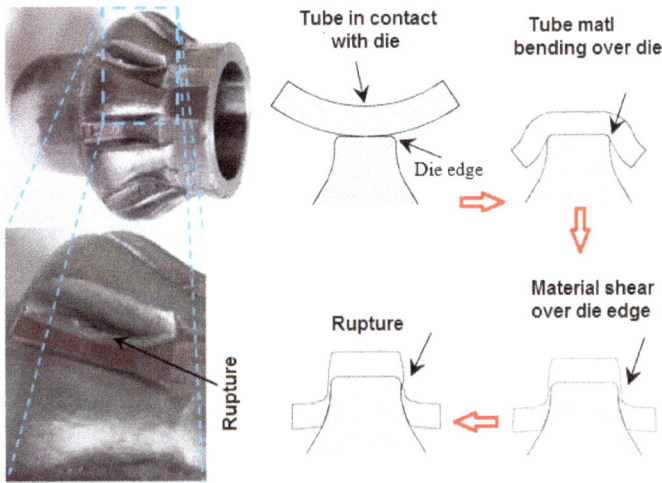

Figure 17. Material failure mode during hydroforging of a bevel gear from AL6061 blank.

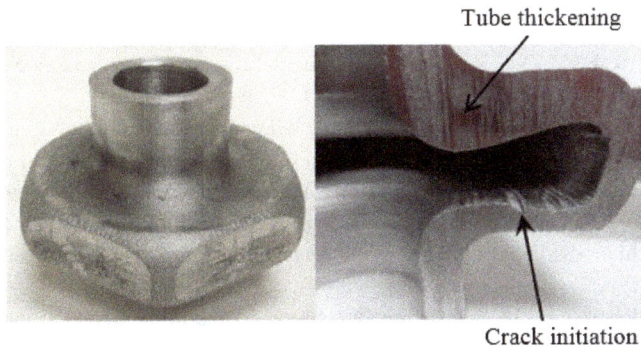

Figure 18. Crack initiation and non-uniform thickness distribution on hexagonal shaped flange.

5. Conclusions

Investigation on the feasibility of hydroforging process where thick tubes are formed by combining forging and hydroforming operations was carried out. Based on initial fluid volume responsible for inducing pressure inside the tube during upsetting of the tube, process windows for determining the required tube length to successfully form a part were established for a few geometries. Preliminary hydroforging experiments were carried out to examine the feasibility and limitations of this process. Hollow flanged vessels, hollow flanged hexagonal parts, and hollow bevel and spur gears were hydroforged from AL6061 and SS306 tubular blanks. The conclusions drawn from this preliminary study are:

- Hydroforging process has a lot of potential for manufacturing hollow components with high-strength-to-weight ratio from thick tubing for applications in power transmission systems.
- The simplicity of the hydroforging tool set up implies that development of hydroforging production machines may not be capital intensive compared to conventional hydroforming machines. This is largely due to the fact that hydroforging does not require a pressure intensifier. Thus, a conventional forging machine could be modified to suit the requirements of the hydroforging process.
- One of the drawbacks of hydroforging process, is that the induced pressure inside the tube relies on the initial volume of fluid in the tube. Thus, the hydroforging process is only applicable for products whose internal tube volume continuously decrease as the tube material is fed toward the die cavity.
- Since the control of fluid pressure for ensuring that the pressure loading path is optimal may be achieved by releasing some of the pressurized fluid inside the tube, much longer tubular blanks to compensate for the release of fluid will be needed.

Acknowledgments: The authors would like to aknowledge Steven Cameron for helping with fabricating the tooling.

Author Contributions: Bandar Alzahrani and Gracious Ngaile conceived and designed the experiments; Bandar Alzahrani performed the experiments; Bandar Alzahrani and Gracious Ngaile analyzed the data; Gracious Ngaile wrote the paper.

Conflicts of Interest: The authors declare no conflict of interest.

References

1. Debin, S.; Jie, X.; Chunju, W.; Bin, G. Hybrid forging processes of micro-double gear using micro-forming technology. *Int. J. Adv. Manuf. Technol.* **2009**, *44*, 238–243. [CrossRef]
2. Araghi, B.T.; Gottmann, A.; Bambach, M.; Hirt, G.; Bergweiler, G.; Diettrich, J.; Saeed-Akbari, A. Review on the development of a hybrid incremental sheet forming system for small batch sizes and individualized production. *Prod. Eng. Res. Dev.* **2011**, *5*, 393–404. [CrossRef]
3. Meng, Z.; Huang, S.; Hu, J.; Xia, Z.; Xia, X. Experimental research on warm and electromagnetic hybrid forming of magnesium alloy sheet. *J. Mech. Eng.* **2011**, *47*, 38–42. [CrossRef]
4. Pender, K.; Ngaile, G. A novel hybrid process for drawing operations. In Proceedings of the ASME 2013 International Manufacturing Science and Engineering Conference Collocated with the 41st North American Manufacturing Research Conference (MSEC 2013), Madison, WI, USA, 10–14 June 2013.
5. Jager, A.; Selvaggio, A.; Hanisch, S.; Haase, M.; Becker, C.; Kolbe, J.; Tekkaya, A.E. Innovative hybrid process in metal forming. In Proceedings of the 41st North American Manufacturing Research Conference (NAMRC 2013), Madison, WI, USA, 10–14 June 2013; Volume 41, pp. 205–214.
6. Ahmetoglu, M.; Altan, T. Tube hydroforming: State-of-the-art and future trends. *J. Mater. Proc. Technol.* **2000**, *98*, 25–33. [CrossRef]
7. Koc, M.; Altan, T. Overall review of the tube hydroforming (THF) technology. *J. Mater. Proc. Technol.* **2001**, *108*, 384–393. [CrossRef]

8. Fujikawa, S.; Yoshioka, H.; Shimamura, S. Cold- and warm-forging applications in the automotive industry. *J. Mater. Proc. Technol.* **1992**, *35*, 317–342. [CrossRef]

9. Balendra, R.; Petersen, S.B.; Colla, D. Preforming considerations for the injection forging of tubes. *Proc. IMechE Conf. Trans.* **1995**, *3*, 79–84.

10. Balendra, R.; Qin, Y. Material-flow considerations for the design of injection forging. *J. Manuf. Sci. Eng. Trans. ASME* **1997**, *119*, 350–357. [CrossRef]

11. Balendra, R.; Qin, Y. Injection forging: Engineering and research. *J. Mater. Proc. Technol.* **2004**, *145*, 189–206. [CrossRef]

12. Ma, Y.; Qin, Y.; Balendra, R. Forming of hollow gear-shafts with pressure-assisted injection forging (PAIF). *J. Mater. Proc. Technol.* **2005**, *167*, 294–301. [CrossRef]

13. Ma, Y.; Qin, Y.; Balendra, R. Upper-bound analysis of the pressure-assisted injection forging of thick-walled tubular components with hollow flanges. *Int. J. Mech. Sci.* **2006**, *48*, 1172–1185. [CrossRef]

14. Qin, Y.; Ma, Y.; Balendra, R. Mechanics of pressure-assisted injection forging of tubular components. *J. Mech. Eng. Sci.* **2004**, *218*, 1195–1212. [CrossRef]

15. Qin, Y.; Ma, Y.; Balendra, R. Pressurising materials and process design considerations of the pressure-assisted injection forging of thick-walled tubular components. In Proceedings of the 9th ISPE International Conference on Concurrent Engineering, Cranfield, UK, 27–31 July 2002; Volume 150, pp. 30–39.

16. Qin, Y.; Balendra, R. Computer-aided design of nett-forming by injection forging of engineering components. *J. Mater. Proc. Technol.* **1998**, *76*, 62–68. [CrossRef]

17. Muller, K.; Stonis, M.; Lucke, M.; Behrens, B. Hydroforging of thick-walled hollow aluminum profiles. In Proceedings of the 15th Conference of the European Scientific Association on Material Forming, (ESAFORM 2012), Erlangen, Germany, 14–16 March 2012.

18. Roeper, M.; Reinsch, S. Hydroforging—A new manufacturing technology for forged lightweight products of aluminum. In Proceedings of the 2005 ASME International Mechanical Engineering Congress and Exposition, (IMECE 2005), Orlando, FL, USA, 5–11 November 2005; pp. 297–304.

13

Improved Sectional Image Analysis Technique for Evaluating Fiber Orientations in Fiber-Reinforced Cement-Based Materials

Bang Yeon Lee [1], Su-Tae Kang [2], Hae-Bum Yun [3] and Yun Yong Kim [4],*

Academic Editor: Hong Wong

[1]   School of Architecture, Chonnam National University, 77 Yongbong-ro, Buk-gu, Gwangju 61186, Korea; bylee@jnu.ac.kr

[2]   Department of Civil Engineering, Daegu University, 201 Daegudae-ro, Jillyang, Gyeongsan, Gyeongbuk 38453, Korea; stkang@daegu.ac.kr

[3]   Department of Civil, Environmental, and Construction Engineering, University of Central Florida, Orlando, FL 32816, USA; Hae-Bum.Yun@ucf.edu

[4]   Department of Civil Engineering, Chungnam National University, 99 Daehak-ro, Yuseong-gu, Daejeon 34134, Korea

*   Correspondence: yunkim@cnu.ac.kr

Abstract: The distribution of fiber orientation is an important factor in determining the mechanical properties of fiber-reinforced concrete. This study proposes a new image analysis technique for improving the evaluation accuracy of fiber orientation distribution in the sectional image of fiber-reinforced concrete. A series of tests on the accuracy of fiber detection and the estimation performance of fiber orientation was performed on artificial fiber images to assess the validity of the proposed technique. The validation test results showed that the proposed technique estimates the distribution of fiber orientation more accurately than the direct measurement of fiber orientation by image analysis.

Keywords: fiber; fiber-reinforced concrete; image analysis; orientation

1. Introduction

The first major investigation was made to evaluate the potential of steel fibers as a reinforcement for concrete in the United States during the early 1960s [1]. Since then, a substantial amount of research, development, experimentation and industrial application of steel or synthetic fiber-reinforced concrete has been carried out [2]. The major role of adding fiber is to bridge microcracks and, thus, to improve tensile resistance [3]. Thus, the distribution of fibers strongly influences the resulting mechanical performance of the composite [4–10]. Short fibers with lengths of 6 to 40 mm, which are randomly distributed in all directions, so as to have isotropic behavior, are commonly used in fiber-reinforced concrete. However, the real fiber distribution is strongly influenced by various factors, such as the fiber characteristics, including the diameter, length and volume fraction, the rheological properties of the matrix, the placing method, the shape of the form, *etc.* Non-uniform fiber distribution decreases the effect of fibers on strengthening the matrix [11,12]. Therefore, it is not reasonable to estimate the uniaxial tensile strength or flexural strength of fiber-reinforced concrete from the assumption of uniformly two-dimensional or three-dimensional distributed fibers. Furthermore, the number of fibers in the sectional image in fiber-reinforced concrete under three-dimensional distribution and two-dimensional distribution are 1/2 and $2/\pi$, respectively, of that under one-dimensional distribution [13]. A directional efficiency coefficient was adopted to consider the effect of fiber orientation distribution on the tensile behavior of fiber-reinforced concrete [14–16].

Micromechanically, the fiber orientation, which is the angle of the fiber inclined to the crack plane, influences the fiber pullout load and fiber strength in the matrix. The fiber pullout load increases when the fiber orientation is increased due to the increase of the normal force between the fiber and the matrix, which increases the frictional bond between the fiber and the matrix. This phenomenon is known as the snubbing effect, and currently, an empirical equation between the pullout load of inclined fiber and the pullout load of fiber without an inclination angle is being adopted [17,18]. On the other hand, the fiber strength in a matrix decreases when the fiber orientation is increased due to an additional stress at the exit point of the crack plane by bending [19]. The effect of fiber orientation on the multiple fibers in the composite is taken into consideration in the form of a probability density function for fiber orientation and single fiber pullout load [20].

Several techniques, including image analysis [21–26], transmission X-ray imaging [27–31] and alternating current impedance spectroscopy (AC-IS) [32,33], are available for evaluating the fiber dispersion and orientation in a composite made of a cement-based matrix and steel, carbon, glass or synthetic fibers [34]. Among the various techniques, image analysis provides direct information on fiber dispersion and orientation. However, previous studies reported that a two-dimensional image analysis technique may induce a significant systematic error in orientation measurements according to image resolution [21,35]. Therefore, we present a new image analysis technique to improve the evaluation accuracy of fiber orientation distribution in the sectional image of fiber-reinforced cement-based material. The proposed image analysis technique estimates the distribution of fiber orientation from the number of fibers in the sectional images because the number of fibers is dependent on the distribution of the fiber orientation.

2. Fiber Distribution Evaluation Method

2.1. Image Analysis for the Evaluation of the Fiber Orientation Distribution

The distribution characteristics of fiber can be quantitatively evaluated by calculating the coefficient based on the coordinates of the fibers and the shape of the fibers in the cutting plane. To detect the fiber in the fiber images, the color image is converted to a grayscale image. The grayscale image is then converted to a binary image based on a set threshold object detection method, which in turn is based on a thresholding algorithm [36]. In this process, other parts aside from the fibers can be detected as fibers due to having similar brightness to fibers. These misdetected objects are classified on the basis of the threshold of the object's area, which is determined by the minimum area of randomly-selected fibers. Misdetected objects with a smaller area than the threshold area are deleted. In addition, aggregate fiber images (otherwise known as misdetected fiber images) can be correctly detected by means of the watershed segmentation algorithm and morphological reconstruction [8,37,38]. Over-segmentation by the watershed segmentation algorithm can be minimized by applying a morphological reconstruction [8]. Fiber orientation was defined as the angle between the fiber axis and the normal direction of cutting plane. This is simply calculated by Equation (1) [13].

$$\theta = \arccos\left(\frac{d}{l}\right) = \arccos\left(\frac{d}{d/\cos\theta}\right) \tag{1}$$

where θ, d and l are the inclined angle of the fiber (out-of-plane fiber orientation), the diameter of the fiber and the major axis length of the fiber image, respectively. (Figure 1) The major axis length and diameter, which is the same as the minor axis length, of the fibers were measured by specifying the length (in pixels) of the major axis and minor axis lengths of the ellipse that had the same normalized second central moments as the region.

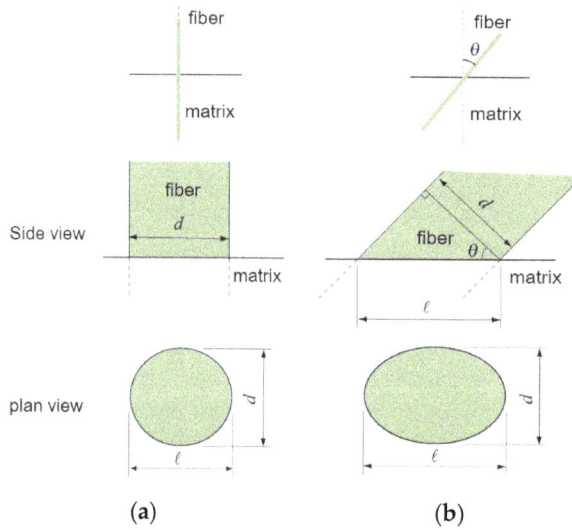

Figure 1. Dimensions of fibers in the side view and plan view on the cutting plane, which is the horizontal line in side view: (**a**) 0° orientation and (**b**) θ orientation.

2.2. Effect of the Number of Pixels on the Detection Accuracy

Table 1 shows artificial fiber images in the cutting plane according to the fiber orientation, and Figure 2 shows the ratio of the major axis length (l) to the minor axis length (d) of the fiber image with the fiber orientation. As seen in Figure 2, it seems that l/d is constant below a fiber orientation (θ) of 30°. The difference in l/d between the fiber images with θ of 0° and 30° is 15.3%. This indicates that an estimation error is unavoidable when we calculate θ based on the l/d of a fiber image with an actual θ smaller than 30°. In contrast, the difference in l/d between the fiber images with θ of 0° and 45° is 41.2%; furthermore, the l/d of a fiber image with θ of 85° is 8.08-times larger than that of a fiber with θ of 45°. This simple numerical simulation shows that the sensitivity of l/d should sharply increase with a decreasing fiber orientation. This is because l/d theoretically equals the inverse of cosθ.

Table 1. Artificial fiber images in the cutting plane (diameter of 150 pixels).

Fiber Orientation (°)	Fiber Image
0	
15	
30	
45	
60	
75	
85	

Table 2 lists the fiber orientation measured using an image analysis in an artificial fiber image with a certain orientation according to the number of pixels in the diameter of the fiber. A high number of pixels indicates a high resolution. Figure 3 shows the error of the measured orientation of an artificial fiber image according to the number of pixels in the diameter. The error increases with a decreasing θ and the number of pixels in the diameter of the fiber. This is attributed to the increase in sensitivity with a decreasing θ and the increase of detection error with a decreasing number of pixels in the diameter of the fiber, *i.e.*, a decreasing resolution. The diameter of the synthetic fibers used in a high ductile fiber-reinforced cementitious composite ranges from 10 to 40 μm, and the diameter of steel fibers used in ultra-high performance concrete is about 200 μm [39–41]. Table 3 gives the unit pixel length according to the number of pixels and the real diameter of the fiber for three types of fibers with different size diameters. The unit pixel length increases with a decreasing number of pixels in the diameter of the fiber and an increasing real fiber size. If there are five pixels, they represent 200 μm in diameter of steel fiber, and the unit pixel represents 40 μm, which means that one pixel image falsely detected during acquisition or processing may induce an error of 40 μm in measuring the fiber diameter. Therefore, a high enough resolution in relation to the size of the fiber should be employed to prevent false detection. However, a higher resolution requires a larger processing time and memory.

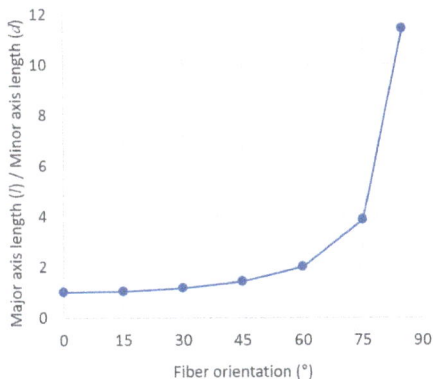

Figure 2. l/d ratio of the fiber image plotted as a function of fiber orientation, θ.

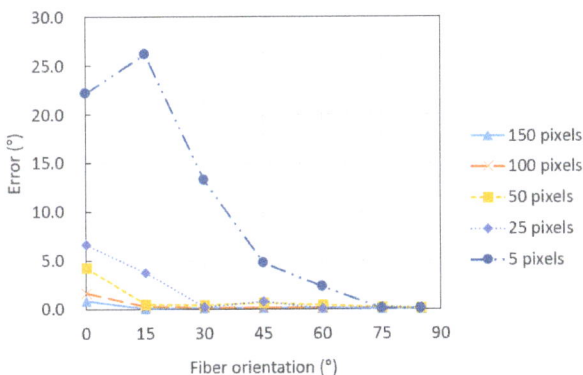

Figure 3. Error of the measured orientation of an artificial fiber image according to the number of pixels in the diameter.

Table 2. Measured orientation of the artificial fiber image according to the number of pixels in the diameter.

Number of Pixels in the Diameter of the Fiber	Fiber Orientation (°)						
	0	15	30	45	60	75	85
150	0.8	14.9	29.9	44.9	59.9	74.9	85.0
100	1.7	14.7	29.9	44.9	59.9	74.9	85.0
50	4.3	14.5	29.6	44.4	59.6	74.8	84.9
25	6.6	18.7	30.2	45.7	59.9	74.8	84.9
5	22.2	41.2	43.2	49.8	62.3	75.1	84.9

Table 3. Unit pixel length with the number of pixels and the real diameter of the fiber.

Number of Pixels in the Diameter of the Fiber	Real Diameter of the Fiber (μm)		
	10	40	200
150	0.07	0.27	1.33
100	0.10	0.40	2.00
50	0.20	0.80	4.00
25	0.40	1.60	8.00
5	2.00	8.00	40.00

3. Image Analysis for Enhancing the Evaluation Accuracy of the Fiber Orientation Distribution

The number of fibers in the sectional images is dependent on the distribution of the fiber orientation. This study suggests that the distribution of the fiber orientation is derived from the number of fibers inversely when enough resolution to ensure the accuracy of fiber detection can be obtained. In this study, a two-parameter exponential function proposed by Xia *et al.* [42] was adopted to express the distribution of fiber orientation, given as follows:

$$g(\theta) = \frac{\sin\theta^{2p-1}\cos\theta^{2q-1}}{\int_{\theta_{min}}^{\theta_{max}} \sin\theta^{2p-1}\cos\theta^{2q-1}d\theta} \tag{2}$$

where p and q are the shape parameters, which can be used to determine the shape of the probability density function. The parameters p and q should be more than 0.5, and θ is in a range from 0 to $\pi/2$. $g(\theta)$ with p of one and q of 0.5 is the same as $\sin\theta$, which is the probabilistic density function for a perfect three-dimensional distribution of fiber orientation. When p and q are one and one, respectively, $g(\theta)$ is $1/\pi$, which is the probabilistic density function for perfect two-dimensional distribution of the fiber orientation. In this study, the parameters p and q are determined by applying an optimization technique for minimizing the error between the number of fibers measured by image analysis, which is described in Section 2.1, and the number of fibers calculated theoretically. The theoretical number of fibers $N_{f,t}$ is calculated from the volume and diameter of the fiber, the area of the sectional image and the assumed fiber distribution, given as follows:

$$N_{f,t} = \frac{4V_f A_s}{\pi d^2} \int_{\theta_{min}}^{\theta_{max}} g(\theta)\cos\theta d\theta \tag{3}$$

where V_f and A_s are the fiber content in terms of the volume fraction and the area of the section of fiber-reinforced concrete, respectively.

The optimum technique of the proposed method adopts a direct search, since it does not require any information about the gradient of the objective function and is easy to implement. In this study, a real-valued genetic algorithm is applied to genetic operations for finding the optimal values of p and q and then estimating the distribution of the fiber orientation. The genetic algorithm is initiated with a set of solutions called populations. Solutions from one population are used to form a new population. This procedure is motivated by the expectation that the new population will be better than the old one. Solutions that are selected to form new solutions are chosen according to their fitness; the more suitable they are, the more opportunities they have to reproduce. This is done by three major processes: selection, crossover and mutation [43–45].

The selection is a process in which the best-fit solutions in the population are fit enough to survive and possibly reproduce new solutions for the next generation. The crossover process creates new solutions by combining pairs of old solutions in the current population. This enables the algorithm to extract the best new solutions from different individuals and recombine them into potentially superior new solutions. The mutation process creates new solutions by randomly changing individual old solutions. This prevents all solutions in the population from falling into a local optimum of solved problems. In this study, the population size was set to 200. In order to create a new generation, the roulette wheel selection, the combination of genes with randomly-selected genes from parents' genes and the addition of a random number taken from a Gaussian distribution with mean zero were adopted for the selection, crossover and mutation process, respectively. When there is no improvement in the objective function for a sequence of consecutive generations of length 50, the process is stopped.

The fitness function is expressed in the form of Equation (4).

$$f_{\text{fitness}}(p,q) = \left| N_{f,t}(p,q) - N_{f,m} \right| \tag{4}$$

where $N_{f,m}$ is the measured number of fibers by image analysis.

4. Validation of the Proposed Technique

To assess the validity of the proposed technique, a series of tests was performed on artificial fiber images. Artificial section fiber images with three sizes of fibers and two- and three-dimensional random distributions of fiber orientation were tested (Table 4). The total fiber volume was assumed to be 2.0 vol%. The number of fibers was calculated from the fiber volume fraction, the area of the images, the diameter of the fibers and the distribution characteristics of the fiber orientation. Figures 4–9 show the sectional fiber images made artificially for the test of the validation. As shown in Figures 4b and 5b, with the diameter of five pixels representing the diameter of the fiber, the surface of the fiber image is not a smooth curve, but a series of discontinuous lines. On the other hand, the surfaces of the fiber images are smoother with an increasing resolution and the number of pixels representing the diameter of the fiber from five to twenty five.

Table 4. Images to assess the performance of the proposed technique.

Image ID	Number of Pixels in the Diameter of the Fiber	Area (Pixel²)	Dimension of the Fiber Orientation Distribution	Number of Fibers
I05-3	5	2000 × 2000	3	2038
I05-2			2	2594
I15-3	15	6000 × 6000	3	2038
I15-2			2	2594
I25-3	25	5000 × 5000	3	510
I25-2			2	649

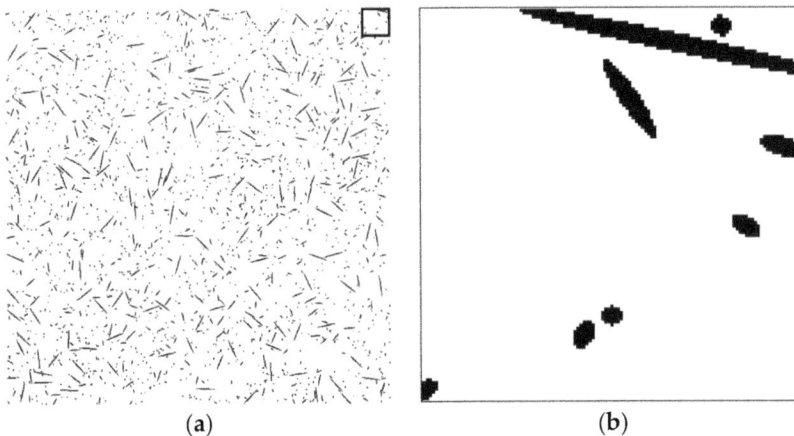

(a)   (b)

Figure 4. Artificial fiber images with a diameter of five pixels in a three-dimensional random distribution in the area of 2000 pixels by 2000 pixels (I05-3): (a) section image and (b) magnified image.

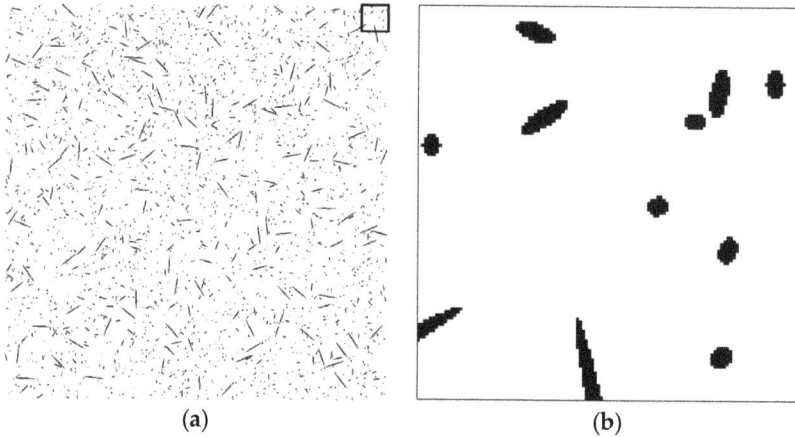

Figure 5. Artificial fiber images with a diameter of five pixels in a two-dimensional random distribution in the area of 2000 pixels by 2000 pixels (I05-2): (**a**) section image and (**b**) magnified image.

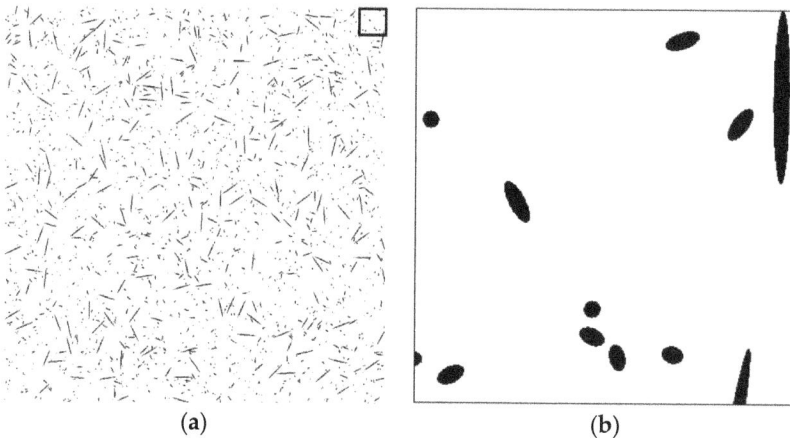

Figure 6. Artificial fiber images with a diameter of 15 pixels in a three-dimensional random distribution in the area of 6000 pixels by 6000 pixels (I15-3): (**a**) section image and (**b**) magnified image.

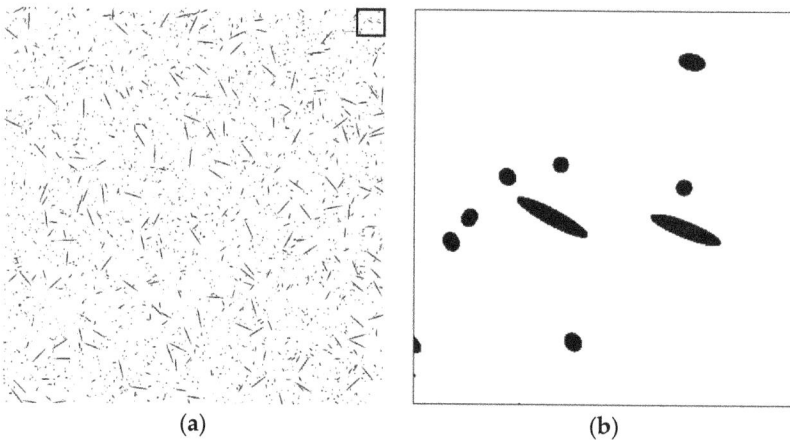

Figure 7. Artificial fiber images with a diameter of 15 pixels in a two-dimensional random distribution in the area of 6000 pixels by 6000 pixels (I15-2): (**a**) section image and (**b**) magnified image.

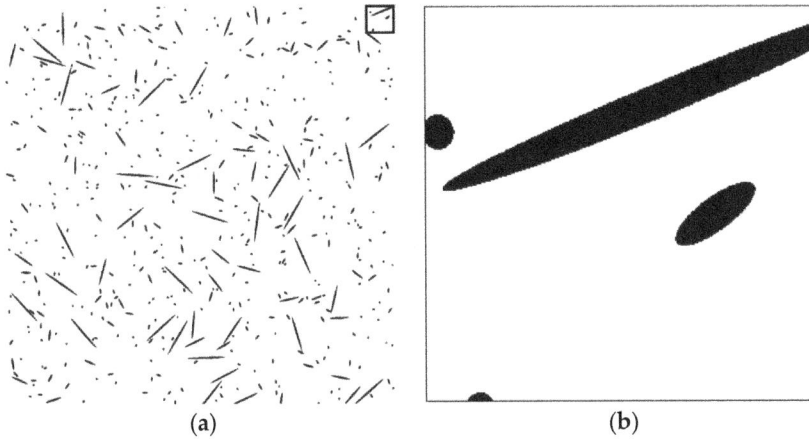

Figure 8. Artificial fiber images with a diameter of 25 pixels in a three-dimensional random distribution in the area of 5000 pixels by 5000 pixels (I25-3): (**a**) section image and (**b**) magnified image.

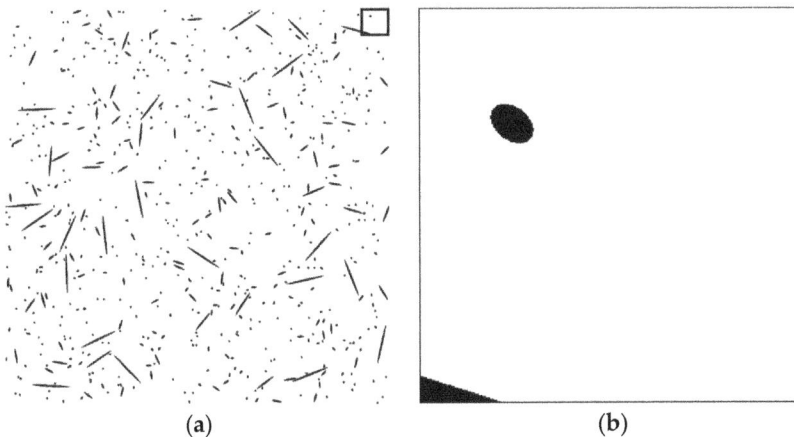

Figure 9. Artificial fiber images with a diameter of 25 pixels in a two-dimensional random distribution in the area of 5000 pixels by 5000 pixels (I25-2): (**a**) section image and (**b**) magnified image.

Figure 10 shows the probability density functions of the fiber orientation according to the number of pixels representing the diameter of the fiber and the distribution characteristics (two- or three-dimensional distribution). The fiber orientation was measured using the image processing technique described in Section 2.1 based on Lee's technique [8]. The randomly-generated artificial fibers show probability density functions similar to those of perfectly uniform two- or three-dimensional distributions. However, the probability density functions from the direct measurement of fiber orientation by image analysis are considerably different from those of real distributions, especially in the region of low θ. The error of the two-dimensional distribution is larger than that of the three-dimensional distribution. This is attributed to the fact that two-dimensional distribution images have a higher probability for low θ fibers than that of three-dimensional distribution images. The error decreased with an increasing number of pixels, which represents the diameter of the fiber. This can be expected from the investigation in Section 2.2. Figure 11 shows the average errors per fiber of the measured orientation of the artificial fiber images from Figures 4–9 according to the number of pixels representing the diameter of the fibers. The average error per fiber in the three-dimensional distribution was 32.4% lower than that in the two-dimensional distribution. In contrast, the average error per fiber in the two-dimensional distribution more sharply decreased with an increasing number of pixels compared to that in the three-dimensional distribution. Average errors per fiber of the I15-2 and I25-2 images decreased by 70.0% and 80.8% compared to I05-2, respectively, while average errors per fiber of the I15-3 and I25-3 images decreased by 60.1% and 79.3% compared to I05-3, respectively. This is also due to the higher proportion of low θ fibers in a two-dimensional distribution image compared to that of a three-dimensional distribution image.

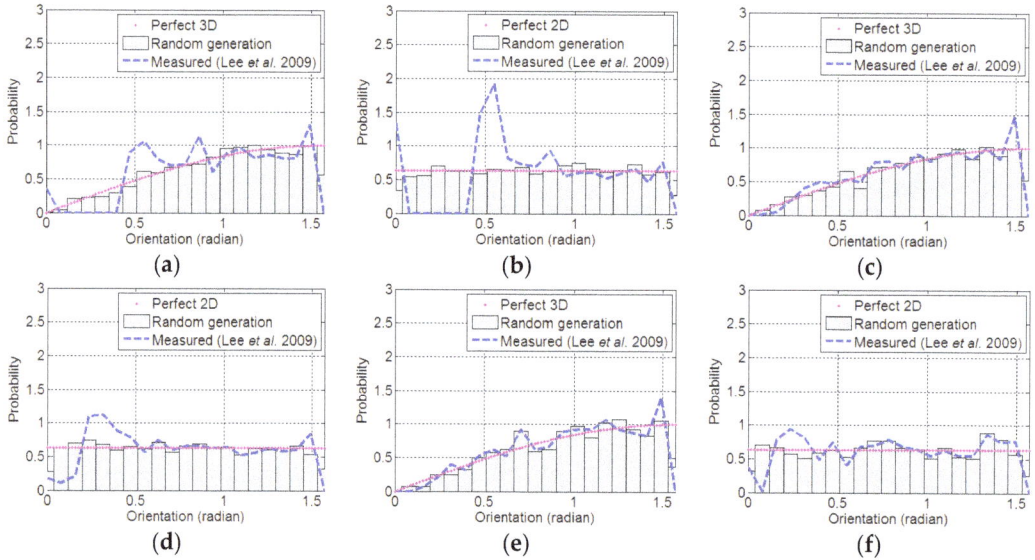

Figure 10. Probability density functions of the orientation of randomly-generated artificial fibers and the orientation measured by the image analysis technique proposed by Lee *et al.* (2009): (**a**) I05-3; (**b**) I05-2; (**c**) I15-3; (**d**) I15-2; (**e**) I25-3; (**f**) I25-2.

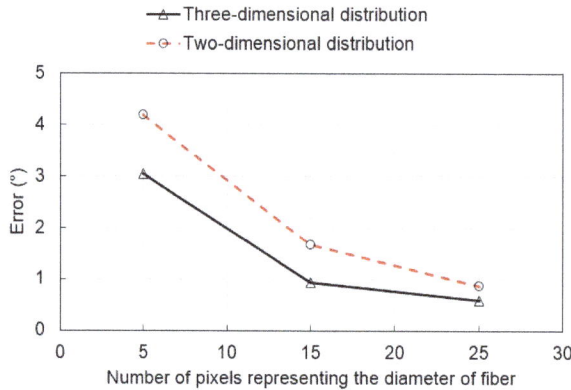

Figure 11. Average error per fiber of the measured orientation of the artificial fiber images according to the number of pixels in the diameter.

Figure 12 shows the probability density functions of the real orientation of the fibers, which was randomly generated artificially, the direct measurement of the fiber orientation by the image analysis proposed by Lee *et al.* [8] and the estimated probability density functions from the technique proposed in this study. Figure 13 shows the difference in the probability densities from the direct measurement of the fiber orientation by image analysis and the estimation by the technique proposed in this study compared to that of the real orientation. No difference indicates that the fiber orientations of all of the fibers are exactly measured or estimated with the real fiber orientation by the techniques. With the three-dimensional distribution, the differences in the probability density for the I05-03 and

I15-03 images decreased by 72.7% and 43.7%, respectively, by the proposed technique compared to the previous technique (direct measurement). However, the differences in the probability density for the I25-03 image was increased by 21.9% with the proposed technique compared to the previous technique. This may be attributed to the larger variation of the probability densities of the fiber orientation of the I25-03 image compared to those of I05-03 and I15-03. With the two-dimensional distribution, the differences in the probability density for the I05-02, I15-02 and I25-02 images decreased by 80.6%, 56.1% and 17.1%, respectively, with the proposed technique compared to the previous technique. The differences in the probability density converged with an increasing number of pixels representing the diameter of the fiber regardless of the dimensions of the distribution of the fiber orientation. The test results confirmed that the technique proposed in this study provided better estimation performance than the previous technique, especially when there were two-dimensional distributions and a small number of pixels representing the diameter of the fiber. The image analysis technique proposed in this study can be used to assess or analyze with more accuracy the effects of the fiber orientation on the mechanical properties of fiber-reinforced concrete when the number of pixels representing the diameter of the fiber is limited in the process of image acquisition.

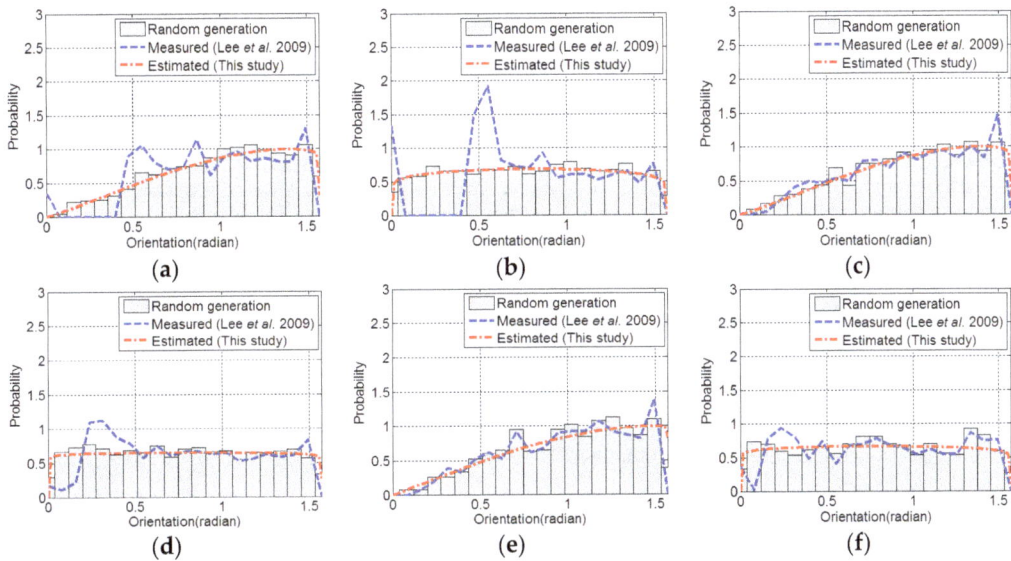

Figure 12. Comparison between the measured distribution and the estimated distribution by the technique proposed in this study: (a) I05-3; (b) I05-2; (c) I15-3; (d) I15-2; (e) I25-3; (f) I25-2.

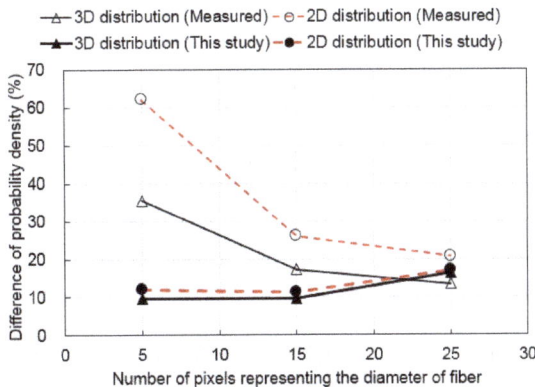

Figure 13. Comparison between the measured distribution and the estimated distribution by the technique proposed in this study.

5. Conclusions

We proposed a new image analysis technique to estimate the distribution of fiber orientation in a sectional image of fiber-reinforced concrete. A series of experimental and analytical investigations with artificial fiber images was carried out to assess the validity of this technique. The following conclusions can be drawn from the results:

(1) We investigated the effect of the number of pixels representing the diameter of the fiber and fiber orientation on the detection accuracy. The error increased with a decreasing fiber orientation and the number of pixels in the diameter of the fiber. We attributed this to an increase in sensitivity with a decreasing fiber orientation and an increase in the detection error with a decreasing number of pixels in the diameter of the fiber, *i.e.*, the decreasing resolution.

(2) The proposed technique estimates the distribution of fiber orientation by finding optimal distribution functions matching the measured number of fibers by an image analysis with the theoretical number of fibers calculated from the volume and diameter of the fiber, the area of the sectional image and the assumed fiber distribution.

(3) Validation tests using artificial fiber images according to the size of the fiber images and the dimensions of the fiber orientation confirmed that the technique proposed in this study ensures better evaluation performance than that by direct measurement of the fiber orientation from image analysis, especially when there is a two-dimensional distribution and a small number of pixels representing the diameter of the fiber.

Acknowledgments: This research was supported by a grant (15CTAP-C097490-01) from the Technology Advancement Research Program funded by the Ministry of Land, Infrastructure and Transport Affairs of the Korean Government and also supported by the Basic Science Research Program through the National Research Foundation of Korea (NRF) funded by the Ministry of Science, ICT & Future Planning (No. 2015R1A5A1037548).

Author Contributions: All authors have equally contributed to the authorship of the article.

Conflicts of Interest: The authors declare no conflict of interest.

References

1. Romualdi, J.P.; Batson, G.B. Mechanics of crack arrest in concrete. *J. Eng. Mech. Division* **1963**, *89*, 147–168.
2. American Concrete Institute (ACI) Committee. *544.1r-96: Report on Fiber Reinforced Concrete*; ACI: Farmington, MI, USA, 1996.
3. Li, V.C. Tailoring ECC for special attributes: A review. *Int. J. Concr. Struct. Mater.* **2012**, *6*, 135–144. [CrossRef]
4. Dupont, D.; Vandewalle, L. Distribution of steel fibres in rectangular sections. *Cem. Concr. Compos.* **2005**, *27*, 391–398. [CrossRef]
5. Jain, L.; Wetherhold, R. Effect of fiber orientation on the fracture toughness of brittle matrix composites. *Acta Metall. Mater.* **1992**, *40*, 1135–1143. [CrossRef]
6. Kwon, S.H.; Kang, S.-T.; Lee, B.Y.; Kim, J.-K. The variation of flow-dependent tensile behavior in radial flow dominant placing of ultra high performance fiber reinforced cementitious composites (UHPFRCC). *Constr. Build. Mater.* **2012**, *33*, 109–121. [CrossRef]
7. Lee, B.Y.; Kim, J.K.; Kim, Y.Y. Prediction of ecc tensile stress-strain curves based on modified fiber bridging relations considering fiber distribution characteristics. *Comput Concr.* **2010**, *7*, 455–468. [CrossRef]
8. Lee, B.Y.; Kim, J.-K.; Kim, J.-S.; Kim, Y.Y. Quantitative evaluation technique of polyvinyl alcohol (pva) fiber dispersion in engineered cementitious composites. *Cem. Concr. Compos.* **2009**, *31*, 408–417. [CrossRef]
9. Tosun-Felekoğlu, K.; Felekoğlu, B.; Ranade, R.; Lee, B.Y.; Li, V.C. The role of flaw size and fiber distribution on tensile ductility of PVA-ECC. *Compos. B Eng.* **2014**, *56*, 536–545. [CrossRef]
10. Zerbino, R.; Tobes, J.; Bossio, M.; Giaccio, G. On the orientation of fibres in structural members fabricated with self compacting fibre reinforced concrete. *Cem. Concr. Compos.* **2012**, *34*, 191–200. [CrossRef]
11. Mobasher, B.; Stang, H.; Shah, S. Microcracking in fiber reinforced concrete. *Cem. Concr. Res.* **1990**, *20*, 665–676. [CrossRef]
12. Poitou, A.; Chinesta, F.; Bernier, G. Orienting fibers by extrusion in reinforced reactive powder concrete. *J. Eng. Mech.* **2001**, *127*, 593–598. [CrossRef]

13. Kang, S.T.; Lee, B.Y.; Kim, J.-K.; Kim, Y.Y. The effect of fibre distribution characteristics on the flexural strength of steel fibre-reinforced ultra high strength concrete. *Constr. Build. Mater.* **2011**, *25*, 2450–2457. [CrossRef]

14. Krenchel, H. *Fibre Reinforcement*; Akademisk Forlag: Copenhagen, Denmark, 1964.

15. Aveston, J.; Kelly, A. Theory of multiple fracture of fibrous composites. *J. Mater. Sci.* **1973**, *8*, 352–362. [CrossRef]

16. Laws, V. The efficiency of fibrous reinforcement of brittle matrices. *J. Phys. D Appl. Phys.* **1971**, *4*, 1737. [CrossRef]

17. Li, V.C. Postcrack scaling relations for fiber reinforced cementitious composites. *J. Mater. Civ. Eng.* **1992**, *4*, 41–57. [CrossRef]

18. Morton, J.; Groves, G. The effect of metal wires on the fracture of a brittle-matrix composite. *J. Mater. Sci.* **1976**, *11*, 617–622. [CrossRef]

19. Redon, C.; Li, V.C.; Wu, C.; Hoshiro, H.; Saito, T.; Ogawa, A. Measuring and modifying interface properties of pva fibers in ECC matrix. *J. Mater. Civ. Eng.* **2001**, *13*, 399–406. [CrossRef]

20. Lin, Z.; Kanda, T.; Li, V.C. On interface property characterization and performance of fiber reinforced cementitious composites. *Concr. Sci. Eng.* **1999**, *1*, 173–184.

21. Eik, M.; Herrmann, H. Raytraced images for testing the reconstruction of fibre orientation distributions. *Proc. Est. Acad. Sci.* **2012**, *61*, 128–136. [CrossRef]

22. Guild, F.J.; Summerscales, J. Microstructural image analysis applied to fibre composite materials: A review. *Composites* **1993**, *24*, 383–393. [CrossRef]

23. Liu, J.; Sun, W.; Miao, C.; Liu, J.; Li, C. Assessment of fiber distribution in steel fiber mortar using image analysis. *J. Wuhan Univ. Technol. Mater. Sci. Ed.* **2012**, *27*, 166–171. [CrossRef]

24. Mlekusch, B. Fibre orientation in short-fibre-reinforced thermoplastics ii. Quantitative measurements by image analysis. *Compos. Sci. Technol.* **1999**, *59*, 547–560. [CrossRef]

25. Mlekusch, B.; Lehner, E.A.; Geymayer, W. Fibre orientation in short-fibre-reinforced thermoplastics i. Contrast enhancement for image analysis. *Compos. Sci. Technol.* **1999**, *59*, 543–545. [CrossRef]

26. Sebaibi, N.; Benzerzour, M.; Abriak, N.E. Influence of the distribution and orientation of fibres in a reinforced concrete with waste fibres and powders. *Constr. Build. Mater.* **2014**, *65*, 254–263. [CrossRef]

27. Benson, S.D.; Karihaloo, B.L. CARDIFRC: Manufacture and constitutive behaviour. In Proceedings of the International Symposium Dedicated to Professor Surendra Shah, Dundee, Scotland, UK, 3–4 September 2003; pp. 233–244.

28. Bordelon, A.C.; Roesler, J.R. Spatial distribution of synthetic fibers in concrete with X-ray computed tomography. *Cem. Concr. Compos.* **2014**, *53*, 35–43. [CrossRef]

29. Pujadas, P.; Blanco, A.; Cavalaro, S.; de la Fuente, A.; Aguado, A. Fibre distribution in macro-plastic fibre reinforced concrete slab-panels. *Constr. Build. Mater.* **2014**, *64*, 496–503. [CrossRef]

30. Stähli, P.; Custer, R.; van Mier, J.G. On flow properties, fibre distribution, fibre orientation and flexural behaviour of FRC. *Mater. Struct.* **2008**, *41*, 189–196. [CrossRef]

31. Stroeven, P. Morphometry of fibre reinforced cementitious materials. *Matér. Constr.* **1979**, *12*, 9–20. [CrossRef]

32. Chung, D. Dispersion of short fibers in cement. *J. Mater. Civ. Eng.* **2005**, *17*, 379–383. [CrossRef]

33. Ozyurt, N.; Woo, L.Y.; Mason, T.O.; Shah, S.P. Monitoring fiber dispersion in fiber-reinforced cementitious materials: Comparison of ac-impedance spectroscopy and image analysis. *ACI Mater. J.* **2006**, *103*, 340–347.

34. Lee, B.Y.; Kim, T.; Kim, Y.Y. Fluorescence characteristic analysis for discriminating fibers in cementitious composites. *J. Adv. Concr. Technol.* **2010**, *8*, 337–344.

35. Eberhardt, C.; Clarke, A.; Vincent, M.; Giroud, T.; Flouret, S. Fibre-orientation measurements in short-glass-fibre composites—II: A quantitative error estimate of the 2d image analysis technique. *Compos. Sci. Technol.* **2001**, *61*, 1961–1974. [CrossRef]

36. Otsu, N. A threshold selection method from gray-level histograms. *Automatica* **1975**, *11*, 23–27.

37. Vincent, L. Morphological grayscale reconstruction in image analysis: Applications and efficient algorithms. *IEEE Trans. Image Process.* **1993**, *2*, 176–201. [CrossRef] [PubMed]

38. Vincent, L.; Soille, P. Watersheds in digital spaces: An efficient algorithm based on immersion simulations. *IEEE Trans. Pattern Anal. Mach. Intell.* **1991**, *6*, 583–598. [CrossRef]

39. Choi, J.-I.; Lee, B.Y. Bonding properties of basalt fiber and strength reduction according to fiber orientation. *Materials* **2015**, *8*, 6719–6727. [CrossRef]

40. Lee, B.Y.; Cho, C.G.; Lim, H.-J.; Song, J.-K.; Yang, K.-H.; Li, V.C. Strain hardening fiber reinforced alkali-activated mortar—A feasibility study. *Constr. Build. Mater.* **2012**, *37*, 15–20. [CrossRef]

41. Park, S.H.; Kim, D.J.; Ryu, G.S.; Koh, K.T. Tensile behavior of ultra high performance hybrid fiber reinforced concrete. *Cem. Concr. Compos.* **2012**, *34*, 172–184. [CrossRef]

42. Xia, M.; Hamada, H.; Maekawa, Z. Flexural stiffness of injection molded glass fiber reinforced thermoplastics. *Int. Polym. Process.* **1995**, *10*, 74–81. [CrossRef]

43. Golberg, D.E. *Genetic Algorithms in Search, Optimization, and Machine Learning*; Addison-Wesley: Boston, MA, USA, 1989.

44. Haupt, R.L.; Haupt, S.E. *Practical Genetic Algorithms*, 2nd ed.; John Wiley & Sons, Inc.: Hoboken, NJ, USA, 2004.

45. Lee, B.Y.; Kim, J.H.; Kim, J.K. Optimum concrete mixture proportion based on a database considering regional characteristics. *J. Comput. Civil. Eng.* **2009**, *23*, 258–265. [CrossRef]

14

Correlation of High Magnetoelectric Coupling with Oxygen Vacancy Superstructure in Epitaxial Multiferroic BaTiO$_3$-BiFeO$_3$ Composite Thin Films

Michael Lorenz [1,*], Gerald Wagner [2], Vera Lazenka [3], Peter Schwinkendorf [1],
Michael Bonholzer [1], Margriet J. Van Bael [4], André Vantomme [3], Kristiaan Temst [3],
Oliver Oeckler [2] and Marius Grundmann [1]

Academic Editor: Jan Ingo Flege

[1] Institut für Experimentelle Physik II, Universität Leipzig, Leipzig D-04103, Germany;
    schwinkendorf@physik.uni-leipzig.de (P.S.); bonholzer@physik.uni-leipzig.de (M.B.);
    grundmann@physik.uni-leipzig.de (M.G.)
[2] Institut für Mineralogie, Kristallographie und Materialwissenschaft, Universität Leipzig, Leipzig D-04103,
    Germany; wagner@chemie.uni-leipzig.de (G.W.); oliver.oeckler@uni-leipzig.de (O.O.)
[3] Instituut voor Kern- en Stralingsfysica, KU Leuven, Leuven B-3001, Belgium;
    vera.lazenka@fys.kuleuven.be (V.L.); andre.vantomme@fys.kuleuven.be (A.V.);
    kristiaan.temst@fys.kuleuven.be (K.T.)
[4] Laboratorium voor Vaste-Stoffysica en Magnetisme, KU Leuven, Leuven B-3001, Belgium;
    margriet.vanbael@fys.kuleuven.be
*   Correspondence: mlorenz@physik.uni-leipzig.de

Abstract: Epitaxial multiferroic BaTiO$_3$-BiFeO$_3$ composite thin films exhibit a correlation between the magnetoelectric (ME) voltage coefficient α_{ME} and the oxygen partial pressure during growth. The ME coefficient α_{ME} reaches high values up to 43 V/(cm·Oe) at 300 K and at 0.25 mbar oxygen growth pressure. The temperature dependence of α_{ME} of the composite films is opposite that of recently-reported BaTiO$_3$-BiFeO$_3$ superlattices, indicating that strain-mediated ME coupling alone cannot explain its origin. Probably, charge-mediated ME coupling may play a role in the composite films. Furthermore, the chemically-homogeneous composite films show an oxygen vacancy superstructure, which arises from vacancy ordering on the {111} planes of the pseudocubic BaTiO$_3$-type structure. This work contributes to the understanding of magnetoelectric coupling as a complex and sensitive interplay of chemical, structural and geometrical issues of the BaTiO$_3$-BiFeO$_3$ composite system and, thus, paves the way to practical exploitation of magnetoelectric composites.

Keywords: oxide thin films; multiferroic composites; magnetoelectric coupling; magnetoelectric voltage coefficient; oxygen vacancy superstructure; pulsed laser deposition

1. Introduction

Multiferroic composites consisting of two different chemical compounds offer unique flexibility in geometrical and structural design to achieve desired functional properties, in particular a high magnetoelectric coupling [1,2]. Correlations at the interfaces and novel approaches in combining magnetism and ferroelectricity are mentioned by Fiebig and Spaldin as outlooks in the research on novel magnetoelectrics [3]. Progress reports highlight the impressive variability of multiferroic composite materials, which in the future should also allow implementation of electrical transport into the multiferroic concept [4,5]. Kleemann *et al.* discussed the complex magnetoelectric (ME) effects of single-phase Type I (such as BiFeO$_3$) and Type II (such as TbMnO$_3$) multiferroics in comparison to multiphase composites and higher-order ME effects in disordered Type III multiferroics [6]. For a discussion of possible units of the ME coefficient $\alpha_{MEij} = E_i/H_j$, see [7]. Giant magnetoelectric coefficients α_{ME} are reported in two-two composite multilayers consisting of AlN and amorphous $(Fe_{90}Co_{10})_{78}Si_{12}B_{10}$ on Si(100). The α_{ME} values reach 737 V/(cm·Oe) at a mechanical resonance of 753 Hz and 3.1 V/(cm·Oe) out of a resonance at 100 Hz [8]. These AlN-$(Fe_{90}Co_{10})_{78}Si_{12}B_{10}$ composites are applied as magnetic field sensors. The consideration of exchange bias coupling allows a tuning of the detection limit and ME coefficient of these sensors [9].

Little work is reported on composites containing BiFeO$_3$; some papers mostly on bulk solid solutions are discussed in the review [10] and in [11]. The coexistence of ferroelectricity and ferromagnetism in BiFeO$_3$-BaTiO$_3$ thin films at room temperature was reported already in 1999 [12]. More recently, the group around Ramesh combined the multiferroic BiFeO$_3$ with ferrimagnetic CoFe$_2$O$_4$ in self-organized nanostructures with a 300-nm total thickness and obtained a transverse ME susceptibility of 60 mV/(cm·Oe) [13]. Muragavel et al. have grown multiferroic epitaxial $(BiFeO_3)_x$-$(BaTiO_3)_{(1-x)}$ composite films (x = 0.3–0.9) on SrTiO$_3$:Nb and found compressive strain for $x < 0.7$ and relaxed growth for $x > 0.7$ and a corresponding phase change from rhombohedral to tetragonal [14]. Epitaxial Co- and Fe-substituted BiFeO$_3$ films for spin-filter applications are reported in [15]. Mößbauer spectroscopy was applied to bulk $(BiFeO_3)_x$-$(BaTiO_3)_{(1-x)}$ ceramics (x = 0.9–0.7), and relatively low α_{ME} values up to 0.6 mV/(cm·Oe) were obtained (1 kHz, H_{DC} = 0) [16]. Nearly a complete mixing ratio of polycrystalline bulk $(BiFeO_3)_x$-$(BaTiO_3)_{(1-x)}$ ceramics (x = 0.025–1) was investigated, and a maximum α_{ME} of 0.87 mV/(cm·Oe) was found for x = 0.725 [17]. Compositionally-modulated $(Co/Mg/Ni)Fe_2O_4$ spinel nanopillars embedded in a BiFeO$_3$ matrix film have been reported recently together with their magnetic response [18]. Reviews on the properties and device applications of the single-phase multiferroic compound BiFeO$_3$ are published in [19,20]. We reported the effect of rare-earth doping on the multiferroic properties of BiFeO$_3$ thin films [21].

In continuation of this latter work on BiFeO$_3$, we found that the ME coefficient of single-phase BiFeO$_3$ thin films can be considerably enhanced by combination with the ferroelectric BaTiO$_3$ in both superlattices and in chemically-homogeneous composite thin films [11,22,23]. In particular, while a BiFeO$_3$ film showed an α_{ME} of about 2 V/(cm·Oe), the corresponding (BaTiO$_3$-BiFeO$_3$) × 15 superlattice showed 9 V/(cm·Oe) at 300 K, clearly demonstrating the interface effect on the magnetic moment direction and magnetoelectric coupling, as published in [23]. In our first chemically-homogeneous composite films published in [11], we combined the ferroelectric BaTiO$_3$ and multiferroic BiFeO$_3$ phase into a thin film nanocomposite structure, and the detailed connectivity scheme and coupling mechanism in terms of the microscopic origin of the measured high ME coefficients α_{ME} up to 21 V/(cm·Oe) have not been clear up to now.

Therefore, we present in this paper a more detailed investigation of the microstructure of the chemically-homogeneous BaTiO$_3$-BiFeO$_3$ composite films based on scanning transmission electron microscopy (STEM) and selected area electron diffraction (SAED). Furthermore, we discuss the optimum oxygen supply during the growth of the epitaxial composite films to achieve the maximum magnetoelectric voltage coefficient and the corresponding film structure, as well as probable magnetoelectric coupling mechanisms. In the following, we provide results on the solid solution BaTiO$_3$-BiFeO$_3$ composite films. The BaTiO$_3$-BiFeO$_3$ superlattice heterostructures with clear spatial separation of both phases as published in [11,22,23] are mentioned in the discussion only for a comparison.

2. Results and Discussion

2.1. Out-of-Plane Strain and Crystalline Structure

In agreement with the modeling of the magnetoelectric voltage coefficient in dependence of the volume fraction of the piezoelectric component in various lead-based and lead-free oxide composites [1,24], we have chosen the source target composition for our composite film growth by pulsed laser deposition (PLD) to be 67 wt % BaTiO$_3$ and 33 wt % BiFeO$_3$. The oxygen partial pressure during PLD was varied between 0.01 mbar and 0.5 mbar. As the growth rate of the composite films depends strongly on the oxygen partial pressure, it is natural that the film thickness of the samples is not uniform. We used several sample series for the results presented here (see the running sample numbers in Table 1). In the last series (G5556–G5562), we tried to compensate for the changing growth rate by using different total numbers of laser pulses for the growth, resulting in reduced thickness variation from 208–388 nm only, instead of up to 1000 nm in the previous series. The substrate material of the films intended for X-ray diffraction (XRD) and ferroelectric and magnetoelectric measurements was $SrTiO_3$:Nb(001). For the STEM and SAED investigations only, films grown simultaneously on MgO(001) were used; see the Experimental Section for further insight into our growth regime. For oxygen-deficient weakly-Mn-doped ZnO thin films, we have recently measured the trigonal distortion of oxygen tetrahedra; see [25]. Thus, lower oxygen partial pressure in PLD growth is clearly correlated to increasing density of oxygen vacancies. With that, we are able to correlate the ME coefficient to the oxygen deficiency of the composites and related structural properties.

Table 1. X-ray diffraction (XRD) lattice parameters of $SrTiO_3$ substrates and $BaTiO_3$-$BiFeO_3$ composite films on $SrTiO_3$:Nb(001) at the indicated growth pressures $p(O_2)$. Composite film thickness was determined at focused ion beam (FIB) cross-sections by scanning transmission electron microscopy (STEM) for one sample of each growth series. c-lattice parameters were calculated from wide-angle 004 peaks. The bulk lattice parameters are as follows (nm): $SrTiO_3$: $a = 0.3905$ (JCPDS 84-0444); $BaTiO_3$: $a = 0.39945$, $c = 0.40335$ (JCPDS 83–1880); $BiFeO_3$: $a = 0.3962$ (JCPDS 73–0548). The out-of-plane strain values $\Delta c/c_0$ are calculated assuming an averaged bulk lattice parameter of the 67 wt % $BaTiO_3$-33 wt % $BiFeO_3$ composite of $c_0 = 0.40099$ nm. For visualization of the c_{film} and FWHM(ω) values, see Figures 1b, 2 and 3, respectively. The accuracy of the lattice parameters is about ± 0.0001 nm. Several samples for each growth pressure are included in the table to get an impression about the reproducibility of pulsed laser deposition (PLD) and the accuracy of lattice parameters. The sample numbers show which films were grown simultaneously (identical No.) or in consecutive PLD runs (successive No.), respectively.

$p(O_2)$ (mbar)	Sample No.	Thickness (nm)	$c_{substrate}$ (nm)	c_{film} (nm)	Strain $(c - c_0)/c_0$	FWHM(ω) $\bar{1}03$ (°)
0.01	G 5084b	730	0.39055	0.41030	2.322%	0.127
0.01	G 5084c	730	0.39066	0.40991	2.224%	0.125
0.1	G 5085c	488	0.39066	0.40708	1.519%	0.109
0.1	G 5562b	208	0.39067	0.40767	1.666%	0.734 *
0.1	G 5562d	208	0.39081	0.40618	1.294%	0.157
0.15	G 5561b	347	0.39067	0.40593	1.232%	0.162
0.15	G 5561d	347	0.39069	0.40554	1.135%	0.320 *
0.2	G 5560b	365	0.39068	0.40403	0.758%	0.131
0.25	G 4728d	285	0.39069	0.40103	0.010%	0.237
0.25	G 5559b	388	0.39067	0.40168	0.172%	0.272
0.325	G 5558b	352	0.39068	0.40148	0.122%	0.303
0.4	G 5557d	311	0.39070	0.40148	0.122%	0.354
0.5	G 5086b	1000	0.39056	0.40133	0.085%	0.346
0.5	G 5556b	256	0.39068	0.40150	0.127%	0.354

* Extra peak broadening due to substrate domains; see Figure S1.

Figure 1a shows XRD 2θ-ω scans of the investigated $BaTiO_3$-$BiFeO_3$ composite films grown on $SrTiO_3$:Nb(001) single crystalline substrates around the 002 peaks. Figure 1a, as well as the 001 reciprocal space maps (RSMs) in Figure 2 demonstrate the film peak shift to lower 2θ angles, *i.e.*, increasing out-of-plane lattice parameter with decreasing oxygen partial pressure. In Figure 1b, the corresponding c-lattice parameters as calculated from the positions of the 004 reflections of the films are plotted as depicted in Figure S2 in the Supplementary Materials. We found that using 004 peaks for the calculation of c values results in a smoother dependence on oxygen partial pressure in comparison to $\cos^2\theta$ extrapolations with omission of 001 or 001 + 002 peaks. Due to the goniometer height error, the accuracy of lattice parameters calculated from low-θ peaks, such as 001 and 002, is not sufficient for a precise analysis. In addition to Figure 2, the RSMs around the 001 peak in the Supplementary Materials show the impact of the tilt mosaicity of the $SrTiO_3$ substrate on the composite film structure (Figure S1); and the impact of a changed 33 wt %-67 wt % $BaTiO_3$-$BiFeO_3$ composite film composition instead of the generally used 67 wt %-33 wt % on the film mosaicity (Figure S3).

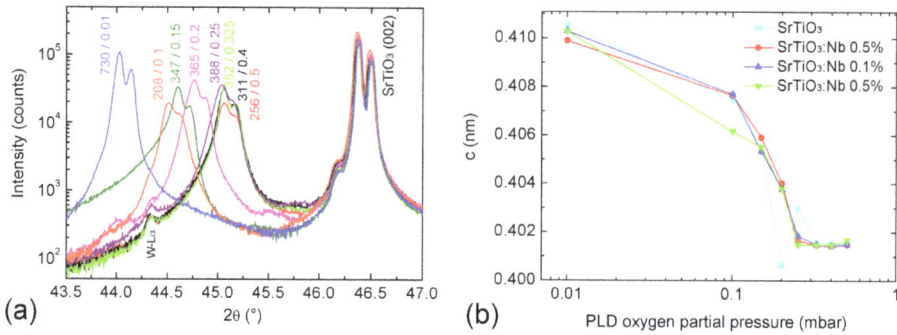

Figure 1. (a) X-ray diffraction (XRD) 2θ-ω scans of $BaTiO_3$-$BiFeO_3$ composite thin films grown on $SrTiO_3$:Nb(001) at the indicated oxygen partial pressures. The two numbers at the film peaks indicate the total composite film thickness in nm, and the oxygen partial pressure during pulsed laser deposition (PLD) growth, respectively. Note the $K\alpha_{1/2}$ splitting of each peak. W-Lα is a spectral line from the X-ray tube. The single-phase contributions of $BaTO_3$ and $BiFeO_3$ cannot be resolved here; see Table 1. (b) PLD oxygen pressure evolution of the c-axis lattice parameters calculated from 004 peaks of four $BaTiO_3$-$BiFeO_3$ composite thin films grown simultaneously for each growth pressure at the indicated substrates. See Table 1 for the values of the c-lattice parameters and more structural details.

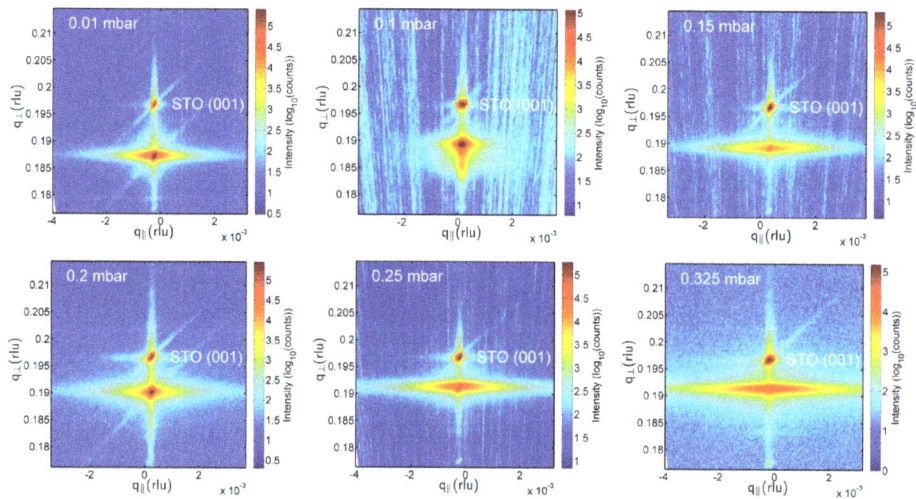

Figure 2. XRD reciprocal space maps around the symmetric $SrTiO_3$ 001 peaks of $BaTiO_3$ (67 wt %)-$BiFeO_3$ (33 wt %) composite thin films grown at the indicated PLD oxygen partial pressures. The separation of a point-like substrate and a broadened composite film peak decreases with increasing oxygen partial pressure due to decreasing out-of-plane strain; see Table 1. The horizontal broadening of film peaks is a measure of the tilt mosaicity of the composites. STO stands for $SrTiO_3$.

As the lattice parameters of $BaTiO_3$ and $BiFeO_3$ as bulk and thin films are very close together (see the Joint Committee on Powder Diffraction Standards (JCPDS) data in the caption of Table 1), the XRD peaks of the $BaTiO_3$ and $BiFeO_3$ phases in our composites cannot be resolved. Even closer agreement of $BaTiO_3$ and $BiFeO_3$ lattice parameters is supported by [26], reporting for $BiFeO_3$ in tetragonal $P4mm$ symmetry $a = 0.3935$ and $c = 0.3998$ nm. Therefore, we assume an averaged out-of-plane lattice constant

c_0 to get estimates about the average out-of-plane strain of the composite films; see Table 1. A higher oxygen deficiency of the composite films results in increased out-of-plane strain. For a comparison, in our $(BaTiO_3\text{-}BiFeO_3) \times 15$ superlattices, both phases could be separately detected by Rutherford backscattering spectrometry, XRD and STEM with nearly no intermixing at the interfaces [11,22].

Figure 3 shows typical RSMs around the asymmetric $\bar{1}03$ peaks of the composite films. As already visible in Figure 2, the film peaks broaden both horizontally and vertically with increasing oxygen pressure, *i.e.*, the film mosaicity and the variation of film lattice parameter increases. This broadening is quantitatively expressed in Table 1 by the FWHM(ω) values. From the increasing vertical misalignment of the film and substrate peaks with increasing growth pressure, we conclude an increasing film relaxation. That means that none of the grown composite films can be considered as in-plane lattice matched to the $SrTiO_3$:Nb substrates. The in-plane a-lattice parameters as calculated using c_{004} (see Table 1) and the d-values extracted from the $\bar{1}03$ RSM peaks scatter around 0.404 ± 0.005 nm. Because of the limited accuracy of these in-plane lattice parameters, statements about in-plane strain cannot be derived from these values.

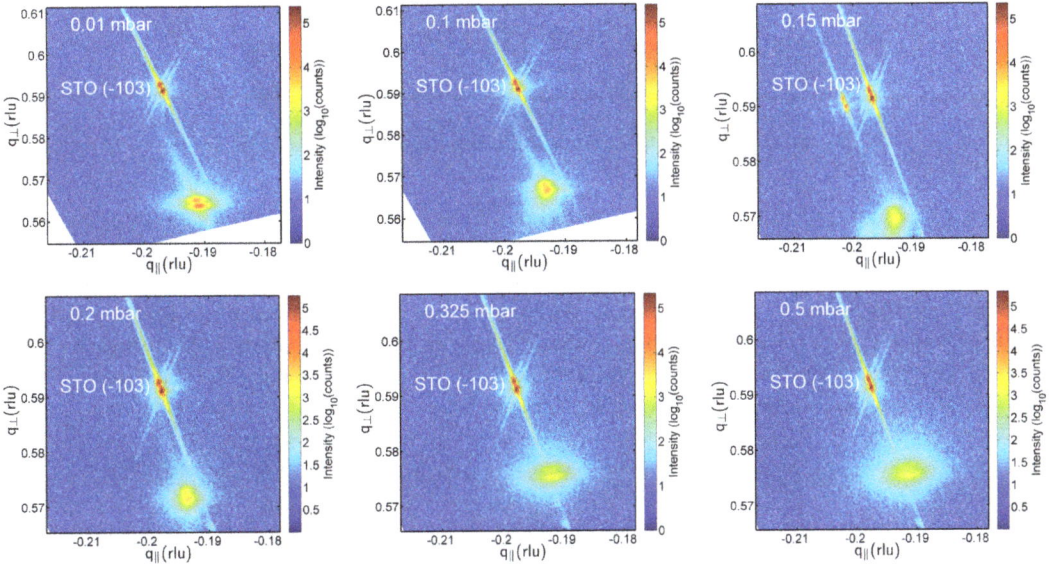

Figure 3. XRD reciprocal space maps around the asymmetric $\bar{1}03$ $SrTiO_3$ substrate peaks (in the Figures referred to as STO (-103)) of the $BaTiO_3$-$BiFeO_3$ composite films grown at the indicated oxygen partial pressures. The lower intensity film peaks are at the bottom. With increasing oxygen pressure, the film peaks broaden, *i.e.*, the composite mosaicity increases. However, film-to-substrate peak separation decreases. Note that the vertical $q^\perp$ axis for the two lowest pressures is enlarged. As is visible from the increasing vertical misalignment of film and substrate peaks with increasing growth pressure, the film relaxation increases. The 0.15-mbar sample shows substrate and corresponding crystalline film domains. The peak splitting is due to the $K\alpha_{1/2}$ radiation used.

2.2. STEM and Oxygen Vacancy Superstructure

Two composite film samples grown on MgO(001) were investigated by STEM and SAED, namely one sample grown at 0.01 mbar (see Figure 4 and Figures S4–S7) and the other at 0.25 mbar (see Figure 5 and Figures S7–S10). Because these films on MgO were grown simultaneously with the samples on $SrTiO_3$ and $SrTiO_3$:Nb, we expect the microstructural features found here to be representative of all of our $BaTiO_3$-$BiFeO_3$ composite thin films. Figure 4 shows a bright-field STEM micrograph of a $BaTiO_3$-$BiFeO_3$ composite thin film grown at the lowest pressure of 0.01 mbar. It shows a high density of dislocation lines in the film. Energy dispersive X-ray spectroscopy (EDX) maps taken at the STEM cross-sections indicate a homogeneous elemental distribution of Ba, Ti, Bi and Fe in the composite film.

On top of the composite film is a thin gold and a platinum layer (Figure 4); see also Figures S4 and S8 for STEM dark field images and EDX maps of the 0.25-mbar and the 0.01-mbar samples, respectively.

In the dark-field STEM image of a cross-section of a $BaTiO_3$-$BiFeO_3$ composite film grown at 0.25 mbar in Figure 5, the columnar domains revealed by the grey scale contrast modulation reflect the different orientation of ordered oxygen vacancy layers, i.e., $(\bar{1}11)$ and/or $(1\bar{1}1)$. Probably, the small microstrain differences between these domains are responsible for the high magnetoelectric coefficients α_{ME} reported below.

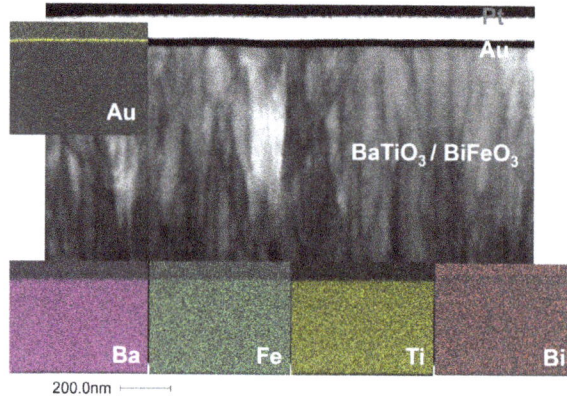

Figure 4. Scanning transmission electron microscopy (STEM) bright-field image of the $BaTiO_3$-$BiFeO_3$ composite film grown at 0.01 mbar oxygen pressure shown as the (110) cross-section. Clearly visible is the high density of dislocation lines in this particular composite film. The colored energy dispersive X-ray spectroscopy (EDX) maps demonstrate the homogeneous distribution of elements Ba, Fe, Ti and Bi in the plane of the cross-section. The gold film results from the extraction of the TEM cross-section out of the planar film sample using a focused ion beam. For more EDX maps and elemental analyses, see the Supplementary Materials, Figures S4, S6, S8 and S9.

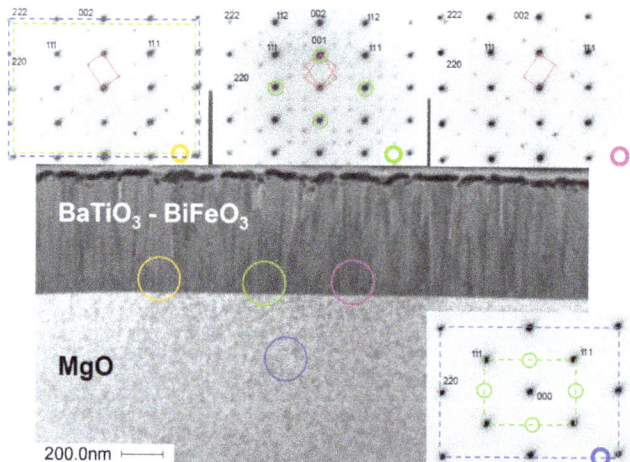

Figure 5. STEM dark-field image of the $BaTiO_3$-$BiFeO_3$ composite film grown at 0.25 mbar on MgO(001) taken from the (110) cross-section. The four selected area electron diffraction (SAED) patterns have been taken from the encircled regions at the interface or the substrate. The main spots from the composite (green circles) confirm the $BaTiO_3$-type structure of the composite film. These are forbidden for MgO, as seen bottom right. The additional weak spots in the composite indicate superstructure reflections that are probably due to oxygen vacancy ordering. The red dotted lines show the two possible orientations of oxygen vacancy ordering.

In Figure 5 (top), in the SAED patterns, the main reflections indicate the $BaTiO_3$-type film structure of the 67 wt % $BaTiO_3$-33 wt % $BiFeO_3$ composite and the MgO structure of the substrate, which overlap. However, far away from the 000 reflection, a splitting appears due to small differences of the lattice parameters of MgO and $BaTiO_3$-$BiFeO_3$ composite film (see the blue and green frame in Figure 5, top left). Upon closer inspection of the film patterns in Figure 5, additional weaker reflections at $\pm^1/_3(hkl)_{pseudocubic}$ become visible, which indicate a superstructure with three-times the $BaTiO_3$-$BiFeO_3$ composite lattice parameter (Table 1). The superstructure clearly appears in two orientations, tilted left and right, and at some position, both orientations are observed simultaneously, such as in Figure 5, top center. The origin of the superstructure may be the same oxygen vacancy ordering, which is known from $BaTiO_3$ grown in a reduced environment [27]. If the structural data (space group $P3m1$; see [27]) for such a "reduced $BaTiO_3$" were used for SAED simulation, exactly the same pattern (also the weak reflections) appears at the same positions as seen experimentally. In "reduced $BaTiO_3$", every third of the hexagonal anion layers stacked along the pseudocubic <111> direction corresponds to a plane of ordered oxygen vacancies, i.e., it remains "empty" [27].

Several other types of perovskite superstructures have been discussed by Glazer [28] in order to explain additional weak reflections. Those in the SAED patterns shown in Figure 5 are consistent with the structure model shown in Figure 6. In addition, the oxygen vacancies involve the reduction of Ti^{4+} to Ti^{3+} as a consequence of charge neutrality, which is corroborated by electron energy loss spectroscopy [27]. Consistent with simple bond-valence considerations, the Ti^{3+} ions should be located in the incomplete oxygen atom octahedra. Thus, oxygen defects lead to ordered dipoles. Figures S5 and S7 in the Supplementary Materials show STEM and SAED images of the corresponding 0.01 mbar sample. There seems to be an indication of visually more intense oxygen superstructure reflections of the sample grown with higher oxygen deficiency, i.e., at a lower pressure of 0.01 mbar. Figure S7 provides a direct comparison of these SAED images of both samples.

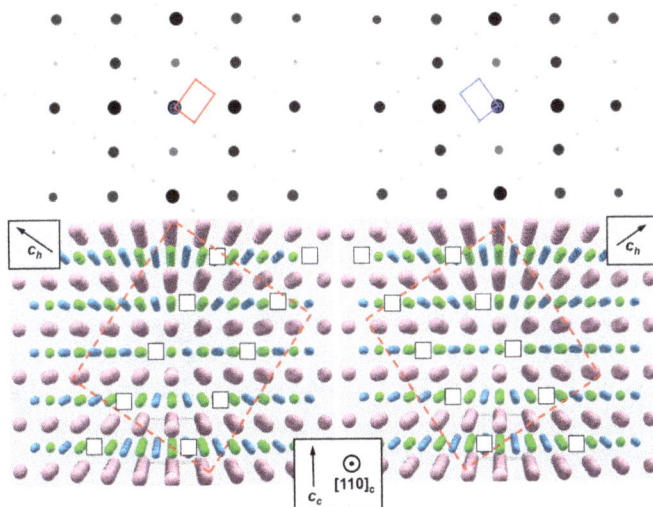

Figure 6. Structure model of oxygen vacancy ordering in planes parallel $(\bar{1}11)$ (**left**) and $(1\bar{1}1)$ (**right**), respectively; projection along [110] (directions according to cubic setting, subscript c; except the additionally indicated [001] direction with respect to the hexagonal setting, subscript h). The indication of atoms is as follows: blue, oxygen; pink, barium; green, titanium.

In order to confirm this kind of vacancy ordering, kinematical diffraction patterns (cf. Figure 6) were calculated based on the structure model for oxygen-deficient trigonal $Ba(Ti^{4+}_{1/3}Ti^{3+}_{2/3})O_{2.67}$ given by [27]. The data for $BaTiO_3$ in the cubic setting were taken from [29]. Apart from a small deviation of the average lattice parameters of the nanocomposite in comparison to bulk $BaTiO_3$

(splitting of the 222 reflection; *cf.* Figure 5), the simulated SAED patterns (Figure 6) agree well with the experimental ones. This may suggest a close analogy between oxygen vacancy ordering in reduced $BaTiO_3$ and that in our $BaTiO_3$-$BiFeO_3$ composite thin films. Figure S11 compares the calculated oxygen vacancy ordered "defective" $BaTiO_3$ electron diffraction pattern with that of pseudocubic "stoichiometric" $BaTiO_3$. The SAED image taken from the 0.25-mbar composite film depicted in Figure S10 confirms again the $BaTiO_3$-type structure of the entire $BaTiO_3$-$BiFeO_3$ composite film.

2.3. Magnetoelectric Voltage Coefficients

Figure 7a,b shows the magnetoelectric voltage coefficient α_{ME} of the $BaTO_3$-$BiFeO_3$ composite thin films as a function of temperature and direct current (DC) bias magnetic field (H), respectively, for the indicated oxygen partial pressures during PLD growth. In these experiments, the measured electric field (E) is oriented parallel to the applied alternating current (AC) and DC H-fields. During the measurements, the AC H-field was kept constant, and we did not apply any stress nor an additional E-field to the samples. In the temperature dependencies, the DC H-field was also constant. For more details, see the Experimental Section.

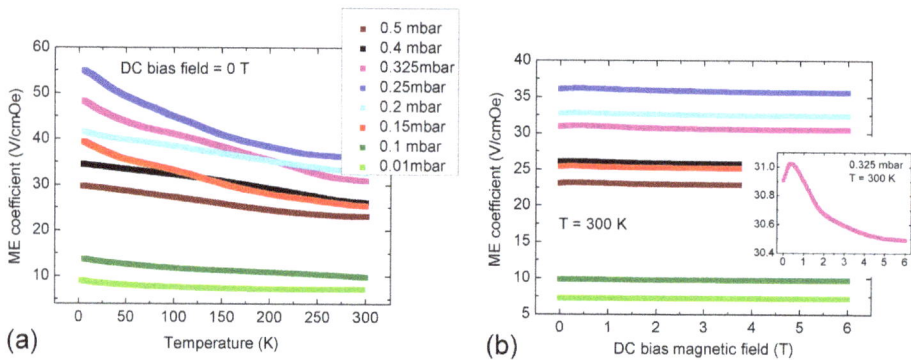

Figure 7. Magnetoelectric voltage coefficient α_{ME} of $BaTiO_3$-$BiFeO_3$ composite thin films in dependence of (**a**) temperature and (**b**) the DC bias magnetic field, for the indicated oxygen partial pressures during the growth of the composites. The legend of (a) is valid for both (a) and (b). The DC bias dependence of α_{ME} in (b) is generally weak with a local maximum around 0.5 T. The inset is an expanded view of $\alpha_{ME}(H_{DC})$ of the 0.325 mbar sample G5558c, compare Table 1. For additional field-dependent α_{ME} graphs, see the Supplementary Materials, Figure S12.

The magnetoelectric coupling of the composite films increases with decreasing temperature, in contrast to that of our recently-reported $BaTO_3$-$BiFeO_3$ superlattices consisting of 15 pairs of stacked thin $BaTiO_3$ and $BiFeO_3$ layers; see [22]. We assume that in superlattice heterostructures with clearly separated phases, as reported in [11,22,23], the magnetoelectric coupling is dominated by magnetostrictive-piezoelectric interaction coupled via strain at the interfaces of both phases. In purely strain-mediated coupling of the piezoelectric (electrostrictive, ferroelectric) phase to the magnetostrictive (magnetoelastic, ferromagnetic) phase, the coupling should decrease with decreasing temperature due to the temperature dependence of magnetostrictive and piezoelectric coefficients. Because the temperature dependence of α_{ME} of the $BaTO_3$-$BiFeO_3$ composite thin films is opposite in tendency (Figure 7a), we expect the participation of an additional coupling mechanism, for example via charge; see the further discussion below. However, the temperature dependence of magnetoelectric coupling is not yet fully understood. The ferroelectric and magnetic phase transitions of undoped and Ba-doped $BiFeO_3$ at high temperatures are reported in [30].

In Figure S13, we show additionally typical ferroelectric hysteresis loops of two series of composite films grown simultaneously to that used for the magnetoelectric measurements in Figures 7 and 8. Almost all samples show ferroelectric switching peaks in the current-voltage (I-V) characteristics used for calculation of the polarization-electrical field (P(E)) hysteresis loops. However, some samples show distorted I–V characteristics and P(E) loops, as $SrTiO_3$:Nb substrates with different Nb contents of 0.1% and 0.5% were partly used for the samples called "b" and "d", respectively; see Figure S13. Too high Nb content might introduce a barrier layer at the interface to the substrate, which results in rectifying, *i.e.*, Schottky behavior concerning the unavoidable leakage current through the films. However, high magnetoelectric coupling seems to correlate clearly with high saturation polarization, as the 0.25 mbar sample shows both the highest α_{ME} and the highest saturation polarization.

In our recent investigation of $BaTiO_3$-$BiFeO_3$ superlattices, we found anomalies in the temperature dependence of α_{ME} around the phase transition from tetragonal to orthorhombic $BaTiO_3$ [22]. Furthermore, it is well known that strain in thin films may shift and remarkably broaden magnetic and ferroelectric transitions. For example, strain-induced shifts of the ferromagnetic Curie temperature of up to 19 K were found in $La_{0.7}Sr_{0.3}MnO_3$ films [31]. The intrinsic piezoelectric coefficient d_{31}, as well as the relative permittivity of $Pb(ZrTi)O_3$ clearly decrease with decreasing temperature, while remanent polarization and the coercive field increase [32]. Because also magnetostriction decreases with decreasing temperature, the generally increasing α_{ME} with decreasing temperature cannot be explained by strain coupling alone. Rather, charge-related coupling mechanisms may play also a role here for our composites. Spurgeon *et al.* report direct local measurements of strain- and charge-mediated magnetization changes in the $La_{0.7}Sr_{0.3}MnO_3/PbZr_{0.2}Ti_{0.8}O_3$ system, which can be tuned by the manganite and ferroelectric layer thicknesses [33]. In contradiction to the chemically almost homogeneous composite films reported here, our superlattices consisting of clearly separated $BaTiO_3$ and $BiFeO_3$ films with a thickness of a few nm show the opposite temperature dependence of α_{ME} down to about 100 K and only below an increasing α_{ME}; compare [22].

Figure 8 shows directly the dependence of α_{ME} on oxygen partial pressure during growth, for 300 K and at the DC-field of maximum α_{ME}, *i.e.*, between 0 and 1 T. We found a reproducible maximum of α_{ME} around a 0.25-mbar growth pressure, and for composite thicknesses of 200–400 nm. Thicker composite films seem to show lower α_{ME} values; see Figure 8. The composite film thicknesses were determined directly and precisely from focused ion beam (FIB)-prepared cross-sections: see Figures S14 and S15. In [25], we have measured the trigonal distortion of oxygen tetrahedra of oxygen-deficient, weakly-Mn-doped ZnO thin films. Lower oxygen partial pressure in PLD growth is clearly correlated to increasing density of oxygen vacancies, which is accompanied by higher structural distortions, expressed locally by increasing variation of bond distances and rocking curve widths, *i.e.*, film mosaicity [25].

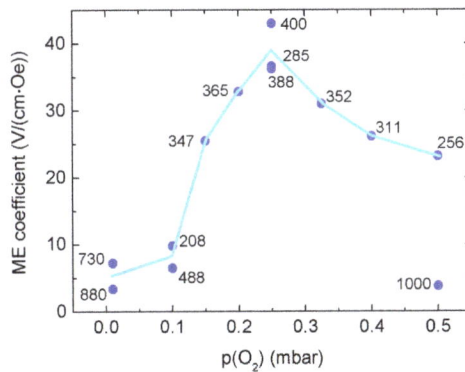

Figure 8. Magnetoelectric voltage coefficients α_{ME} of $BaTO_3$-$BiFeO_3$ composite thin films on $SrTiO_3$:Nb(001) measured at 300 K and at the DC bias magnetic field with the maximum α_{ME}. The number at each data point is the composite film thickness as determined from STEM images of cross-sections prepared by focused ion beam (see the Supplementary Materials, Figures S14 and S15). The line is drawn to guide the eye only. Using the film thicknesses, structural details of the samples (Table 1) can be assigned to the α_{ME} values.

However, in contrast to the $BaTiO_3$-$BiFeO_3$ superlattices [22], we do not find a clear correlation of α_{ME} and the ω-related widths of the $\bar{1}03$ RSM lattice points; see Table 1 and Figure 3. The lowest growth pressures result in the narrowest XRD peak widths, *i.e.*, the smallest tilt mosaicity of the composite films. This behavior is again contradictory to the $(BaTO_3$-$BiFeO_3) \times 15$ superlattices, where we found strongly increasing (001) ω-broadening with both decreasing growth pressure and α_{ME} [22]. The contradictory temperature dependencies of the α_{ME} of composites and superlattices and their sensitivity to external growth parameters, such as oxygen supply and layer geometry, indicate a complex interplay of strain- and charge-mediated magnetoelectric coupling, which requires further extensive efforts to become clearer in the future.

3. Experimental Section

The 67 wt % $BaTiO_3$-33 wt % $BiFeO_3$ composite films were grown by PLD from mixed phase targets with the above-mentioned composition, using a KrF excimer laser and a 10-cm target-to-substrate distance. The optimum growth temperature was 680 °C; there, the films are highly crystalline. The PLD oxygen partial pressure was controlled in between 0.01 and 0.5 mbar. In our PLD approach, four very similar films were grown simultaneously using a multi-substrate holder for epi-polished $SrTiO_3$(001) and $SrTiO_3$:Nb(001) with 0.1% and 0.5% Nb content and MgO(001). The $SrTiO_3$ and $SrTiO_3$:Nb substrates were HF etched and annealed prior to deposition to achieve a monolayer-terraced Ti-terminated surface. By XRD 2θ-ω scans, minor changes of the out-of-plane lattice parameter in dependence of the used substrate were found; see Ref. [11]. The conducting $SrTiO_3$:Nb substrates are required for the determination of ME voltage coefficients together with Pt top contacts of various diameters to get a thin film capacitor structure. The film thickness was determined directly by STEM on embedded cross-sections exposed by focused ion beam (FIB); see the Supplementary Materials, Figures S14 and S15. For these STEM images with lower resolution, a field emission scanning electron microscope (FEM) FEI NOVA Nanolab 200 (FEI Europe Nano Port, Eindhoven, The Netherlands) was used. We adjusted the number of laser pulses in PLD to compensate partially the effect of background pressure on growth rate and to achieve a thickness of 250–400 nm for all applied oxygen partial pressures; see Figure 8. Thicker films stem from earlier growth runs. With that, we monitor film thickness effects on magnetoelectric coupling. More details of the growth process, as well as the structural, ferroelectric and magnetic response of the samples can be found in [11]. For a recent description of the state of the art of PLD growth, see also the Special Issue "25 years of pulsed laser deposition" [34].

XRD wide-angle 2θ-ω scans, reciprocal space maps around the symmetric 001 and 002 and the asymmetric $\bar{1}03$ lattice points of the $SrTiO_3$:Nb(001) substrates were measured using a PANalytical X'pert PRO MRD (PANalytical B.V., Almelo, The Netherlands) with Cu $K\alpha_{1/2}$ radiation from a parabolic mirror and a PIXcel^{3D} array detector (PANalytical B.V.) with an electronically controlled receiving slit width.

Free-standing cross-sections for STEM were prepared from two samples, *i.e.*, the 67 wt % $BaTiO_3$-33 wt % $BiFeO_3$ composite films grown at either 0.25 mbar or 0.01 mbar oxygen partial pressure on MgO(001). Cross-sections about 100 nm thick were extracted using the FIB of the above-mentioned FEM Nanolab 200. Further thinning up to 200 kV electron transparency was done by Ar^+ ion milling in a Gatan PIPS instrument (Gatan, Inc., Pleasanton, CA, USA). STEM was carried out in a Philips CM-200 STEM (FEI Europe Nano Port) with a super-twin objective lens (point resolution of 0.23 nm). SAED patterns were taken from selected regions of the cross-sections. The distribution of elements was determined by EDX mapping (EDAX detector system). A weak Bi-deficit (more pronounced for the 0.01 mbar sample) was found in the composite films due to the use of a stoichiometric PLD source target and the lower sticking of Bi-species at the heated substrate surface; see Figures S6 and S9 for quantitative EDX analyses of the two investigated samples in the Supplementary Materials. Kinematical electron diffraction patterns were simulated using the JEMS software package

(École Polytechnique Fédérale de Lausanne, Villigen, Switzerland) [35]. The structure data (not refined) for "reduced" $BaTiO_3$ required for simulation were taken from [27] with data code ICSD 54785.

The magnetoelectric voltage coefficients α_{ME} were measured in a Quantum Design physical property measurement system (PPMS). The AC voltage induced inside the capacitor structure with the composite film as the medium in between the two electrodes is measured with a lock-in amplifier SR 830 (Stanford Research Systems, Inc., Sunnyvale, CA, USA) in response to a small AC magnetic field with a fixed frequency of 1 kHz and a field strength H_{AC} of 10 Oe (1 Oe = $10^3/4\pi \cdot A \cdot m^{-1}$). Both AC and possible additional DC magnetic fields are applied in the out-of-plane direction of the capacitor, *i.e.*, along the ME voltage in longitudinal geometry. For more details and corresponding ME measurements on $BiFeO_3$-based bulk samples and $BiFeO_3$ films, see [36] and [23], respectively.

Ferroelectric hysteresis loops were measured using a thin film analyzer TF 2000 HS (aixACCT Systems GmbH, Aachen, Germany) in dynamic hysteresis mode with frequencies of 100 Hz up to 1 kHz, without or with leakage current compensation. Triangular excitation pulses were applied in the dynamic sequences. The diameter of Pt top contacts was usually 225 μm.

4. Conclusions

$BaTiO_3$-$BiFeO_3$ composite thin films with a mixing ratio of the PLD source target of 67 wt % $BaTiO_3$ and 33 wt % $BiFeO_3$ and a total film thickness of typically 350 nm show a clear dependence of magnetoelectric voltage coefficient α_{ME} on oxygen partial pressure during growth. Three composite films grown at 0.25 mbar show the highest α_{ME} values of 36–43 V/(cm·Oe) at 300 K. α_{ME} decreases for both lower and higher growth pressure, due to increasing oxygen deficiency and increasing crystallite size, respectively. The composite films are grown epitaxially on $SrTiO_3$:Nb(001) substrates and exhibit increasing out-of-plane strain with decreasing PLD growth pressure in the range from 0.5 mbar down to 0.01 mbar. Because the out-of-plane lattice parameters of $BaTiO_3$ and $BiFeO_3$ are very close together, the corresponding XRD peaks of the two phases could not be resolved. Therefore, the detailed microstructure of the composite films on the atomic length scale has been unknown up to now.

However, STEM micrographs and SAED patterns taken from different regions of cross-sections of the composite films indicate an oxygen vacancy superstructure, which arises from vacancy ordering on the {111} planes of the pseudocubic $BaTiO_3$-type structure of the composite films. The intensity of the additional superstructure reflections seems to correlate to the oxygen growth pressure of the two investigated samples. This means that the sample grown at lower (0.01 mbar) pressure seems to show a more pronounced oxygen vacancy superstructure in comparison to the sample grown at higher (0.25 mbar) $p(O_2)$. In our previous work on oxygen-deficient ZnO thin films [25], we found a clear correlation of the PLD growth pressure and oxygen deficiency of films, which goes along with the structural distortions of the cation-anion octahedra.

Contrary to our recently investigated ($BaTiO_3$-$BiFeO_3$) × 15 superlattices [11,22,23], α_{ME} shows an increasing behavior with decreasing temperature, which cannot be explained by strain-mediated ME coupling of piezoelectric and magnetostrictive phases in the composite films alone. Rather, charge-mediated ME coupling may play a role here. Further research is planned to understand the observed temperature and DC magnetic field dependencies of α_{ME} more clearly to be able to design magnetoelectric composites with clear application perspectives.

Acknowledgments: We are indebted to Jörg Lenzner for the thickness determination at FIB-extracted cross-sections and for the extraction of free-standing cross-sections for TEM. We thank Holger Hochmuth for growing the PLD thin films and Gabriele Ramm for preparing the PLD target. Financial support from the Deutsche Forschungsgemeinschaft within Sonderforschungsbereich SFB 762 "Functionality of oxide interfaces" is kindly acknowledged. Work at KU Leuven was supported by the Research Foundation Flanders (FWO) and the Concerted Research Actions GOA/09/006 and GOA/14/007.

Author Contributions: Michael Lorenz and Vera Lazenka are the principal investigators. Michael Lorenz analyzed all samples by XRD and mainly wrote the manuscript with support from all coauthors. Gerald Wagner and Oliver Oeckler performed the STEM and SAED investigations and provided their interpretation. Peter Schwinkendorf has done the ferroelectric measurements. Michael Bonholzer developed the RSM presentation software. Vera Lazenka, Margriet J. Van Bael and André Vantomme and Kristiaan Temst performed or supported the magnetoelectric measurements, respectively. Marius Grundmann has conceptualized parts of the study.

Conflicts of Interest: The authors declare no conflict of interest.

References

1. Nan, C.-W.; Bichurin, M.I.; Dong, S.; Viehland, D.; Srinivasan, G. Multiferroic magnetoelectric composites: Historical perspective, status, and future directions. *J. Appl. Phys.* **2008**, *103*, 031101. [CrossRef]
2. Vaz, C.A.F. Electric field control of magnetism in multiferroic heterostructures. *J. Phys. Condens. Matter* **2012**, *24*, 333201. [CrossRef] [PubMed]
3. Fiebig, M.; Spaldin, N.A. Current trends of the magnetoelectric effect. *Eur. Phys. J. B* **2009**, *71*, 293–297. [CrossRef]
4. Vaz, C.A.F.; Hoffman, J.; Ahn, C.H.; Ramesh, R. Magnetoelectric coupling effects in multiferroic complex oxide composite structures. *Adv. Mater.* **2010**, *22*, 2900–2918. [CrossRef] [PubMed]
5. Ma, J.; Hu, J.; Li, Z.; Nan, C.-W. Recent progress in multiferroic magnetoelectric composites: From bulk to thin films. *Adv. Mater.* **2011**, *23*, 1062–1087. [CrossRef] [PubMed]
6. Kleemann, W.; Borisov, P.V.; Shvartsman, V.; Bedanta, S. Multiferroic and magnetoelectric materials—Developments and perspectives. *EPJ Web Conf.* **2012**, *29*, 00046. [CrossRef]
7. Rivera, J.-P. A short review of the magnetoelectric effect and related experimental techniques on single phase (multi-) ferroics. *Eur. Phys. J. B* **2009**, *71*, 299–313. [CrossRef]
8. Greve, H.; Woltermann, E.; Quenzer, H.-J.; Wagner, B.; Quandt, E. Giant magnetoelectric coefficients in $(Fe_{90}Co_{10})_{78}Si_{12}B_{10}$-AlN thin film composites. *Appl. Phys. Lett.* **2010**, *96*, 182501. [CrossRef]
9. Röbisch, V.; Yarar, E.; Urs, N.O.; Teliban, I.; Knöchel, R.; McCord, J.; Quandt, E.; Meyners, D. Exchange biased magnetoelectric composites for magnetic field sensor application by frequency conversion. *J. Appl. Phys.* **2015**, *117*, 17B513. [CrossRef]
10. Freitas, V.F.; Dias, G.S.; Protzek, O.A.; Montanher, D.Z.; Catellani, I.B.; Silva, D.M.; Cótica, L.F.; dos Santos, I.A. Structural phase relations in perovskite-structured $BiFeO_3$-based multiferroic compounds. *J. Adv. Ceram.* **2013**, *2*, 103–111. [CrossRef]
11. Lorenz, M.; Lazenka, V.; Schwinkendorf, P.; Bern, F.; Ziese, M.; Modarresi, H.; Volodin, A.; Van Bael, M.J.; Temst, K.; Vantomme, A.; *et al.* Multiferroic $BaTiO_3$-$BiFeO_3$ composite thin films and multilayers: Strain engineering and magnetoelectric coupling. *J. Phys. D Appl. Phys.* **2014**, *47*, 135303. [CrossRef]
12. Ueda, K.; Tabata, H.; Kawai, T. Coexistence of ferroelectricity and ferromagnetism in $BiFeO_3$–$BaTiO_3$ thin films at room temperature. *Appl. Phys. Lett.* **1999**, *75*, 555. [CrossRef]
13. Oh, Y.S.; Crane, S.; Zheng, H.; Chu, Y.H.; Ramesh, R.; Kim, K.H. Quantitative determination of anisotropic magnetoelectric coupling in $BiFeO_3$-$CoFe_2O_4$ nanostructures. *Appl. Phys. Lett.* **2010**, *97*, 052902. [CrossRef]
14. Murugavel, P.; Lee, J.-H.; Jo, J.Y.; Sim, H.Y.; Chung, J.-S.; Jo, Y.; Jung, M.-H. Structure and ferroelectric properties of epitaxial $_{(1-x)}BiFeO_{3-x}BaTiO_3$ solid solution films. *J. Phys. Condens. Matter* **2008**, *20*, 415208. [CrossRef]
15. Begum, H.A.; Naganuma, H.; Oogane, M.; Ando, Y. Fabrication of Multiferroic Co-Substituted $BiFeO_3$ Epitaxial Films on $SrTiO_3$ (100) Substrates by Radio Frequency Magnetron Sputtering. *Materials* **2011**, *4*, 1087–1095. [CrossRef]
16. Kowal, K.; Jartych, E.; Guzdek, P.; Stoch, P.; Wodecka-Duś, B.; Lisińska-Czekaj, A.; Czekaj, D. X-ray diffraction, Mössbauer spectroscopy, and magnetoelectric effect studies of $(BiFeO_3)_x$-$(BaTiO_3)_{1-x}$ solid solutions. *Nukleonika* **2013**, *58*, 57–61.

17. Yang, S.-C.; Kumar, A.; Petkov, V.; Priya, S. Room-temperature magnetoelectric coupling in single-phase $BaTiO_3$-$BiFeO_3$ system. *J. Appl. Phys.* **2013**, *113*, 144101. [CrossRef]

18. Kim, D.H.; Aimon, N.M.; Sun, X.; Ross, C.A. Compositionally modulated magnetic epitaxial spinel/perovskite nanocomposite thin films. *Adv. Funct. Mater.* **2014**, *24*, 2334–2342. [CrossRef]

19. Lu, J.; Günther, A.; Schrettle, F.; Mayr, F.; Krohns, S.; Lunkenheimer, P.; Pimenov, A.; Travkin, V.D.; Mukhin, A.A.; Loidl, A. On the room temperature multiferroic $BiFeO_3$: Magnetic, dielectric and thermal properties. *Eur. Phys. J. B* **2010**, *75*, 451–460. [CrossRef]

20. Sando, D.; Barthélémy, A.; Bibes, M. $BiFeO_3$ epitaxial thin films and devices: Past, present and future. *J. Phys. Condens. Matter* **2014**, *26*, 473201. [CrossRef] [PubMed]

21. Lazenka, V.V.; Lorenz, M.; Modarresi, H.; Brachwitz, K.; Schwinkendorf, P.; Böntgen, T.; Vanacken, J.; Ziese, M.; Grundmann, M.; Moshchalkov, V.V. Effect of rare-earth ion doping on the multiferroic properties of $BiFeO_3$ thin films grown epitaxially on $SrTiO_3$(100). *J. Phys. D Appl. Phys.* **2013**, *46*, 175006. [CrossRef]

22. Lorenz, M.; Wagner, G.; Lazenka, V.; Schwinkendorf, P.; Modarresi, H.; Van Bael, M.J.; Vantomme, A.; Temst, K.; Oeckler, O.; Grundmann, M. Correlation of magnetoelectric coupling in multiferroic $BaTiO_3$-$BiFeO_3$ superlattices with oxygen vacancies and antiphase octahedral rotations. *Appl. Phys. Lett.* **2015**, *106*, 012905. [CrossRef]

23. Lazenka, V.; Lorenz, M.; Modarresi, H.; Bisht, M.; Rüffer, R.; Bonholzer, M.; Grundmann, M.; Van Bael, M.J.; Vantomme, A.; Temst, K. Magnetic spin structure and magnetoelectric coupling in $BiFeO_3$-$BaTiO_3$ multilayer. *Appl. Phys. Lett.* **2015**, *106*, 082904. [CrossRef]

24. Bichurin, M.; Petrov, V.; Zakharov, A.; Kovalenko, D.; Yang, S.C.; Maurya, D.; Bedekar, V.; Priya, S. Magnetoelectric interactions in lead-based and lead-free composites. *Materials* **2011**, *4*, 651–702. [CrossRef]

25. Lorenz, M.; Böttcher, R.; Friedländer, S.; Pöppl, A.; Spemann, D.; Grundmann, M. Local lattice distortions in oxygen deficient Mn-doped ZnO thin films, probed by electron paramagnetic resonance. *J. Mater. Chem. C* **2014**, *2*, 4947–4956. [CrossRef]

26. Wang, J.; Neaton, J.B.; Zheng, H.; Nagarajan, V.; Ogale, S.B.; Liu, B.; Viehland, D.; Vaithyanathan, V.; Schlom, D.G.; Waghmare, U.V.; *et al.* Epitaxial $BiFeO_3$ multiferroic thin film heterostructures. *Science* **2003**, *299*, 1719–1722. [CrossRef] [PubMed]

27. Woodward, D.I.; Reaney, I.M.; Yang, G.Y.; Dickey, E.C.; Randall, C.A. Vacancy ordering in reduced barium titanate. *Appl. Phys. Lett.* **2004**, *84*, 4650. [CrossRef]

28. Glazer, A.M. The classification of tilted octahedra in perovskites. *Acta Cryst. B* **1972**, *28*, 3384–3392. [CrossRef]

29. Buttner, R.H.; Maslen, E.N. Structural parameters and electron difference density in $BaTiO_3$. *Acta Cryst. B* **1992**, *48*, 764–769. [CrossRef]

30. Das, R.; Mandal, K. Magnetic, ferroelectric and magnetoelectric properties of Ba-doped $BiFeO_3$. *J. Magn. Magn. Mater.* **2012**, *324*, 1913–1918. [CrossRef]

31. Thiele, C.; Dörr, K.; Bilani, O.; Rödel, J.; Schultz, L. Influence of strain on the magnetization and magnetoelectric effect in $La_{0.7}A_{0.3}MnO_3$ /PMN-PT(001) (A = Sr,Ca). *Phys. Rev. B* **2007**, *75*, 054408. [CrossRef]

32. Wolf, R.A.; Trolier-McKinstry, S. Temperature dependence of the piezoelectric response in lead zirconate titanate films. *J. Appl. Phys.* **2004**, *95*, 1397. [CrossRef]

33. Spurgeon, S.R.; Sloppy, J.D.; Kepaptsoglou, D.M.; Balachandran, P.V.; Nejati, S.; Karthik, J.; Damodaran, A.R.; Johnson, C.L.; Ambaye, H.; Goyette, R.; *et al.* Thickness-dependent crossover from charge- to strain-mediated magnetoelectric coupling in ferromagnetic/piezoelectric oxide heterostructures. *ACS Nano* **2014**, *8*, 894–903. [CrossRef] [PubMed]

34. Lorenz, M.; Rao, R. Special issue 25 years of pulsed laser deposition. *J. Phys. D. Appl. Phys.* **2014**, *47*, 030301. [CrossRef]

35. Stadelmann, P. JEMS software package, version 2008. *Ultramicroscopy* **1987**, *21*, 131–146. [CrossRef]

36. Lazenka, V.V.; Zhang, G.; Vanacken, J.; Makoed, I.I.; Ravinski, A.F.; Moshchalkov, V.V. Structural transformation and magnetoelectric behaviour in $Bi_{1-x}Gd_xFeO_3$ multiferroics. *J. Phys. D Appl. Phys.* **2012**, *45*, 125002. [CrossRef]

15

Effect of Rare Earth Metals on the Microstructure of Al-Si Based Alloys

Saleh A. Alkahtani [1], Emad M. Elgallad [2], Mahmoud M. Tash [1], Agnes M. Samuel [2] and Fawzy H. Samuel [2,*]

Academic Editors: Mark T. Whittaker and Robert Lancaster

[1] Industrial Engineering Program, Mechanical Engineering Department, College of Engineering, Prince Sattam bin AbdulAziz University, Al Kharj 11942, Saudi Arabia; salqahtany@hotmail.com (S.A.A.); mahmoud_tash1@yahoo.com (M.M.T.)
[2] Département des Sciences Appliquées, Université du Québec à Chicoutimi, Chicoutimi, QC G7H 2B1, Canada; eelgalla@uqac.ca (E.M.E.); amsamuel@uqac.ca (A.M.S.)
* Correspondence: fhsamuel@uqac.ca

Abstract: The present study was performed on A356 alloy [Al-7 wt %Si 0.0.35 wt %Mg]. To that La and Ce were added individually or combined up to 1.5 wt % each. The results show that these rare earth elements affect only the alloy melting temperature with no marked change in the temperature of Al-Si eutectic precipitation. Additionally, rare earth metals have no modification effect up to 1.5 wt %. In addition, La and Ce tend to react with Sr leading to modification degradation. In order to achieve noticeable modification of eutectic Si particles, the concentration of rare earth metals should exceed 1.5 wt %, which simultaneously results in the precipitation of a fairly large volume fraction of insoluble intermetallics. The precipitation of these complex intermetallics is expected to have a negative effect on the alloy performance.

Keywords: aluminum alloys; La and Ce addition; rare earth metals; modification; intermetallics

1. Introduction

The discovery of Pacz [1] in 1921, that Al-Si alloys containing between 5% and 15% silicon could be treated with alkali fluoride fluxes (preferably sodium fluoride) to yield alloys of improved ductility and machinability, made it possible to extensively use aluminum foundry alloys in various applications. This treatment is commonly referred to as *eutectic modification* by which the morphology of eutectic silicon may be changed from acicular-like to a finer and more interconnected fibrous structure [2,3].

Chemical modification is the most commonly applied method of modification which produces a fine fibrous silicon structure through the addition of trace levels of certain modifying elements. The modification mechanism induced by such chemical modifiers has been related to the nucleation and growth of Si particles [4–7]. The modifier addition deactivates the heterogeneous inoculants AlP, oxides, and other melt substrates that act as nucleation sites for the eutectic phase, resulting in an increase of eutectic undercooling and a refined morphology of eutectic silicon. From the point of view of the growth theory, the modifier atoms segregate into favored growth surfaces of the Si crystals, poisoning the attaching mechanism of Si atoms. As a result, the silicon growth is halted and a refined eutectic morphology is obtained.

The most common chemical modifiers include Sr, Na and Sb. Strontium is preferred as a modifying agent in the light of the rapid fading of Na and the toxic effect associated with Sb [2]. The addition of Sr in the range of 150 to 250 ppm was reported to significantly improve the mechanical properties of Al-Si alloys [8,9]. However, if present in larger amounts, Sr can result in the formation of undesirable intermetallic compounds, such as Al_2SrSi_2 and Al_4SrSi_2, and cause over-modification of the eutectic silicon [10,11].

Rare earth (RE) elements and mischmetal (mixture of RE elements) were also reported to cause modification in both hypo- and hypereutectic Al-Si alloys [12–17]. It was found that the modification efficiency of 1.0 wt % La in the microstructures and mechanical properties of A356 alloy is similar to that of 0.01 wt % Sr [12]. Li *et al.* [13] indicated that the addition of 1.0 wt % La or 1.0 wt % Ce to Al-11.7%Si-1.8%Cu alloy could refine the Si particles, but without any obvious modification effect. The effect of different concentrations of individual additions of RE metals including La, Ce, Pr, Nd, Sm, Eu, Gd, Tb, Dy, Ho, Er, Tm, Yb and Lu on the eutectic modification in Al-10%Si was studied [14]. The results showed that all of the RE elements caused a depression of the eutectic growth temperature, but only Eu produced fully-modified, fine-fibrous silicon. The remaining elements resulted in only a small degree of refinement of the plate-like silicon morphology.

Mousavi *et al.* [15] demonstrated that the optimum levels of mischmetal (combination of Ce, La, Pr and Nd) addition to cause modification in A357 Al-Si casting alloy were 0.1 and 0.3 wt % for thin and thick section castings, respectively. The addition of La was reported to optimize the modification effect of P in the Al-18%Si alloy [16]. Zhang *et al.* [17] reported that the addition of Ce to the Al-18%Si alloy changed the primary Si from branched shape to fine facetted shape, and that the modification effect was more obvious for the eutectic Si away from the primary Si. To arrive at a better understanding of the role of RE elements in the eutectic modification, the present study was undertaken to investigate the effect of individual and combined additions of La and Ce in the presence and absence of Sr on the eutectic microstructure of A356 casting alloy. This alloy was specifically selected in this study because it represents one of the most important hypoeutectic Al-Si alloys based on its wide use in the automotive industry which is the largest consumer of Al-Si casting alloys. The modification effect of RE elements was examined using optical and scanning electron microscopy along with thermal analysis technique.

2. Experimental Procedures

The chemical compositions of the A356 ingots used in the present study are listed in Table 1, whereas the aimed and actual modifier additions along with the codes of the resulting alloys are listed in Table 2.

Table 1. Composition of A356 alloy used in the present study.

Alloy	Elements (wt %)					
	Si	Cu	Mg	Fe	Zn	Al
A356	7	<0.20	0.35	<0.20	<0.10	Bal

The 12-Kg ingots of A356 alloy were cut into smaller pieces, cleaned, dried, and then melted in a SiC crucible, using an electrical resistance furnace. The melt temperature was kept at $750 \pm 5\,°C$. The holding time before pouring was around 40 s. The modifiers used in the present study, namely Sr, La and Ce, were added in the form of Al-10%Sr, Al-15%La and Al-15%Ce master alloys, respectively. The melt was degassed for 15 min using pure, dry argon injected by using a graphite impeller rotating at 120 rpm. The melt was poured into a graphite mold, a schematic of which is shown in Figure 1. The mold was preheated to $600\,°C$ to obtain cooling rates close to equilibrium conditions. Samplings for spectrochemical analysis were simultaneously taken for each casting poured from the different melts.

Thermal analysis was performed simultaneously using the graphite mold casting setup by attaching a high-sensitive K-type thermocouple to the mold system, passing through the bottom of the mold, and extending half way up into the mold cavity, along the mold centerline. The temperature-time data from the fully liquid state, through the solidification range, to the solid state were recorded using a high-speed PC-based data acquisition system connected to the thermocouple. The part of the thermocouple within the mold was protected using double-walled ceramic tubing. From the thermal analysis data, the cooling curves and their first derivatives were plotted and analyzed.

Table 2. The aimed and actual modifier additions and the codes of the resulting alloys.

Alloy	Mold Type	Preheating Mold Temperature (°C)	Alloy Code	Modifier Addition (wt %)					
				Aimed			Actual		
				Sr	La	Ce	Sr	La	Ce
A356	Graphite	600	TB	0	0	0	0	0	0
			TBS	0.01	0	0	0.011	0	0
			T10	0	0.2	0	0	0.17	0
			T1	0	0.5	0	0	0.40	0
			T2	0	1	0	0	0.85	0
			T3	0	1.5	0	0	1.30	0
			T11	0	0	0.2	0	0	0.18
			T4	0	0	0.5	0	0	0.38
			T5	0	0	1	0	0	0.82
			T6	0	0	1.5	0	0	1.38
			T7	0	0.5	0.5	0	0.44	0.38
			T8	0	1	1	0	0.78	0.87
			T9	0	1.5	1.5	0	1.37	1.53
			T10S	0.01	0.2	0	0.009	0.17	0
			T1S	0.01	0.5	0	0.011	0.40	0
			T2S	0.01	1	0	0.008	0.85	0
			T3S	0.01	1.5	0	0.010	1.30	0
			T11S	0.01	0	0.2	0.008	0	0.18
			T4S	0.01	0	0.5	0.009	0	0.38
			T5S	0.01	0	1	0.008	0	0.82
			T6S	0.01	0	1.5	0.010	0	1.38
			T7S	0.01	0.5	0.5	0.009	0.44	0.38
			T8S	0.01	1	1	0.009	0.78	0.87
			T9S	0.01	1.5	1.5	0.011	1.37	1.53

TB: Base A356 alloy; TBS: Sr-modified base A356 alloy.

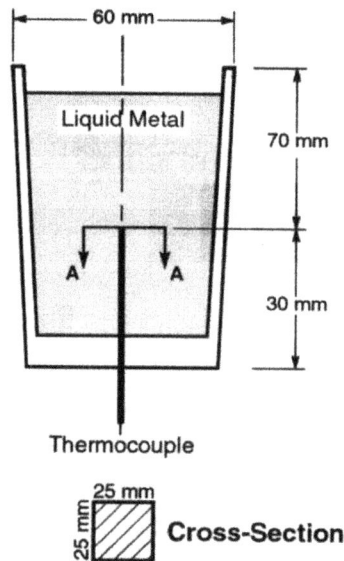

Figure 1. Schematic sketch showing the graphite mold used in this study.

For metallographic observations, samples were sectioned from each casting, mounted, and then polished. The polished samples were examined using an optical microscope and a scanning electron microscope (Hitachi-SU-8000, Hitachi High-Technologies Corporation, Tokyo, Japan). The quantitative analysis of the silicon particle characteristics was carried out using a Clemex image analyzer system (Clemex Technologies Inc., Longueuil, QC, Canada) in conjunction with the optical microscope. In each case, the measurements were carried out over fifty fields at 200× magnification such that the entire sample surface was traversed in a regular, systematic fashion, to obtain the average values of the measured parameters.

3. Results and Discussion

3.1. Thermal Analysis

The solidification curve for the base alloy, obtained from its time-temperature data, and its first derivative plot are provided in Figure 2a. The temperatures at which the main reactions take place during solidification were determined from the first derivative curve and are marked on the diagram. Figure 2b displays the micro-constituents of the microstructure after solidification at the rate of approximately 0.8 °C/s.

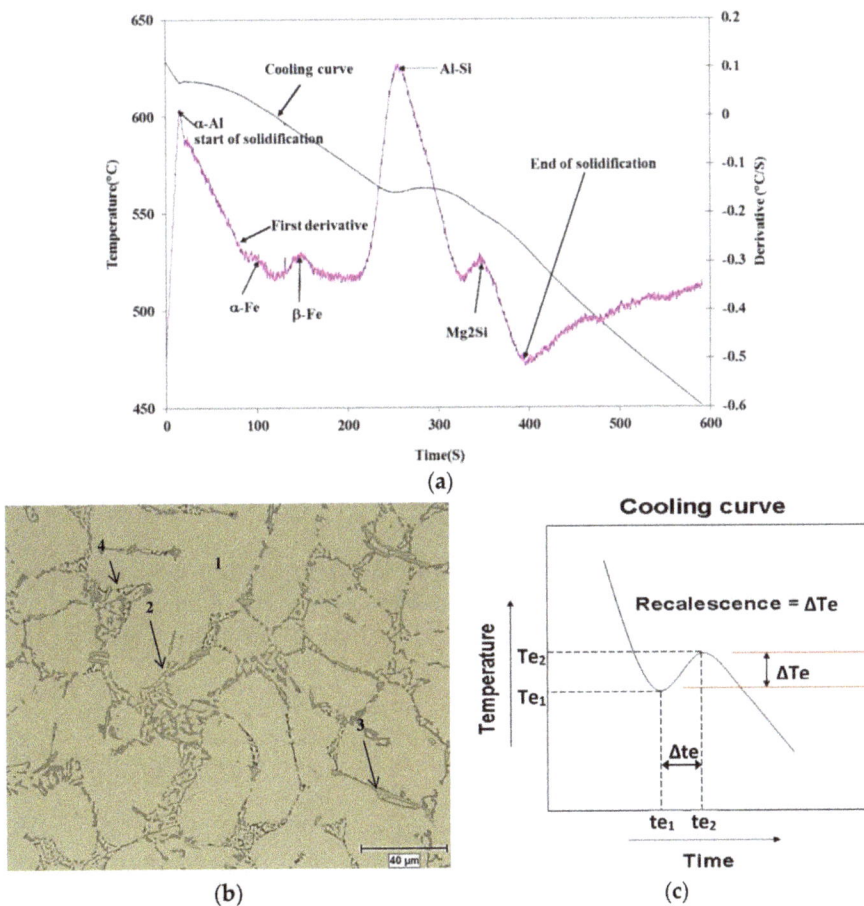

Figure 2. (a) Solidification curve and its first derivative of A356 alloy; (b) Optical microstructure of A356 alloy (coded TB) following solidification at the rate of ~0.8 °C/s: 1—α-Al, 2—eutectic Si, 3—Fe-intermetallic, 4—Mg$_2$Si phase; (c) Schematic diagram of a hypothetical cooling curve showing recalescence.

Figure 2c shows a diagram representing recalescence (ΔT) which is defined as the heat generated during solidification relating to the nucleation and growth of a new phase [18]. Without intending to make an in-depth approach to the mathematics of solidification, but rather to understand the constituent parts governing the solidification process itself, Equation (1) provides a basic mathematical formula which introduces the principal parameters of recalescence:

$$\Delta T = [(\Delta hf/cp)(\delta fs/\delta t)] - [(q_e/cp)(A/V)] \tag{1}$$

As will be observed from Equation (1), Δhf is the latent heat of fusion denoting a decrease in enthalpy due to the transformation from liquid to solid; cp is the specific heat per unit volume; q_e is the heat flux; A and V are area and volume, respectively, of the solidifying sample; and fs is the solid fraction. The first term on the right-hand side of Equation (1) takes into account the continuing evolution of the latent heat of fusion during solidification. The second term reflects principally the effects of casting geometry, area, and volume of the solidifying samples. Regarding this equation, recalescence will occur when the first term on the right-hand side of Equation (1) becomes greater than the second one.

As may be deduced from Equation (1), the heat extraction of the samples in thermal analysis plays an essential role in perceiving recalescence based on solidification data, in the eventuality that there is a high discrepancy between both terms of this equation; for example, if the second term is much higher than the first term, the perception of the recalescence may remain concealed, and as a result, the thermal analysis data will not reveal any noticeable reaction. Generally speaking, throughout the course of thermal analysis experiments at slow cooling rates, the heat loss from the surrounding environment exceeds the heat generated by nascent, or incipient, phase reactions to a great extent, specifically during the initial solidification period, making it impossible to detect their appearance on the cooling curve, *i.e.*, to perceive the recalescence [19].

Figures 3 and 4 show the solidification curves obtained from the present alloys with compositions listed in Table 2, whereas Tables 3 and 4 list the variation in the eutectic recalescence in alloys containing La (as an example). It is evident from these data that;

1. Introduction of La or Ce increased the start of solidification temperature of A356 alloy by about 11 °C at 1.5 wt % La or Ce with a marginal effect on the Al-Si eutectic temperature.

2. Introduction of 100 ppm Sr to A356 alloy reduced the eutectic temperature by about 7 °C. Addition of La, Ce, or La + Ce up to 1.5 wt % each has no further effect on the eutectic temperature.

3. No explicit peak corresponding to precipitation of rare earth (RE)—containing phases could be detected using the thermal analysis technique.

4. Although the introduction of Sr has no noticeable effect on the Al-Si eutectic recalescence temperature (ΔT), the recalescence time increased from 20 s to 40 s as inferred from Tables 3 and 4. In this case as well, the presence of La or Ce did not have an effect on either ΔT or Δt.

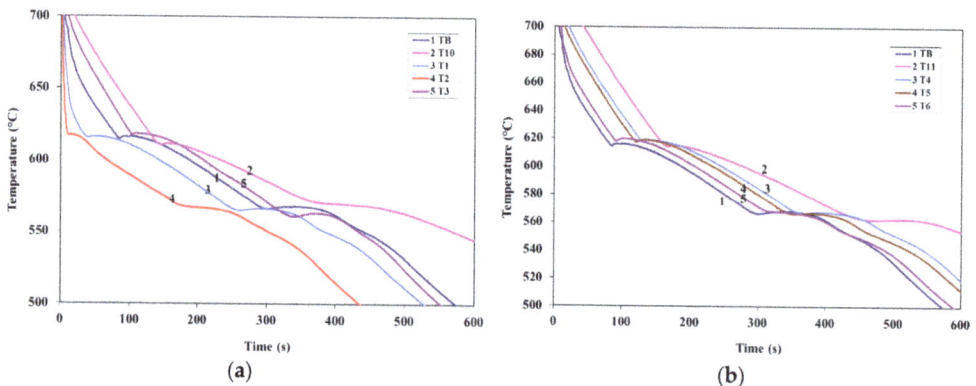

Figure 3.   (a) Solidification curves of La-containing alloys; (b) solidification curves of Ce-containing alloys.

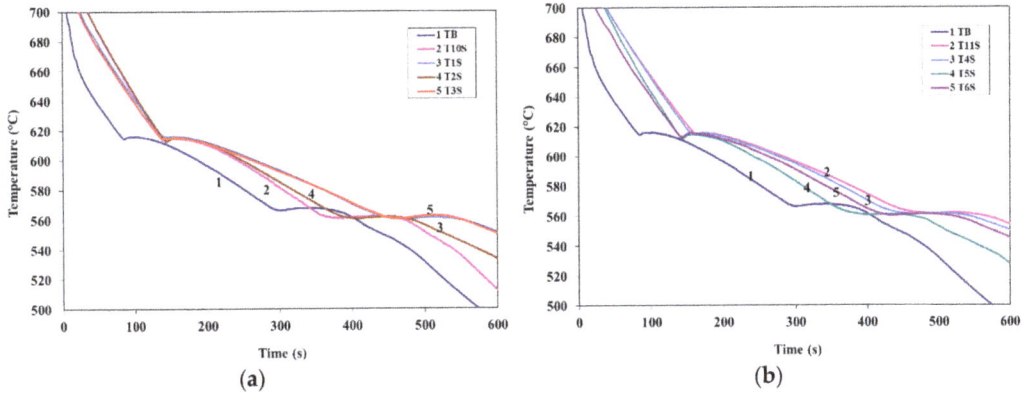

Figure 4. (a) Solidification curves of La+Sr-containing alloys; (b) solidification curves of Ce+Sr-containing alloys.

Table 3. Effect of La addition on the recalescence of the eutectic Si.

Alloy Code	Recalescence of Eutectic Si					
	Te_1 *	Te_2	ΔTe	te_1 **	te_2	Δte
T10	565.1	566.2	1.1	242.5	252.3	9.8
T1	565.2	566.4	1.2	239.8	260.2	20.4
T2	563.4	564.8	1.4	254.3	270.3	16.0
T3	560.8	563.2	2.4	257.0	281.2	24.2

* Temperature (°C); ** time (s).

Table 4. Effect of La and Sr additions on the recalescence of the eutectic Si.

Alloy Code	Recalescence of Eutectic Si					
	Te_1	Te_2	ΔTe	te_1	te_2	Δte
T10S	560.7	562.6	1.9	264.6	308	43.4
T1S	560.4	561.8	1.4	339.6	388.8	49.2
T2S	560.6	562.2	1.6	279.6	322.2	42.6
T3S	560.7	563.0	2.3	347.8	389.4	41.6

3.2. Microstructural Characterization

The coarsening process of eutectic Si particles, called Ostwald ripening, holds that larger particles grow at the expense of smaller ones. Solution treatment tends to spheroidize constituents which cannot be fully dissolved, as in the case of casting Al-Si alloys. At the beginning of the solution treatment, the acicular silicon platelets in the unmodified structure begin to break down into smaller fragments and gradually spheroidize. In modified structures the spheroidization takes place at an early stage [20]. Figure 5 shows a schematic representation of the spheroidization and coarsening process.

Figure 6 show the microstructure of the A356 alloy in the non-modified and different modified conditions. In the non-modified condition (Figure 6a), the eutectic silicon typically displayed a coarse plate-like structure. The addition of 0.01% Sr (Figure 6b) significantly modified the eutectic silicon, changing its morphology from a coarse plate-like structure to a fine fibrous one. The addition of 0.2% La or 0.2% Ce (Figure 6c,d, respectively) resulted in a relative refinement in the plate-like silicon morphology when compared to the non-modified TB alloy (Figure 6a), but without any obvious modification effect. Such a refining effect of La and Ce was also reported by other researchers [13,14] for eutectic and hypoeutectic Al-Si alloys, respectively. However, neither refining nor modifying effect was observed for La or Ce additions beyond 0.2%, *i.e.*, 0.5% and 1%. This can be discerned from the micrographs shown in Figure 6e,f, which were obtained from A356 alloy samples modified by 1% La and 1% Ce, respectively. It is interesting to note that the addition of 1% La or 1% Ce caused a noticeable coarsening of eutectic silicon particles. It was also observed that the combined addition of 0.5% La and 0.5% Ce or 1% La and 1% Ce did not bring about any substantial change in the eutectic silicon morphology.

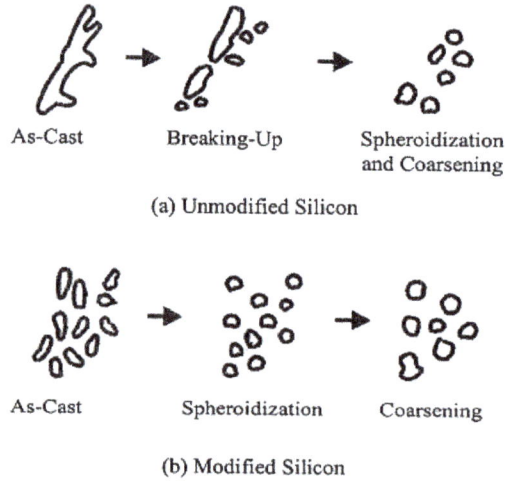

As-Cast   Breaking-Up   Spheroidization
                                                    and Coarsening

(a) Unmodified Silicon

As-Cast   Spheroidization   Coarsening

(b) Modified Silicon

Figure 5. Schematic characterization of the three stages of spheroidization and coarsening of the eutectic silicon phase in the case of (**a**) unmodified; and (**b**) modified Al-Si alloy [20].

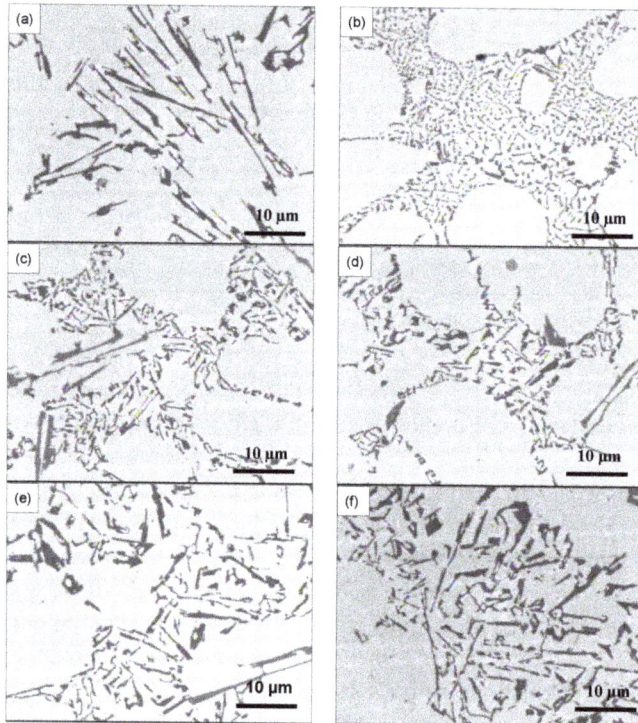

Figure 6. Microstructures of different A356-based alloys: (**a**) base TB alloy (non-modified); (**b**) TBS alloy (0.01% Sr); (**c**) T10 alloy (0.2% La); (**d**) T11 alloy (0.2% Ce); (**e**) T2 alloy (1% La); and (**f**) T5 alloy (1% Ce).

To better understand these observations, X-ray mapping of intermetallic phases found in an A356 alloy sample modified by the combination of 1% La and 1% Ce (T8 alloy) was carried out to determine the distribution of these elements, as shown in Figure 7. The corresponding enregy dispersive X-ray spectroscopic [EDS] spectra are shown in Figure 8. It is evident that the addition of La and Ce resulted in the formation of two intermetallic phases which appeared as coarse white and gray particles in the backscattered image shown in Figure 7. In addition to La and Ce, these phase also contained different contents of Al, Si and Ti. It can therefore be suggested that high levels of La and/or Ce can result in the formation of coarse intermetallic compounds instead of modifying or even refining the eutectic silicon. The formation of La- and Ce-rich intermetallic phases in Al-Si alloys was also reported in several studies [15,21,22]. A new AlSiLa intermetallic phase was detected in an A357 alloy when increasing the La-based mischmetal level beyond 0.3% [15]. Elsebaie *et al.* [21], and Hosseinifar and Malakhov [22] reported the formation of complex phases such as AlTiLa(Ce)Mg, AlSiLa(Ce) and La(Al,Si)$_2$ phases. Figures 9 and 10 display the affinity of RE metals to react with Sr leading to the observed loss of modification in the present alloys. Moreover, La/Ce revealed a clear tendency to react with Si forming a complex intermetallic as shown in Figure 11, which is in good agreement with the abovementioned results [21,22].

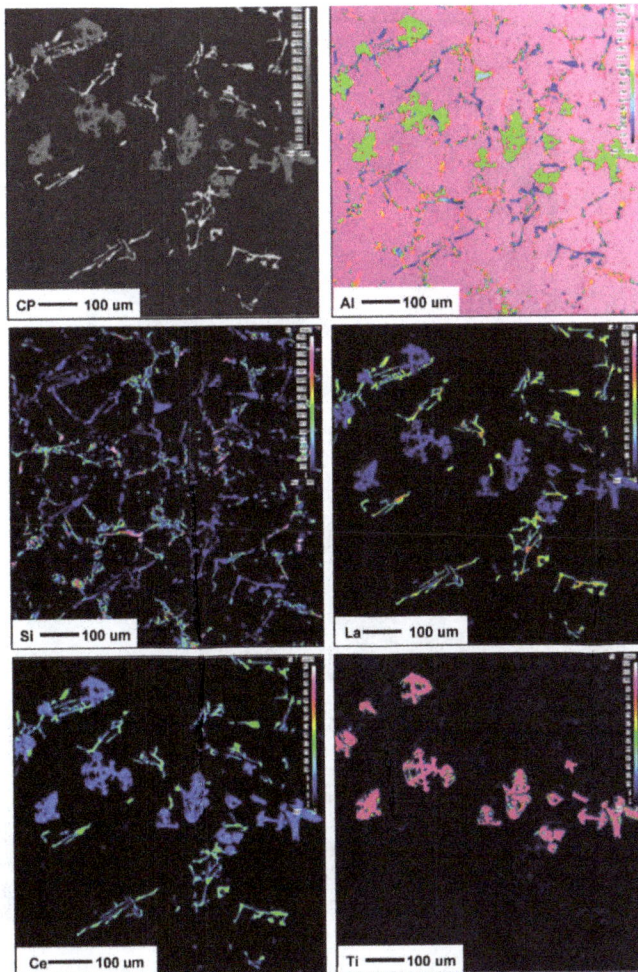

Figure 7. Backscattered electron (BSE) images showing the formation of La- and Ce-rich phases in an A356 alloy sample modified by the combination of 1% La and 1% Ce (T8 alloy) and the corresponding X-ray images of Al, Si, La, Ce and Ti.

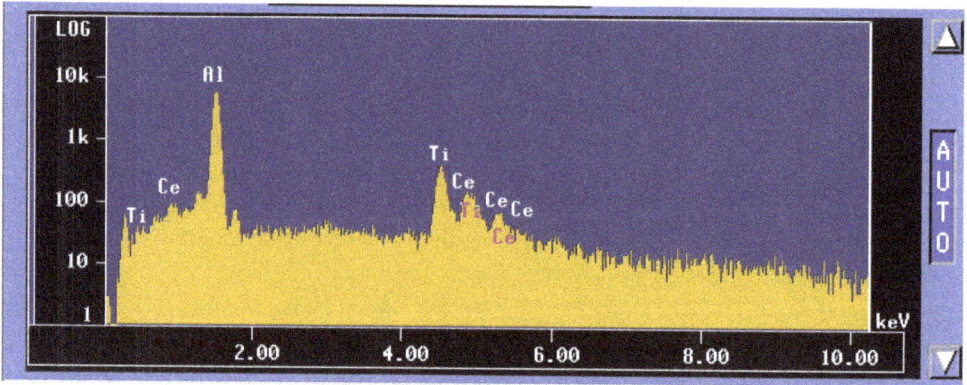

Figure 8. EDS spectrum corresponding to the gray phase observed in the BSE image in Figure 7 (CP).

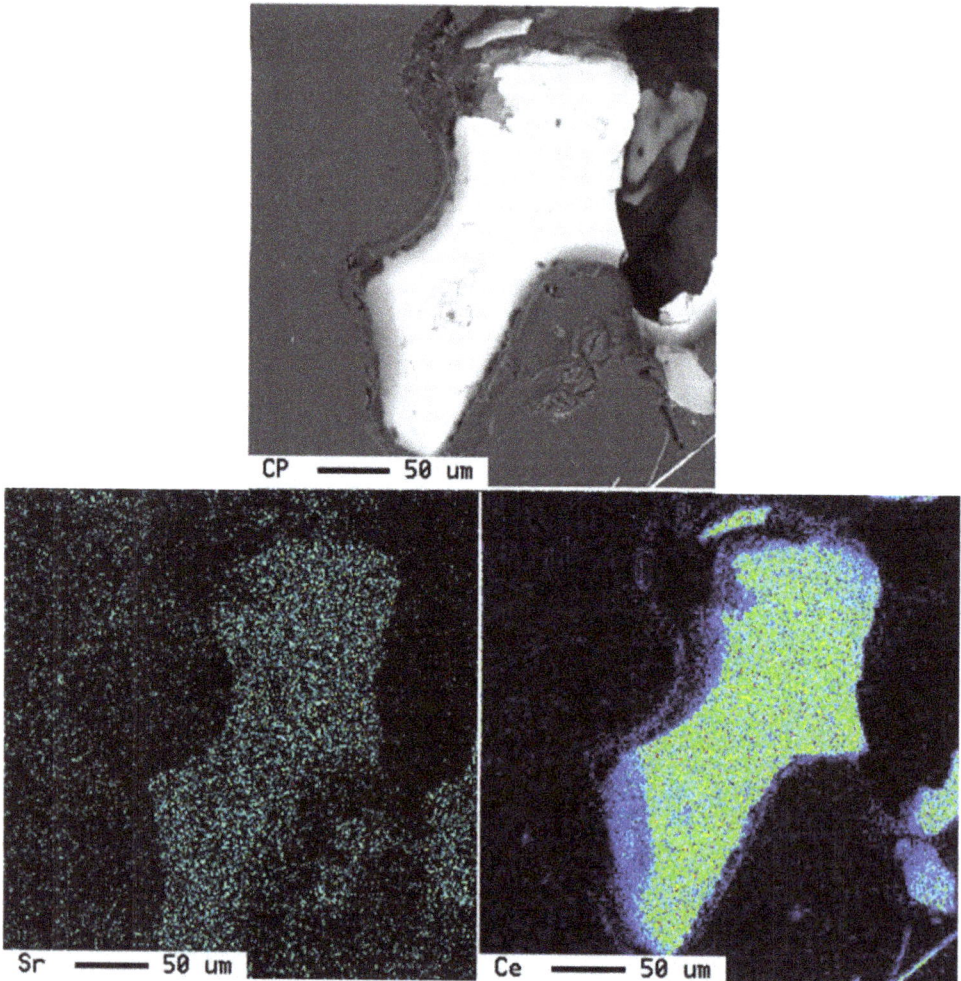

Figure 9. Ce-Sr interactions in A356 alloy modified with 1.0% Ce + 0.01% Sr (T5S alloy).

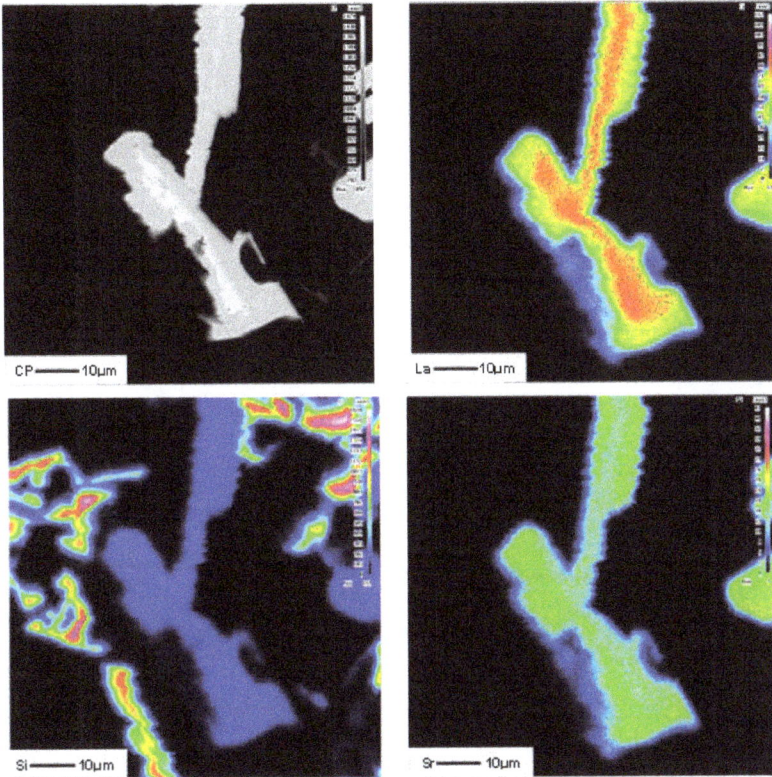

Figure 10. La, Si and Sr distribution in A356 alloy modified with 1.0% La + 0.01% Sr (T2S alloy).

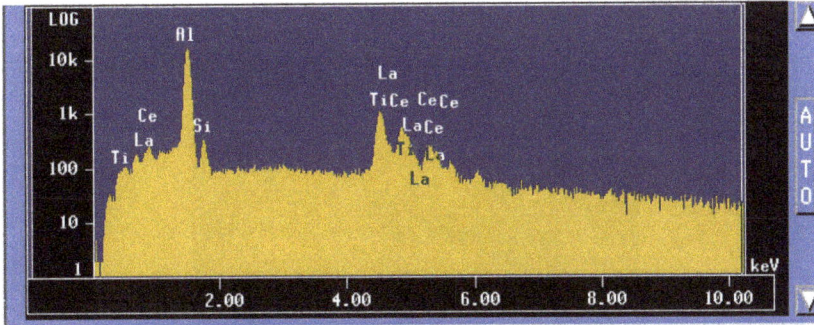

Figure 11. EDS spectrum corresponding to Figure 10, revealing a significant peak due to Si.

Figures 12 and 13 illustrate the variation in Si particle characteristics as a function of the added La or Ce, respectively. It is evident from these diagrams that the affinity of La to react with Sr leading to degradation of modification effectiveness is greater than that of Ce. This observation is limited to 1 wt % of La beyond which a dramatic decrease in the Si particle size is seen to take place (Figure 12). The observed decrease in the size of the eutectic Si particles may interpreted in terms of La-Si interaction forming AlSiLa and La(Al,Si)$_2$ compounds resulting in significant reduction in the amount of free Si. Figure 14 shows the precipitation of insoluble La-rich intermetallics in A356 alloy containing 1.5 wt % La (coded T3 in Table 2). Similar findings have been reported by Hong-Kun and Di [23] who added 3 wt % of La to Al-17 wt %Si. However, the presence of such hard brittle intermetallics would have a negative effect on the alloy mechanical properties [24].

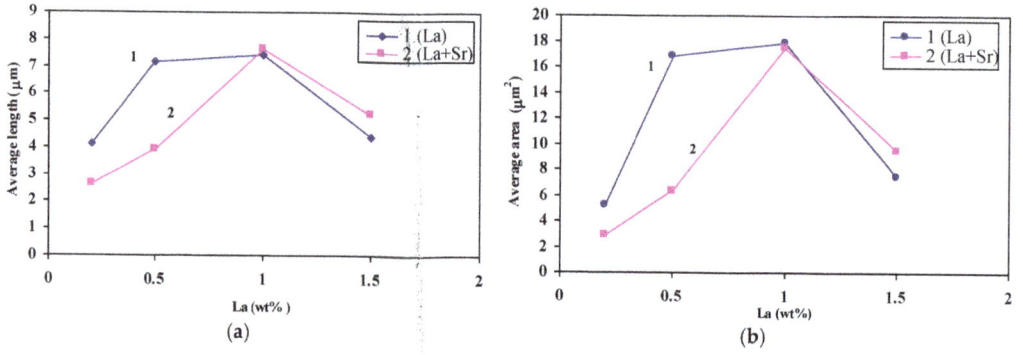

Figure 12. (a) The effect of La and Sr on the average of Si particle length; (b) the effect of La and Sr on the average of Si particle area.

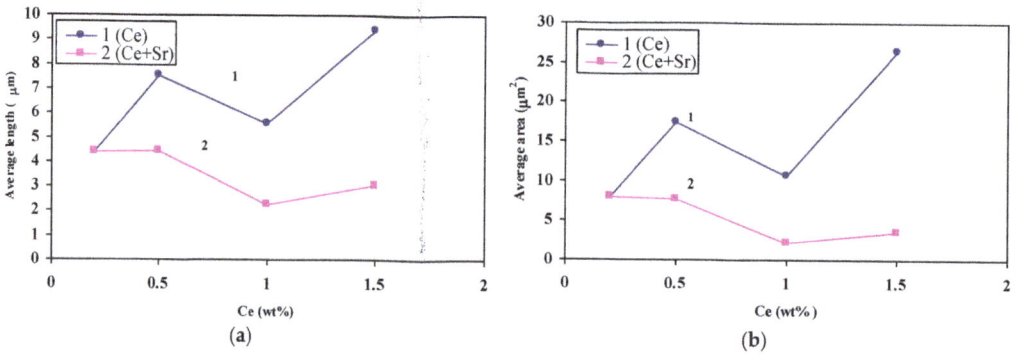

Figure 13. (a) The effect of Ce and Sr on the average of Si particle length; (b) the effect of Ce and Sr on the average of Si particle area.

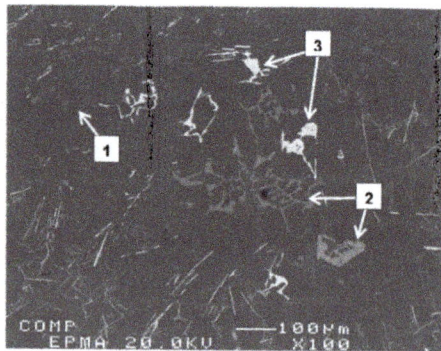

Figure 14. Precipitation of La-rich phases in A356 alloy containing 1.5 wt % La (T3 alloy): (1) α-Al; (2) α-Fe and (3) La-rich phase.

4. Conclusions

Based on the results documented in the present work, the following conclusions may be drawn:

1. Addition of La or Ce individually or combined up to 1.5 wt % increases the melting point of A356 alloy at approximately 1 °C/0.15 wt % RE metals.

2. The addition of RE metals has no significant effect either on the Al-Si eutectic temperature or on the modification of the eutectic Si particles.

3.   Both La and Ce have high affinity to react with Sr resulting in marked reduction in the Sr modification effect.

4.   Modification of Si particles may take place when La concertation exceeds 1.5 wt %.

5.   Increasing the amount of La, however, leads to precipitation of insoluble intermetallics that could negatively affect the alloy performance.

Acknowledgments: Financial and in-kind received support from the Deanship of Scientific Research, Vice Rectorate for Post Graduate and Scientific Research, Prince Sattam Abduaziz University (PSAU) is gratefully acknowledged. This project was supported by the deanship of scientific research at Prince Sattam bin Abdulaziz University under the Saudi Arabia Basic Industries Corporation (SABIC) Research Program Project #.4409/01/2015.

Author Contributions: Saleh A. Alkahtani and Fawzy H. Samuel conceived and designed the experiments with input from Emad M. Elgallad, Mahmoud M. Tash and Agnes M. Samuel; Saleh A. Alkahtani and Mahmoud M. Tash provided financial support for the project; Fawzy H. Samuel, Emad M. Elgallad and Agnes M. Samuel performed the electron microscopic-, thermal analysis experiments, and optical microscopic examinations and analysis of data., with input from Mahmoud M. Tash. Fawzy H. Samuel and Emad M. Elgallad wrote the paper. Agnes M. Samuel and Emad M. Elgallad contributed to the final revisions.

Conflicts of Interest: The authors declare no conflict of interest.

References

1.   Aladar, P. Alloy. U.S. Patent 1,387,900, 16 August 1921.

2.   Closset, B.M.; Gruzleski, J.E. *The Treatment Liquid Aluminum-Silicon Alloys*; American Foundrymen's Society: Des Plaines, IL, USA, 1990.

3.   Elsebaie, O.; Samuel, F.H.; Alkahtani, S.A. Intermetallic phases observed in non-modified and Sr modified Al-Si cast alloys containing mischmetal. *Int. J. Cast Met. Res.* **2013**, *26*, 1–15. [CrossRef]

4.   Elsebaie, O.; Samuel, A.M.; Samuel, F.H.; Doty, H.W. The effects of mischmetal, cooling rate and heat treatment on the eutectic Si particle characteristics of A319.1, A356.2 and A413.1 Al-Si casting alloys. *Mater. Sci. Eng. A* **2008**, *480*, 342–355.

5.   Samuel, A.M.; Garza-Elizondo, G.H.; Doty, H.W.; Samuel, F.H. Role of modification and melt thermal treatment processes on the microstructure and tensile properties of Al-Si alloys. *Mater. Des.* **2015**, *80*, 99–108. [CrossRef]

6.   Hegde, S.; Prabhu, K.N. Modification of eutectic silicon in Al-Si alloys. *J. Mater. Sci.* **2008**, *43*, 3009–3027. [CrossRef]

7.   Onyia, C.W.; Okorie, B.A.; Neife, S.I.; Obayi, C.S. Structural modification of sand cast eutectic Al-Si alloys with sulfur/sodium and its effect on mechanical properties. *World J. Eng. Technol.* **2013**, *1*, 9–16. [CrossRef]

8.   Sarada, B.N.; Srinivasamurthy, P.L. Swetha, microstructural characteristics of Sr and Na modified Al-Mg-Si alloy. *Int. J. Innov. Res. Sci. Eng. Technol.* **2013**, *2*, 3975–3983.

9.   Fat-Halla, N. Structural modification of Al-Si eutectic alloy by Sr and its effect on tensile and fracture characteristics. *J. Mater. Sci.* **1989**, *24*, 2488–2492. [CrossRef]

10.   Barrireroa, J.; Engstlera, M.; Ghafoorb, N.; Jongec, N.; Odénb, M.; Mücklicha, F. Comparison of segregations formed in unmodified and Sr-modified Al-Si alloys studied by atom probe tomography and transmission electron microscopy. *J. Alloys Compd.* **2014**, *611*, 410–421. [CrossRef]

11.   Srirangam, P.; Chattopadhyay, S.; Bhattacharya, A.; Nag, S.; Kaduke, J.; Shankar, S.; Banerjee, R.; Shibata, T. Probing the local atomic structure of Sr-modified Al-Si alloys. *Acta Mater.* **2014**, *65*, 185–193. [CrossRef]

12.   Zhang, J.; Chen, H.; Yu, H.; Jin, Y. Study on dual modification of Al-17%Si alloys by structural heredity. *Metals* **2015**, *5*, 1112–1126. [CrossRef]

13.   Li, H.J.; Shivkumar, S.; Luos, X.J.; Apelian, D. Influence of modification on the solution heat treatment response of cast Al-Si-Mg Alloys. *Cast Met.* **1989**, *1*, 227–234.

14.   Nogita, K.; McDonald, S.D.; Dahle, A.K. Eutectic modification of Al-Si alloys with rare earth metals. *Mater. Trans. JIM* **2004**, *45*, 323–326. [CrossRef]

15.   Mousavi, G.S.; Emamyn, M.; Rassizadehghani, J. The effect of mischmetal and heat treatment on the microstructure and tensile properties of A357 Al-Si casting alloy. *Mater. Sci. Eng. A* **2012**, *556*, 573–581. [CrossRef]

16. Ouyang, Z.-Y.; Mao, X.-M.; Hong, M. Multiplex modification with rare earth elements and P for hypereutectic Al-Si alloys. *J. Shanghai Univ. Engl. Ed.* **2007**, *11*, 400–402. [CrossRef]

17. Zhang, H.; Duan, H.; Shao, G.; Xu, L.; Yin, J.; Yan, B. Modification mechanism of cerium on the Al-18Si alloy. *Rare Met.* **2006**, *25*, 11–15. [CrossRef]

18. Dobrzański, L.A.; Maniara, R.; Sokolowski, J.H. The effect of cast Al-Si-Cu alloy solidification rate on alloy thermal characteristics. *J. Achiev. Mater. Manuf. Eng.* **2006**, *17*, 217–220.

19. Kurz, W.; Fisher, D.J. *Fundamentals of Solidification*, 4th revised ed.; Trans Tech. Publications Ltd.: Uetikon-Zuerich, Switzerland, 1998.

20. Paray, F.; Gruzleski, J.E. Microstructure-mechanical property relationship in A356 alloy. Part 1: Microstructure. *Cast Met.* **1994**, *7*, 29–40.

21. Elsebaie, O. l'Effet de l'Addition du "Mischmetal", du Taux de Refroidissement et du Traitement Thermique sur la Microstucture et la Dureté des Alliages Al-Si de Type 319, 356, et 413. Mater's Thesis, Université du Québec à Chicoutimi (UQAC), Chicoutimi, QC, Canada, 2006.

22. Hosseinifar, M.; Malakhov, D.V. The sequence of intermetallics formation during the solidification of an Al-Mg-Si alloy containing La. *Metall. Mater. Trans. A* **2011**, *42*, 825–833. [CrossRef]

23. Yi, H.-K.; Zhang, D. Modification effect of pure rare earth metal La on the as cast Al-17%Si alloys. *Trans. Nonferr. Met. Soc. China* **2003**, *13*, 1–7.

24. Hosseinifar, M. Aluminum Alloys Containing Cerium and Lanthanum. Ph.D. Thesis, McMaster University, Hamilton, ON, Canada, 2009.

16

Comparison of Cyclic Hysteresis Behavior between Cross-Ply C/SiC and SiC/SiC Ceramic-Matrix Composites

Longbiao Li

Academic Editor: Dinesh Agrawal

College of Civil Aviation, Nanjing University of Aeronautics and Astronautics, No. 29 Yudao St., Nanjing 210016, China; llb451@nuaa.edu.cn

Abstract: In this paper, the comparison of cyclic hysteresis behavior between cross-ply C/SiC and SiC/SiC ceramic-matrix composites (CMCs) has been investigated. The interface slip between fibers and the matrix existed in the matrix cracking mode 3 and mode 5, in which matrix cracking and interface debonding occurred in the $0°$ plies are considered as the major reason for hysteresis loops of cross-ply CMCs. The hysteresis loops of cross-ply C/SiC and SiC/SiC composites corresponding to different peak stresses have been predicted using present analysis. The damage parameter, *i.e.*, the proportion of matrix cracking mode 3 in the entire matrix cracking modes of the composite, and the hysteresis dissipated energy increase with increasing peak stress. The damage parameter and hysteresis dissipated energy of C/SiC composite under low peak stress are higher than that of SiC/SiC composite; However, at high peak stress, the damage extent inside of cross-ply SiC/SiC composite is higher than that of C/SiC composite as more transverse cracks and matrix cracks connect together.

Keywords: ceramic-matrix composites (CMCs); cross-ply; hysteresis loops; matrix cracking; interface debonding

1. Introduction

Nickel-based superalloys with thermal and environmental ceramic coatings are the current load bearing material system that can operate above the metal substrate's melting temperature, *i.e.*, about 1100 °C, at which a combined-cycle gas turbine will operate at 60% fuel efficiency. However, due to demands for reduced fuel consumption, lighter and hotter engines are required, especially in aviation. Ceramic materials can operate at high temperature with creep resistance, at which metals cannot. However, their use as structural components is severely limited because of their brittleness. Continuous fiber-reinforced ceramic-matrix composites (CMCs), by incorporating fibers in ceramic matrices, however, can be made as strong as metal, yet are much lighter and can withstand much higher temperatures exceeding the capability of current nickel alloys typically used in high-pressure turbines, which can increase the efficiency of aero engines [1]. CMC durability has been validated through ground testing or commercial flight testing in demonstrator or customer gas turbine engines accumulating almost 30,000 h of operation. The CMC combustion chamber and high-pressure turbine components were designed and tested in the ground testing of the GEnx aero engine [2]. The CMC rotating low-pressure turbine blades in a F414 turbofan demonstrator engine were successfully tested for 500 grueling cycles to validate the unprecedented temperature and durability capabilities by GE Aviation (Fairfield, CT, USA). The CMC tail nozzles were designed and fabricated by SNECMA (SAFRAN, Paris, France) and completed the first commercial flight on CFM56-5B aero engine (CFM International, Cincinnati, OH, USA) on 2015. CMCs will play a key role in the performance of CFM's

LEAP (Leading Edge Aviation Propulsion) turbofan engine, which would enter into service in 2016 for Airbus A320 and in 2017 for the Boeing 737 max.

Under cyclic loading and unloading, matrix cracking and fiber/matrix interface debonding occur inside of CMCs [3]. The hysteresis loops appear as the fiber slips relative to matrix in the interface debonded region [4]. The shape, location, and area of hysteresis loops can reveal the internal damage evolution of CMCs subjected to cyclic loading [5]. Many researchers investigated characteristics of hysteresis loops. Kotil *et al.* [6] investigated the effect of interface shear stress on the shape and area of hysteresis loops in unidirectional CMCs. Pryce and Smith [7] investigated the effect of interface partially debonding on hysteresis loops of unidirectional CMCs by assuming purely frictional load transfer between fibers and the matrix. Ahn and Curtin [8] investigated the effect of matrix stochastic cracking on hysteresis loops of unidirectional CMCs and compared with the Pryce-Smith model [7]. Solti *et al.* [9] investigated the effect of interface partially and completely debonding on hysteresis loops in unidirectional CMCs using the maximum interface shear strength criterion to determine interface slip lengths. Vagaggini *et al.* [10] investigated the effect of interface debonded energy on hysteresis loops of unidirectional CMCs based on the Hutchinson-Jensen fiber pull-out model [11]. Cho *et al.* [12] investigated the evolution of interface shear stress under cyclic-fatigue loading from frictional heating measurements. Li *et al.* investigated the effect of interface debonding [13], fibers Poisson contraction [14], fiber fracture [15], and interface wear [16] on hysteresis loops of unidirectional CMCs, and developed an approach to estimate interface shear stress in unidirectional CMCs through hysteresis loop area [17]. Kuo and Chou [18] investigated matrix multicracking in cross-ply CMCs and classified the multiple cracking states into five modes, in which cracking mode 3 and mode 5 involve matrix cracking and interface debonding in the 0° plies.

The objective of this paper is to compare the cyclic hysteresis behavior between cross-ply C/SiC and C/SiC CMCs. The interface slip between fibers and the matrix existed in matrix cracking mode 3 and mode 5, in which matrix cracking and interface debonding occurred in the 0° plies, are considered as the major reason for hysteresis loops of cross-ply CMCs. The hysteresis loops of cross-ply C/SiC and SiC/SiC composites corresponding to different peak stresses have been predicted using present analysis. The differences between C/SiC and SiC/SiC composite on damage parameters and hysteresis dissipated energy have been investigated.

2. Materials and Experimental Procedures

2.1. Cross-Ply C/SiC Composite

The T-700™ carbon (Toray Institute Inc., Tokyo, Japan) fiber-reinforced silicon carbide matrix composites (C/SiC CMCs) were provided by Shanghai Institute of Ceramics, People's Republic of China. The fibers have an average diameter of 7 μm and come on a spool as a tow of 12 k fibers. The cross-ply C/SiC composite was manufactured by hot-pressing method, which offered the ability to fabricate dense composite via a liquid phase sintering method at a low temperature. The lay-ups supplied were in the form of (0/90/0/90/0/90/0/90/0). The volume fraction of fibers was about 40%. The void content in the manufactured plates is below 5%. Low pressure chemical vapor infiltration was employed to deposit approximately 5~20 layer PyC/SiC with mean thickness of 0.2 μm in order to enhance the desired non-linear/non-catastrophic tensile behavior.

The dog bone-shaped specimens, with dimensions of 123 mm length, 3.8 mm thickness according to ASTM (American Society for Testing and Materials) standard C 1360-10 [19], and 10 mm width in the gage section of cross-ply C/SiC composite, were cut from 150 mm × 150 mm panels by water cutting. The specimens were further coated with SiC of ~20 μm thick by chemical vapor deposition to prevent oxidation at elevated temperature.

The loading/unloading tensile experiments at room temperature were conducted on an MTS Model 809 servo hydraulic load-frame (MTS System Crop., Minneapolis, MN, USA) equipped with edge-loaded grips, operated at the loading rate of 2.0 MPa/s. The gage-section strains were measured using a clip-on extensometer (Model No. 634.12F-24, MTS Systems Corp.; modified for a 25 mm gage-length). The direct observations of matrix cracking were made using a HiROX optical microscope (Tokyo, Japan). The matrix crack density was determined by counting the number of the cracks in a length of about 15 mm.

2.2. Cross-Ply SiC/SiC Composite

The Hi-Nicalon Type S™ fiber reinforced pre-impregnated melt-infiltrated silicon-carbide matrix composites (SiC/SiC CMC) were provided by GE Aviation (Cincinnati, OH, USA) [20]. The specimens were machined to a dogbone shape with dimensions of 203 mm length, 10.16 mm width, and 1.88 mm thickness. The lay-ups supplied were in the form of $[0/90]_{2s}$. During the final phase of manufacturing the laminates, molten silicon is infiltrated into the pre-impregnated lamina tapes to form a SiC and silicon mixed matrix.

The loading/unloading tensile experiments at room temperature were conducted on an MTS servo hydraulic load-frame (MTS System Crop., Minneapolis, MN, USA) equipped with edge-loaded grips, operated under displacement control with the loading rate of 0.1–0.5 mm/min. The gage-section strains were measured using a 25.4 mm clip-on MTS extensometer with a maximum displacement of 2% strain. The direct observations of matrix cracking were made using Mitutoyo binocular optical microscope (Tokyo, Japan). The matrix crack density was determined by counting the number of the cracks in a length of 5–10 mm.

3. Hysteresis Loops Models Considering Multiple Matrix Cracking Modes

Under cyclic loading, the matrix cracking modes in cross-ply CMCs can be divided into five different modes, i.e., mode 1: transverse cracking in the 90° plies; mode 2: transverse cracking and matrix cracking occurred in the 90° and 0° plies, respectively, with perfect fiber/matrix interface bonding in the 0° plies; mode 3: transverse cracking and matrix cracking occurred in the 90° and 0° plies, respectively, with fiber/matrix interface debonding in the 0° plies; mode 4: matrix cracking in the 0° plies with fiber/matrix interface bonding; and mode 5: matrix cracking in the 0° plies with fiber/matrix interface debonding, as shown in Figure 1. Upon unloading and subsequent tensile reloading, matrix cracking mode 3 and mode 5 both exist within cross-ply C/SiC composite, as shown in Figure 2.

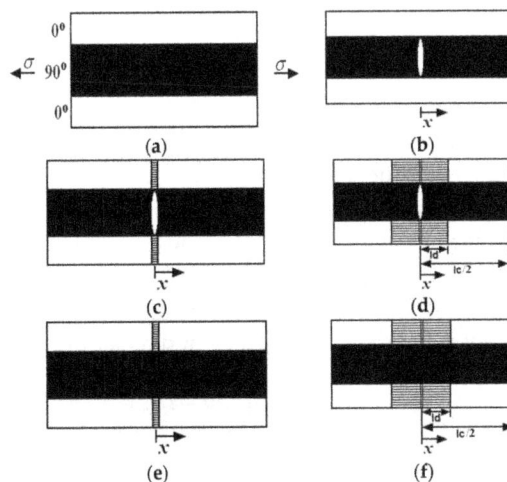

Figure 1. The undamaged state and five damaged modes of cross-ply ceramic composites: (**a**) undamaged composite; (**b**) mode 1: transverse crack; (**c**) mode 2: transverse crack and matrix crack with perfect fiber/matrix bonding; (**d**) mode 3: transverse crack and matrix crack with fiber/matrix interface debonding; (**e**) mode 4: matrix crack with perfect fiber/matrix bonding; and (**f**) mode 5: matrix cracking with fiber/matrix debonding.

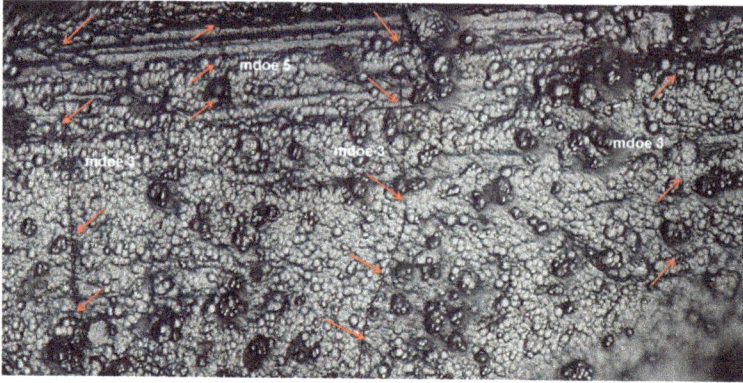

Figure 2. The matrix cracking mode 3 and mode 5 of cross-ply C/SiC composite under cyclic loading/unloading tensile.

Upon unloading and reloading, the frictional slip occurred between fibers and the matrix in the $0°$ plies is the major reason for the hysteresis loops of cross-ply CMCs [5]. In cross-ply laminates, besides the fiber debonding and relative fiber/matrix sliding, other events, *i.e.*, delamination, relative ply sliding, near-tip matrix micro-cracking, and crack surface bridging followed by frictional fiber pull-out may also contribute to the hysteresis behavior. However, in the present analysis, the hysteresis loops models consider only the major factor of interface frictional slip in the matrix cracking mode 3 and mode 5. For matrix cracking mode 3, the hysteresis loops can be divided into four different cases, *i.e.*, case 1: interface partially debonds and fiber slips completely relative to matrix; case 2: interface partially debonds and fiber slips partially relative to matrix; case 3: interface completely debonds and fiber slips partially relative to matrix; and case 4: interface completely debonds and fiber slips completely relative to matrix. The unloading and reloading strains when interface partially debonds are [21]:

$$\varepsilon_{cu} = \frac{\sigma}{V_{f\_axial}E_f} + 4\frac{\tau_i}{E_f}\frac{y^2}{r_f l_c} - 2\frac{\tau_i}{E_f}\frac{(2y - l_d)(2y - l_c + l_d)}{r_f l_c} - (\alpha_c - \alpha_f)\Delta T \tag{1}$$

$$\varepsilon_{cr} = \frac{\sigma}{V_{f\_axial}E_f} - 4\frac{\tau_i}{E_f}\frac{z^2}{r_f l_c} + \frac{4\tau_i}{E_f}\frac{(y - 2z)^2}{r_f l_c} \\ +2\frac{\tau_i}{E_f}\frac{(l_d - 2y + 2z)(l_d + 2y - 2z - l_c)}{r_f l_c} - (\alpha_c - \alpha_f)\Delta T \tag{2}$$

in which V_{f_axial} denotes the fiber volume content in the $0°$ plies; E_f denotes the fiber elastic modulus; r_f denotes the fiber radius; τ_i denotes the fiber/matrix interface shear stress in the $0°$ plies; l_c denotes the matrix crack spacing; l_d denotes the interface debonded length; α_f and α_c denote the fiber and composite thermal expansion coefficient, respectively; ΔT denotes the temperature difference between fabricated temperature T_0 and room temperature T_1 ($\Delta T = T_1 - T_0$); and y and z denote the interface counter-slip length and interface new-slip length, respectively.

When interface completely debonds, the unloading and reloading strains are [21]:

$$\varepsilon_{cu} = \frac{\sigma}{V_{f\_axial}E_f} + 4\frac{\tau_i}{E_f}\frac{y^2}{r_f l_c} - 2\frac{\tau_i}{E_f}\frac{(2y - l_c/2)^2}{r_f l_c} - (\alpha_c - \alpha_f)\Delta T \tag{3}$$

$$\varepsilon_{cr} = \frac{\sigma}{V_{f\_axial}E_f} - 4\frac{\tau_i}{E_f}\frac{z^2}{r_f l_c} + 4\frac{\tau_i}{E_f}\frac{(y - 2z)^2}{r_f l_c} - 2\frac{\tau_i}{E_f}\frac{(l_c/2 - 2y + 2z)^2}{r_f l_c} - (\alpha_c - \alpha_f)\Delta T \tag{4}$$

For matrix cracking mode 5, the hysteresis loops can also be divided into four different cases. The unloading and reloading strains when interface partially debonds are [21]:

$$\varepsilon_{cu} = \frac{1}{V_{f\_axial}E_f}(\sigma - k\sigma_{to}) + 4\frac{\tau_i}{E_f}\frac{y^2}{r_f l_c} - 2\frac{\tau_i}{E_f}\frac{(2y-l_d)(2y+l_d-l_c)}{r_f l_c} - (\alpha_c - \alpha_f)\Delta T \tag{5}$$

$$\varepsilon_{cr} = \frac{1}{V_{f\_axial}E_f}(\sigma - k\sigma_{to}) - 4\frac{\tau_i}{E_f}\frac{z^2}{r_f l_c} + \frac{4\tau_i}{E_f}\frac{(y-2z)^2}{r_f l_c}$$
$$+2\frac{\tau_i}{E_f}\frac{(l_d - 2y + 2z)(l_d + 2y - 2z - l_c)}{r_f l_c} - (\alpha_c - \alpha_f)\Delta T \tag{6}$$

in which k denotes the proportion of transverse plies in the entire composite.

When interface completely debonds, the unloading and reloading strains are [21]:

$$\varepsilon_{cu} = \frac{1}{V_{f\_axial}E_f}(\sigma - k\sigma_{to}) + 4\frac{\tau_i}{E_f}\frac{y^2}{r_f l_c} - 2\frac{\tau_i}{E_f}\frac{(2y-l_c/2)^2}{r_f l_c} - (\alpha_c - \alpha_f)\Delta T \tag{7}$$

$$\varepsilon_{cu} = \frac{1}{V_{f\_axial}E_f}(\sigma - k\sigma_{to}) - 4\frac{\tau_i}{E_f}\frac{z^2}{r_f l_c} + 4\frac{\tau_i}{E_f}\frac{(y-2z)^2}{r_f l_c}$$
$$-2\frac{\tau_i}{E_f}\frac{(l_c/2 - 2y + 2z)^2}{r_f l_c} - (\alpha_c - \alpha_f)\Delta T \tag{8}$$

Considering the effect of multiple matrix cracking modes on hysteresis loops of cross-ply CMCs, the unloading and reloading strains of the composite are [21]:

$$(\varepsilon_u)_c = \eta(\varepsilon_{cu})_3 + (1-\eta)(\varepsilon_{cu})_5 \tag{9}$$

$$(\varepsilon_r)_c = \eta(\varepsilon_{cr})_3 + (1-\eta)(\varepsilon_{cr})_5 \tag{10}$$

in which $(\varepsilon_u)_c$ and $(\varepsilon_r)_c$ denote the unloading and reloading strain of the composite, respectively; $(\varepsilon_{cu})_3$ and $(\varepsilon_{cr})_3$ denote the unloading and reloading strain of the matrix cracking mode 3, respectively; $(\varepsilon_{cu})_5$ and $(\varepsilon_{cr})_5$ denote the unloading and reloading strain of the matrix cracking mode 5, respectively; η is the damage parameter determined by the composite's damage condition, i.e., the proportion of matrix cracking mode 3 in the entire of matrix cracking modes of the composite, $\eta \in [0,1]$.

4. Experimental Comparisons

4.1. Cross-Ply C/SiC Composite

The cyclic loading/unloading tensile behavior of cross-ply C/SiC composite at room temperature has been investigated. The specimen was unloading and subsequent reloading at the peak stress of 20, 40, 60, 80, 100, and 120 MPa, respectively. The peak stress represents the macroscopic stress, i.e., applied loading divided by the specimen cross section. The basic material properties of cross-ply C/SiC composite are given by: $V_f = 40\%$, $E_f = 230$ GPa, $E_m = 350$ GPa, $r_f = 3.5$ μm, $\tau_i = 6$ MPa, $\zeta_d = 0.1$ J/m^2, $\alpha_f = -0.38 \times 10^{-6}/°C$, $\alpha_m = 2.8 \times 10^{-6}/°C$, $\Delta T = -1000$ °C.

For $\sigma_{max} = 60$ MPa, the experimental and theoretical hysteresis loops are shown in Figure 3a, in which the proportion of matrix cracking mode 3 is $\eta = 0.3$. For matrix cracking mode 3, the hysteresis loops correspond to interface slip case 2, as shown in Figure 3b. Upon completely unloading, the interface counter-slip length approaches to 83.8% of interface debonded length, i.e., $y(\sigma_{min})/l_d = 86.5\%$, as shown in Figure 3b; upon reloading to peak stress, the interface new-slip length approaches to 86.5% of interface debonded length, i.e., $z(\sigma_{max})/l_d = 86.5\%$, as shown in Figure 3b. For matrix cracking mode 5, the hysteresis loops correspond to interface slip case 1, as shown in Figure 3b. Upon unloading, the interface counter-slip length approaches to interface debonded length at $\sigma_{tr\_pu} = 45$ MPa, i.e., $y(\sigma_{tr\_pu})/l_d = 1$, as shown in Figure 3b; upon reloading to $\sigma_{tr\_pr} = 15$ MPa, the interface new-slip length approaches to interface debonded length, i.e., $z(\sigma_{tr\_pr})/l_d = 1$, as shown in Figure 3b.

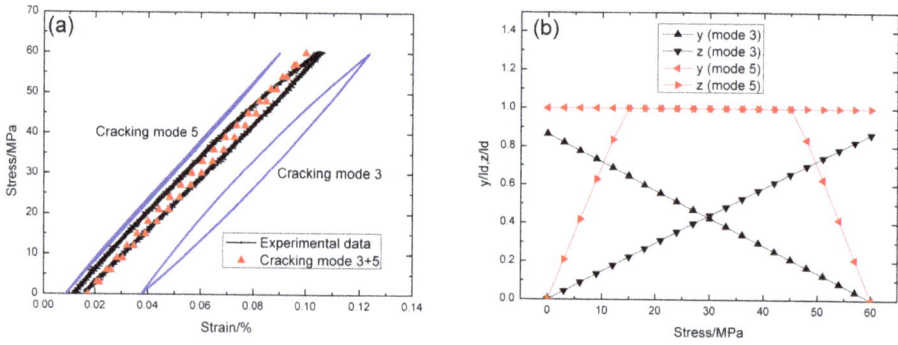

Figure 3. (a) The theoretical and experimental hysteresis loops; and (b) the interface slip lengths, *i.e.*, y/l_d and z/l_d, of matrix cracking mode 3 and mode 5 of cross-ply C/SiC composite when σ_{max} = 60 MPa.

For σ_{max} = 80 MPa, the experimental and theoretical hysteresis loops are shown in Figure 4a, in which the proportion of matrix cracking mode 3 is η = 0.35. For matrix cracking mode 3, the hysteresis loops correspond to interface slip case 2, as shown in Figure 4b. Upon completely unloading, the interface counter-slip length approaches to 73.5% of interface debonded length, *i.e.*, $y(\sigma_{min})/l_d$ = 73.5%, as shown in Figure 4b; upon reloading to peak stress, the interface new-slip length approaches to 73.5% of interface debonded length, *i.e.*, $z(\sigma_{max})/l_d$ = 73.5%, as shown in Figure 4b. For matrix cracking mode 5, the hysteresis loops correspond to interface slip case 1, as shown in Figure 4b. Upon unloading, the interface counter-slip length approaches to interface debonded length at σ_{tr_pu} = 24 MPa, *i.e.*, $y(\sigma_{tr\_pu})/l_d$ = 1, as shown in Figure 4b; upon reloading to σ_{tr_pr} = 56 MPa, the interface new-slip length approaches to interface debonded length, *i.e.*, $z(\sigma_{tr\_pr})/l_d$ = 1, as shown in Figure 4b.

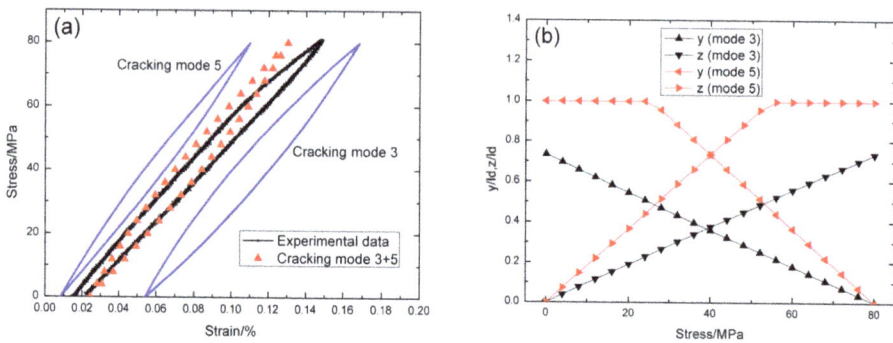

Figure 4. (a) The theoretical and experimental hysteresis loops; and (b) the interface slip lengths, *i.e.*, y/l_d and z/l_d, of matrix cracking mode 3 and mode 5 of cross-ply C/SiC composite when σ_{max} = 80 MPa.

For σ_{max} = 100 MPa, the experimental and theoretical hysteresis loops are shown in Figure 5a, in which the proportion of matrix cracking mode 3 is η = 0.4. For matrix cracking mode 3, the hysteresis loops correspond to interface slip case 4, as shown in Figure 5b. Upon completely unloading, the interface counter-slip length approaches to matrix crack spacing at σ_{tr_fu} = 70 MPa, *i.e.*, $2y(\sigma_{tr\_fu})/l_c$ = 1, as shown in Figure 5b; upon reloading to σ_{tr_fr} = 30 MPa, the interface new-slip length approaches to matrix crack spacing, *i.e.*, $2z(\sigma_{tr\_fr})/l_c$ = 1, as shown in Figure 5b. For matrix cracking mode 5, the hysteresis loops correspond to interface slip case 1, as shown in Figure 5b. Upon unloading, the interface counter-slip length approaches to interface debonded length at σ_{tr_pu} = 95 MPa, *i.e.*, $y(\sigma_{tr\_pu})/l_d$ = 1, as shown in Figure 5b; upon reloading to σ_{tr_pr} = 5 MPa, the interface new-slip length approaches to interface debonded length, *i.e.*, $z(\sigma_{tr\_pr})/l_d$ = 1, as shown in Figure 5b.

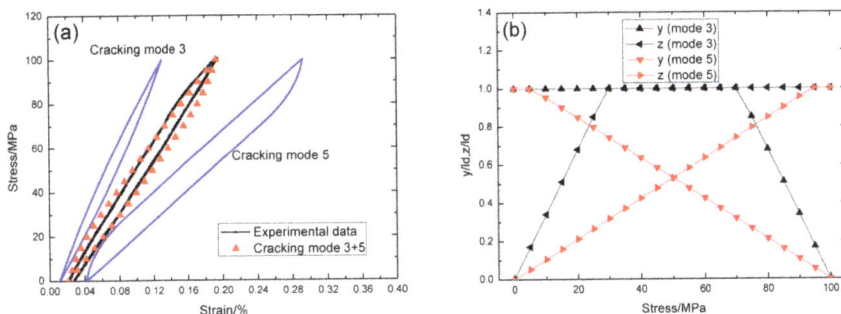

Figure 5. (a) The theoretical and experimental hysteresis loops; and (b) the interface slip lengths, i.e., y/l_d and z/l_d, of matrix cracking mode 3 and mode 5 of cross-ply C/SiC composite when $\sigma_{max} = 100$ MPa.

For $\sigma_{max} = 120$ MPa, the experimental and theoretical hysteresis loops are shown in Figure 6a, in which the proportion of matrix cracking mode 3 is $\eta = 0.42$. For matrix cracking mode 3, the hysteresis loops correspond to interface slip case 4, as shown in Figure 6b. Upon unloading, the interface counter-slip length approaches to matrix crack spacing at $\sigma_{tr\_fu} = 90$ MPa, i.e., $2y(\sigma_{tr\_fu})/l_c = 1$, as shown in Figure 6b; upon reloading to $\sigma_{tr\_fr} = 30$ MPa, the interface new-slip length approaches to matrix crack spacing, i.e., $2z(\sigma_{tr\_fr})/l_c = 1$, as shown in Figure 6b. For matrix cracking mode 5, the hysteresis loops correspond to interface slip case 2, as shown in Figure 6b. Upon completely unloading, the interface counter-slip length approaches to 89.3% of interface debonded length, i.e., $y(\sigma_{min})/l_d = 89.3\%$, as shown in Figure 6b; upon reloading to $\sigma_{max} = 120$ MPa, the interface new-slip length approaches to 89.3% of interface debonded length, i.e., $z(\sigma_{max})/l_d = 89.3\%$, as shown in Figure 6b.

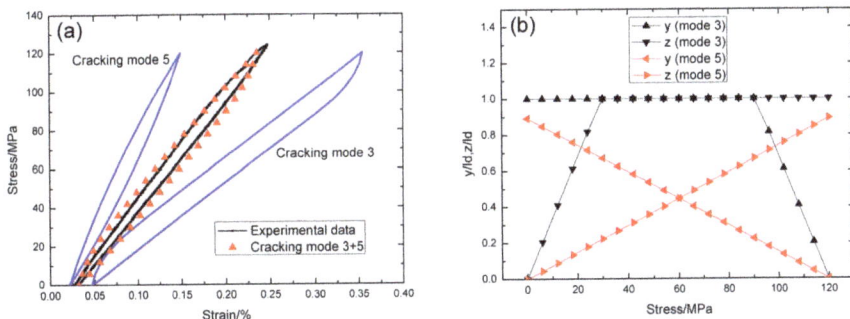

Figure 6. (a) The theoretical and experimental hysteresis loops; and (b) the interface slip lengths, i.e., y/l_d and z/l_d, of matrix cracking mode 3 and mode 5 of cross-ply C/SiC composite when $\sigma_{max} = 120$ MPa.

4.2. Cross-Ply SiC/SiC Composite

Gordon [20] investigated the cyclic loading/unloading hysteresis behavior of cross-ply SiC/SiC composite. The loading/unloading peak stresses are 190, 200 and 210 MPa, respectively. The basic material properties of cross-ply SiC/SiC composite are given by reference [20]: $V_f = 30\%$, $E_f = 420$ GPa, $E_m = 364$ GPa, $r_f = 7.5$ μm, $\tau_i = 15$ MPa, $\zeta_d = 1.5$ J/m^2, $\alpha_f = 4.6 \times 10^{-6}/°C$, $\alpha_m = 4.38 \times 10^{-6}/°C$, and $\Delta T = -1400$ °C.

For $\sigma_{max} = 190$ MPa, the experimental and theoretical hysteresis loops are shown in Figure 7a, in which the proportion of matrix cracking mode 3 is $\eta = 0.35$. For matrix cracking mode 3, the hysteresis loops correspond to interface slip case 1, as shown in Figure 7b. Upon unloading, the interface counter-slip length approaches to interface debonded length at $\sigma_{tr\_pu} = 9.5$ MPa, i.e., $y(\sigma_{tr\_pu})/l_d = 1$, as shown in Figure 7b; upon reloading to $\sigma_{tr\_pr} = 180.5$ MPa, the interface new-slip length approaches to interface debonded length, *i.e.*, $z(\sigma_{\text{tr\_pr}})/l_d = 1$, as shown in Figure 7b. For matrix cracking mode 5, the hysteresis loops correspond to interface slip case 1, as shown in Figure 7b. Upon unloading, the interface counter-slip length approaches to interface debonded length at $\sigma_{\text{tr\_pu}} = 161.5$ MPa, *i.e.*, $y(\sigma_{\text{tr\_pu}})/l_d = 1$, as shown in Figure 7b; upon reloading to $\sigma_{\text{tr\_pr}} = 28.5$ MPa, the interface new-slip length approaches to interface debonded length, *i.e.*, $z(\sigma_{\text{tr\_pr}})/l_d = 1$, as shown in Figure 7b.

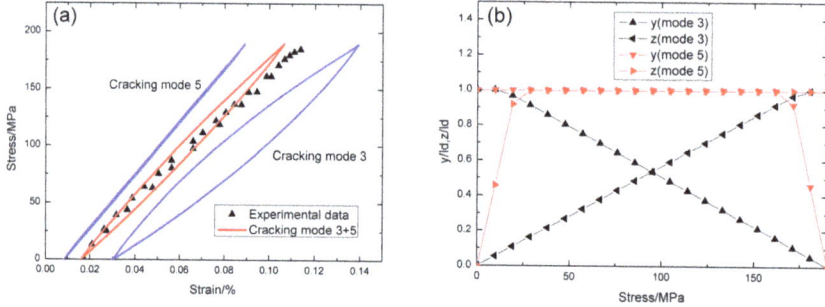

Figure 7. (a) The theoretical and experimental hysteresis loops; and (b) the interface slip lengths, *i.e.*, y/l_d and z/l_d, of matrix cracking mode 3 and mode 5 of cross-ply SiC/SiC composite when $\sigma_{\text{max}} = 190$ MPa.

For $\sigma_{\text{max}} = 200$ MPa, the experimental and theoretical hysteresis loops are shown in Figure 8a, in which the proportion of matrix cracking mode 3 is $\eta = 0.45$. For matrix cracking mode 3, the hysteresis loops correspond to interface slip case 4, as shown in Figure 8b. Upon unloading, the interface counter-slip length approaches to matrix crack spacing at $\sigma_{\text{tr\_fu}} = 120$ MPa, *i.e.*, $2y(\sigma_{\text{tr\_fu}})/l_c = 1$, as shown in Figure 8b; upon reloading to $\sigma_{\text{tr\_fr}} = 80$ MPa, the interface new-slip length approaches to matrix crack spacing, *i.e.*, $2z(\sigma_{\text{tr\_fr}})/l_c = 1$, as shown in Figure 8b. For matrix cracking mode 5, the hysteresis loops correspond to interface slip case 1, as shown in Figure 8b. Upon unloading, the interface counter-slip length approaches to interface debonded length at $\sigma_{\text{tr\_pu}} = 150$ MPa, *i.e.*, $y(\sigma_{\text{tr\_pu}})/l_d = 1$, as shown in Figure 8b; upon reloading to $\sigma_{\text{tr\_pr}} = 50$ MPa, the interface new-slip length approaches to interface debonded length, *i.e.*, $z(\sigma_{\text{tr\_pr}})/l_d = 1$, as shown in Figure 8b.

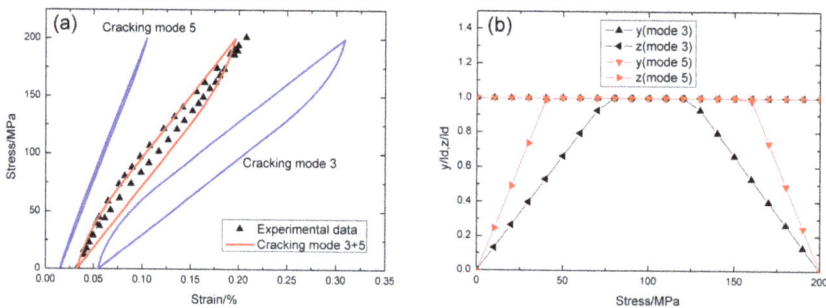

Figure 8. (a) The theoretical and experimental hysteresis loops; and (b) the interface slip lengths, *i.e.*, y/l_d and z/l_d, of matrix cracking mode 3 and mode 5 of cross-ply C/SiC composite when $\sigma_{\text{max}} = 200$ MPa.

For $\sigma_{\text{max}} = 210$ MPa, the experimental and theoretical hysteresis loops are shown in Figure 9a, in which the proportion of matrix cracking mode 3 is $\eta = 0.7$. For matrix cracking mode 3, the hysteresis loops correspond to interface slip case 4, as shown in Figure 9b. Upon unloading, the interface counter-slip length approaches to matrix crack spacing at $\sigma_{\text{tr\_fu}} = 147$ MPa, *i.e.*, $2y(\sigma_{\text{tr\_fu}})/l_c = 1$, as shown in Figure 9b; upon reloading to $\sigma_{\text{tr\_fr}} = 63$ MPa, the interface new-slip length approaches to matrix crack spacing, *i.e.*, $2z(\sigma_{\text{tr\_fr}})/l_c = 1$, as shown in Figure 9b. For matrix cracking mode 5, the hysteresis loops correspond to interface slip case 1, as shown in Figure 9b. Upon unloading, the interface counter-slip length approaches to interface debonded length at $\sigma_{tr\_pu} = 147$ MPa, *i.e.*, $y(\sigma_{tr\_pu})/l_d = 1$, as shown in Figure 9b; upon reloading to $\sigma_{tr\_pr} = 63$ MPa, the interface new-slip length approaches to interface debonded length, *i.e.*, $z(\sigma_{tr\_pr})/l_d = 1$, as shown in Figure 9b.

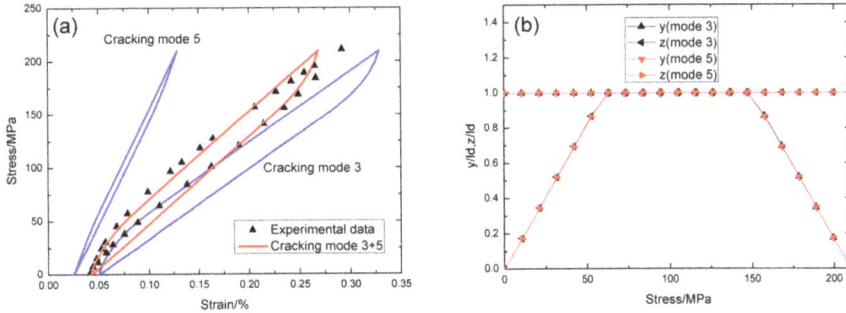

Figure 9. (a) The theoretical and experimental hysteresis loops; and (b) the interface slip lengths, *i.e.*, y/l_d and z/l_d, of matrix cracking mode 3 and mode 5 of cross-ply SiC/SiC composite when $\sigma_{max} = 210$ MPa.

5. Comparison between C/SiC and SiC/SiC Composite

The damage parameter η *vs.* normalized stress $\sigma_{max}/\sigma_{uts}$ curves of cross-ply C/SiC and SiC/SiC composites are illustrated in Figure 10. With increasing peak stress, the damage parameter η increases. At low peak stress, the damage parameter η of cross-ply C/SiC is higher than that of cross-ply SiC/SiC composite, *i.e.*, $\eta = 0.3$ at $\sigma_{max} = 60$ MPa or 48.2% σ_{uts} of C/SiC composite, and $\eta = 0.35$ at $\sigma_{max} = 190$ MPa or 82.6% σ_{uts} of SiC/SiC composite. However, at high peak stress, the damage parameter η of cross-ply SiC/SiC is higher than that of cross-ply C/SiC composite, *i.e.*, $\eta = 0.7$ at $\sigma_{max} = 210$ MPa or 91.3% σ_{uts} of SiC/SiC composite, and $\eta = 0.42$ at $\sigma_{max} = 120$ MPa or 96.3% σ_{uts} of C/SiC composite. The matrix crack density of cracking mode 3 in the 0° plies *vs.* normalized stress σ/σ_{max} curves of cross-ply C/SiC and SiC/SiC composites can be used to show the damage extent inside of composites, as shown in Figure 11. It can be found that the matrix crack density of cracking mode 3 in the C/SiC composite is higher than that of SiC/SiC composite under lower peak stress, *i.e.*, $\sigma/\sigma_{max} < 0.8$, and lower for C/SiC composite than that of SiC/SiC composite under higher peak stress, *i.e.*, $\sigma/\sigma_{max} > 0.8$.

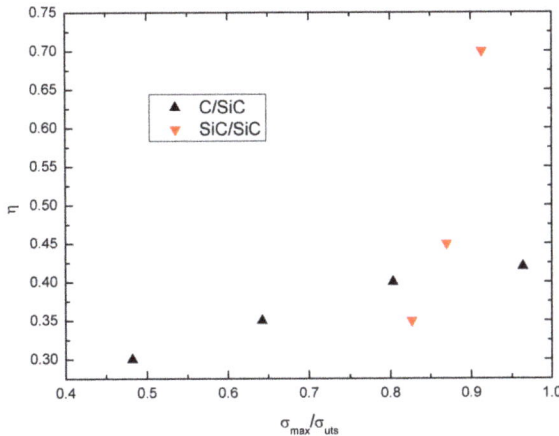

Figure 10. The damage parameter η *vs.* normalized stress $\sigma_{max}/\sigma_{uts}$ curves of cross-ply C/SiC and SiC/SiC composites.

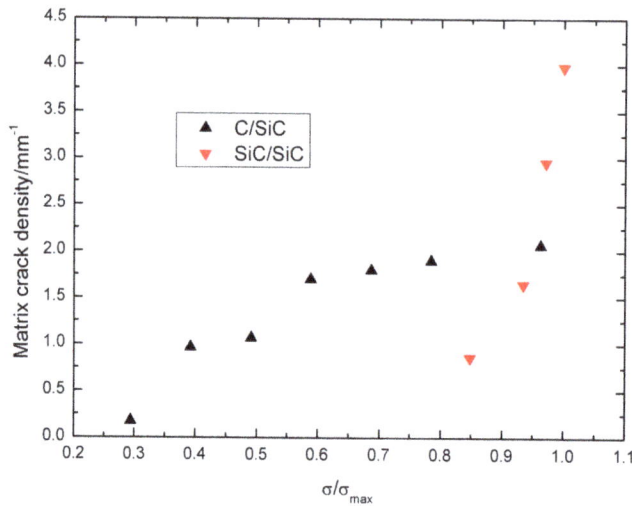

Figure 11. The matrix crack density of matrix cracking mode 3 in the $0°$ plies *vs.* normalized stress σ/σ_{max} of cross-ply C/SiC and SiC/SiC composites.

As the axial thermal residual tensile stress existed in SiC matrix due to the large mismatch of the axial thermal expansion coefficient between carbon fibers and silicon carbide matrix, *i.e.*, $-0.38 \times 10^{-6}/°C$ *vs.* $2.8 \times 10^{-6}/°C$, and the radial thermal residual tensile stress existed in the fiber/matrix interface due to the large mismatch of the radial thermal expansion coefficient between carbon fibers and silicon carbide matrix, *i.e.*, $7 \times 10^{-6}/°C$ *vs.* $2.8 \times 10^{-6}/°C$, there are unavoidable microcracks existing within the SiC matrix in the $90°$ and $0°$ plies when the composite was cooled down from high fabricated temperature to ambient temperature. These processing-induced microcracks propagated and, in conjunction with new microcracks during the loading process, formed mode 5 matrix cracks in the $90°$ plies. With increasing applied stress, some matrix cracks in the $90°$ plies connected with matrix cracks in the $0°$ plies forming mode 3 matrix cracks, which propagate through the $90°$ and $0°$ plies. For cross-ply SiC/SiC composite, the axial thermal residual compressive stress existed in SiC matrix due to the large mismatch of the axial thermal expansion coefficient between silicon carbide fibers and silicon carbide matrix, *i.e.*, $5.1 \times 10^{-6}/°C$ *vs.* $3.5 \times 10^{-6}/°C$, and the radial thermal residual compressive stress existed in the fiber/matrix interface due to the large mismatch of the radial thermal expansion coefficient between silicon carbide fibers and silicon carbide matrix, *i.e.*, $2.9 \times 10^{-6}/°C$ *vs.* $3.5 \times 10^{-6}/°C$, which decreases matrix cracking evolution rate and also the damage parameter η at low peak stress. However, with increasing peak stress, the damage extent inside of cross-ply SiC/SiC composite, *i.e.*, the damage parameter η, is much higher than that of C/SiC composite as more transverse cracks and matrix cracks connecting together.

The hysteresis dissipated energy *vs.* normalized stress $\sigma_{max}/\sigma_{uts}$ curves of cross-ply C/SiC and SiC/SiC composites are illustrated in Figure 12. With increasing peak stress, the hysteresis dissipated energy of C/SiC and SiC/SiC composites increase, *i.e.*, from 3.7 kPa at σ_{max} = 60 MPa or 48.2% σ_{uts}, to 15.2 kPa at σ_{max} = 120 MPa or 96.3% σ_{uts}; and from 5.2 kPa at σ_{max} = 190 MPa or 82.6% σ_{uts}, to 46.6 kPa at σ_{max} = 210 MPa or 91.3% σ_{uts}. The hysteresis dissipated energy of C/SiC composite under low peak stress is higher than that of SiC/SiC composite due to a higher damage parameter η at low peak stress of C/SiC composite compared with that of SiC/SiC composite. However, at high peak stress, the damage parameter η of SiC/SiC composite is higher than that of C/SiC composite, leading to higher hysteresis dissipated energy compared with that of C/SiC composite.

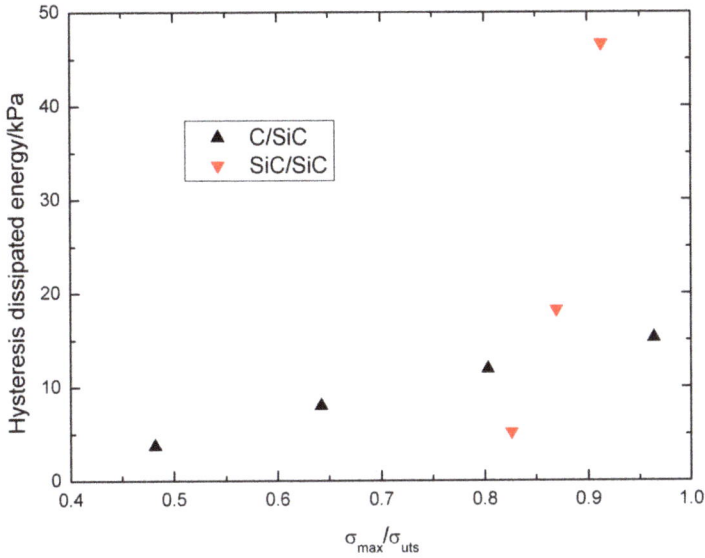

Figure 12. The hysteresis dissipated energy *vs.* normalized stress $\sigma_{max}/\sigma_{uts}$ curves of cross-ply C/SiC and SiC/SiC composites.

For C/SiC composite, the hysteresis loops of matrix cracking mode 3 and mode 5 correspond to interface slip case 2 and case 1, respectively, under low peak stresses of 60 and 80 MPa; when the peak stress is 100 MPa, the hysteresis loops of matrix cracking mode 3 transfers from case 2 to case 4; and when the peak stresses are 120 MPa, the hysteresis loops of matrix cracking mode 5 transfers from case 1 to case 2, as shown in Table 1. For SiC/SiC composite, the hysteresis loops of matrix cracking mode 3 and mode 5 both correspond to interface slip case 1 under low peak stresses of 190 MPa; and when the peak stress increases to 200 MPa, the hysteresis loops of matrix cracking mode 3 transfers from case 1 to case 4, and the hysteresis loops of matrix cracking mode 5 remains to be interface slip case 1, as shown in Table 2.

Table 1. The interface slip type of matrix cracking mode 3 and mode 5 corresponding to different peak stresses of cross-ply C/SiC composite.

Cracking Modes	60 MPa	80 MPa	100 MPa	120 MPa
Matrix cracking mode 3	case 2	case 2	case 4	case 4
Matrix cracking mode 5	case 1	case 1	case 1	case 2

Table 2. The interface slip type of matrix cracking mode 3 and mode 5 corresponding to different peak stresses of cross-ply SiC/SiC composite.

Cracking Modes	190 MPa	200 MPa	210 MPa
Matrix cracking mode 3	case 1	case 4	case 4
Matrix cracking mode 5	case 1	case 1	case 1

6. Conclusions

The comparison of cyclic hysteresis behavior between cross-ply C/SiC and SiC/SiC CMCs has been investigated. The interface slip between fibers and the matrix existed in matrix cracking mode 3 and mode 5 are considered as the major reason for hysteresis loops of cross-ply CMCs. The hysteresis loops of cross-ply C/SiC and SiC/SiC composites corresponding to different peak stresses have been predicted using present analysis. The differences between C/SiC and SiC/SiC composite on damage parameters and hysteresis dissipated energy have been investigated.

(1) The damage parameter, *i.e.*, the proportion of matrix cracking mode 3 in the entire matrix cracking modes of the composite, and the hysteresis dissipated energy both increase with increasing peak stress;

(2) The damage parameter and hysteresis dissipated energy of C/SiC composite under low peak stress are higher than those of SiC/SiC composite; However, with increasing peak stress, the damage extent inside of cross-ply SiC/SiC composite, *i.e.*, the damage paramter η and hysteresis dissipated energy, is much higher than that of C/SiC composite as more transverse cracks and matrix cracks connecting together.

Acknowledgments: The author thanks the Science and Technology Department of Jiangsu Province for the funding that made this research study possible. This study has received the support from the Science and Technology Department of Jiangsu Province through the Natural Science Foundation of Jiangsu Province (Grant No. BK20140813), and the Fundamental Research Funds for the Central Universities (Grant No. NS2016070). The author also wishes to thank three anonymous reviewers and editors for their helpful comments on an earlier version of the paper.

Conflicts of Interest: The author declares that he has no conflict of interest.

References

1. Naslain, R. Design, preparation and properties of non-oxide CMCs for application in engines and nuclear reactors: An overview. *Compos. Sci. Technol.* **2004**, *64*, 155–170. [CrossRef]

2. Bednarcyk, B.A.; Mital, S.K.; Pineda, E.J.; Arnold, S.M. Multiscale modeling of ceramic matrix composites. In Proceedings of the 56th AIAA/ASCE/AHS/ASC Structures Dynamics Materials Conference, Kissimmee, FL, USA, 5–9 January 2015.

3. Gowayed, Y.; Ojard, G.; Santhosh, U.; Jefferso, G. Modeling of crack density in ceramic matrix composites. *J. Compos. Mater.* **2015**, *49*, 2285–2294. [CrossRef]

4. Reynaud, P. Cyclic fatigue of ceramic-matrix composites at ambient and elevated temperatures. *Compos. Sci. Technol.* **1996**, *56*, 809–814. [CrossRef]

5. Fantozzi, G.; Reynaud, P. Mechanical hysteresis in ceramic matrix composites. *Mater. Sci. Eng. Part A Struct.* **2009**, *521–522*, 18–23. [CrossRef]

6. Kotil, T.; Holmes, J.W.; Comninou, M. Origin of hysteresis observed during fatigue of ceramic matrix composites. *J. Am. Ceram. Soc.* **1990**, *73*, 1879–1883. [CrossRef]

7. Pryce, A.W.; Smith, P.A. Matrix cracking in unidirectional ceramic matrix composites under quasi-static and cyclic loading. *Acta Metall. Mater.* **1993**, *41*, 1269–1281. [CrossRef]

8. Ahn, B.K.; Curtin, W.A. Strain and hysteresis by stochastic matrix cracking in ceramic matrix composites. *J. Mech. Phys. Solids* **1997**, *45*, 177–209. [CrossRef]

9. Solti, J.P.; Mall, S.; Robertson, D.D. Modeling damage in unidirectional ceramic-matrix composites. *Compos. Sci. Technol.* **1995**, *54*, 55–66. [CrossRef]

10. Vagaggini, E.; Domergue, J.M.; Evans, A.G. Relationships between hysteresis measurements and the constituent properties of ceramic matrix composites: I, theory. *J. Am. Ceram. Soc.* **1995**, *78*, 2709–2720. [CrossRef]

11. Hutchison, J.W.; Jensen, H.M. Models of fiber debonding and pullout in brittle composites with friction. *Mech. Mater.* **1990**, *9*, 139–163. [CrossRef]

12. Cho, C.D.; Holmes, J.W.; Barber, J.R. Estimation of interfacial shear in ceramic composites from frictional heating measurements. *J. Am. Ceram. Soc.* **1991**, *74*, 2802–2808. [CrossRef]

13. Li, L.B.; Song, Y.D.; Sun, Z.G. Influence of interface de-bonding on the fatigue hysteresis loops of ceramic matrix composites. *Chin. J. Solid. Mech.* **2009**, *30*, 8–14.

14. Li, L.B.; Song, Y.D.; Sun, Z.G. Effect of fiber Poisson contraction on fatigue hysteresis loops of ceramic matrix composites. *J. Nanjing Univ. Aero. Astron.* **2009**, *41*, 181–186.

15. Li, L.B.; Song, Y.D. Influnece of fiber failure on fatigue hysteresis loops of ceramic matrix composites. *J. Reinf. Plast. Compos.* **2011**, *30*, 12–25.

16. Li, L.B. Modeling the effect of interface wear on fatigue hysteresis behavior of carbon fiber-reinforced ceramic-matrix composites. *Appl. Compos. Mater.* **2015**. [CrossRef]

17. Li, L.B.; Song, Y.D.; Sun, Y.C. Estimate interface shear stress of unidirectional C/SiC ceramic matrix composites from hysteresis loops. *Appl. Compos. Mater.* **2013**, *20*, 693–707. [CrossRef]

18. Kuo, W.S.; Chou, T.W. Multiple cracking of unidirectional and cross-ply ceramic matrix composites. *J. Am. Ceram. Soc.* **1995**, *78*, 745–755. [CrossRef]

19. *Standard Practice for Constant-Amplitude, Axial, Tension-Tension Cyclic Fatigue of Continuous Fiber-Reinforced Advanced Ceramics at Ambient Temperatures*; ASTM C 1360-10; ASTM: West Conshohocken, PA, USA, 2015.

20. Gordon, N. Material Health Monitoring of SiC/SiC Laminated Ceramic Matrix Composites With Acoustic Emission And Electrical Resistance. Master Thesis, University of Akron, Akron, OH, USA, December 2014.

21. Li, L.B.; Song, Y.D.; Sun, Y.C. Effect of matrix cracking on hysteresis behavior of cross-ply ceramic matrix composites. *J. Compos. Mater.* **2014**, *48*, 1505–1530. [CrossRef]

17

Influence of Oxygen Concentration on the Performance of Ultra-Thin RF Magnetron Sputter Deposited Indium Tin Oxide Films as a Top Electrode for Photovoltaic Devices

Jephias Gwamuri [1], Murugesan Marikkannan [2,†], Jeyanthinath Mayandi [2,†], Patrick K. Bowen [1,†] and Joshua M. Pearce [1,3,*]

Academic Editor: Lioz Etgar

[1] Department of Materials Science & Engineering, Michigan Technological University, 1400 Townsend, Houghton, MI 49931, USA; jgwamuri@mtu.edu (J.G.); pkbowen@mtu.edu (P.K.B.)

[2] Department of Materials Science, School of Chemistry, Madurai Kamaraj University, Ta mil Nadu, Madurai 625 019, India; marikannan.mku@gmail.com (M.M.); jeyanthinath.mayandi@gmail.com (J.M.)

[3] Department of Electrical & Computer Engineering, Michigan Technological University, 1400 Townsend, Houghton, MI 49931, USA

[*] Correspondance: pearce@mtu.edu

[†] These authors contributed equally to this work.

Abstract: The opportunity for substantial efficiency enhancements of thin film hydrogenated amorphous silicon (a-Si:H) solar photovoltaic (PV) cells using plasmonic absorbers requires ultra-thin transparent conducting oxide top electrodes with low resistivity and high transmittances in the visible range of the electromagnetic spectrum. Fabricating ultra-thin indium tin oxide (ITO) films (sub-50 nm) using conventional methods has presented a number of challenges; however, a novel method involving chemical shaving of thicker (greater than 80 nm) RF sputter deposited high-quality ITO films has been demonstrated. This study investigates the effect of oxygen concentration on the etch rates of RF sputter deposited ITO films to provide a detailed understanding of the interaction of all critical experimental parameters to help create even thinner layers to allow for more finely tune plasmonic resonances. ITO films were deposited on silicon substrates with a 98-nm, thermally grown oxide using RF magnetron sputtering with oxygen concentrations of 0, 0.4 and 1.0 sccm and annealed at 300 °C air ambient. Then the films were etched using a combination of water and hydrochloric and nitric acids for 1, 3, 5 and 8 min at room temperature. In-between each etching process cycle, the films were characterized by X-ray diffraction, atomic force microscopy, Raman Spectroscopy, 4-point probe (electrical conductivity), and variable angle spectroscopic ellipsometry. All the films were polycrystalline in nature and highly oriented along the (222) reflection. Ultra-thin ITO films with record low resistivity values (as low as 5.83×10^{-4} $\Omega \cdot$cm) were obtained and high optical transparency is exhibited in the 300–1000 nm wavelength region for all the ITO films. The etch rate, preferred crystal lattice growth plane, d-spacing and lattice distortion were also observed to be highly dependent on the nature of growth environment for RF sputter deposited ITO films. The structural, electrical, and optical properties of the ITO films are discussed with respect to the oxygen ambient nature and etching time in detail to provide guidance for plasmonic enhanced a-Si:H solar PV cell fabrication.

Keywords: transparent conducting oxide; indium tin oxide; plasmonics; wet etching; photovoltaics; optics

1. Introduction

Solar photovoltaic (PV) based electricity production is one of the significant ecofriendly methods to generate sustainable energy needed to mitigate the looming global energy crisis [1]. Despite technical improvements [2] and scaling [3], which have resulted in a significant reduction in crystalline silicon (c-Si) PV module costs, for continued PV industry growth [4,5], PV costs must continue to decline to reach a levelized cost of electricity [6] low enough to dominate the electricity market. One approach to reduced PV costs further is to transition to thin film PV technology [7]. Hydrogenated amorphous silicon (a-Si:H) based PV [8] have shown great potential for large scale [9] sustainable commercial production due to lower material costs and use of well-established fabrication techniques [10,11]. However, there is need to improve the efficiency of a-Si:H PV devices if they are to become the next dominant technology for solar cells commercialization. One method to improve a-Si:H PV performance is with optical enhancement [12]. Recent developments in plasmonic theory promise new light management methods for thin-film a-Si:H based solar cells [13–23]. However, previous work has shown these plasmonic approaches require the development of ultra-thin, low-loss and low-resistivity transparent conducting oxides (TCOs) [24]. Tin doped indium oxide (ITO), zinc oxide (ZnO) and tin oxide (SnO_2) are the three most important TCOs and are already widely used in the commercial thin film solar cells [25]. In addition, aluminum-dope zinc oxide (AZO) and fluorine-doped tin oxide (FTO) are among the other most dominant TCOs in various technological fields particularly the optoelectronic devices industry where TCOs have proved indispensable for applications such as photo electrochemical devices, light emitting diodes, liquid crystal displays and gas sensors [26,27]. ITOs can be prepared by direct current (DC) and radio frequency (RF) magnetron sputtering, electron beam evaporation, thermal vapor evaporation, spray pyrolysis, chemical solution deposition, and sol gel methods [28–34]. RF magnetron sputtering can be used to control the electrical and optical properties of the ITO thin films and is heavily used in industry [35].

Recent work by Vora et al. has emphasized the need for ultra-thin ITO top electrodes with low resistivity and high transmittances in the visible range of the electromagnetic spectrum as a prerequisite for the commercial realization of plasmonic-enhanced a-Si:H solar cells [36]. However, research by Gwamuri et al. has demonstrated that fabricating ultra-thin ITO films (sub-50 nm) using conversional methods presented a number of challenges since there is a trade-off between electrical and optical properties of the films [37]. It was evidenced from their results that electrical properties of RF sputter deposited sub-50 nm ITO films degraded drastically as their thickness is reduced, while the optical properties of the same films were seen to improve greatly [37]. To solve this problem, a novel method involving chemical shaving of thicker (greater than 80 nm) RF sputter deposited films was proposed and demonstrated [38]. Building on the promise of that technique, this study seeks to further understand the effect of oxygen concentration on the etch rates of RF sputter deposited ITO films and the impact on the TCO quality as a top electrode for PV devices. A detailed understanding of the interaction of all critical parameters, which determines the quality of ultra-thin ITO will help create even thinner layers with good quality to allow more finely tuned plasmonics resonances. ITO films were deposited using four different oxygen concentrations (0 sccm, 0.4 sccm, 1.0 sccm), annealed in air at 300 °C for 30 min and then etched for four different times (1, 3, 5 and 8 min) to establish the effect of oxygen on etch rates. These materials were characterized by X-ray diffraction (XRD), atomic force microscopy (AFM), Raman Spectroscopy, 4-point probe (4PP), and variable angle spectroscopic Ellipsometry (VASE). In addition, the thin films were investigated for candidates as acid-resistant TCOs for encapsulation of PV devices, which may reduce device processing steps and fabrication costs of completed modules in the future. The results are presented and discussed.

2. Materials and Methods

2.1. ITO Fabrication Process

ITO films were grown on (100) prime silicon substrates with a 98 nm thermally grown oxide, and on glass substrates using a 99.99% 100 mm diameter pressed ITO (SnO_2:In_2O_3 10:90 wt%) target.

Before the deposition the substrates were ultrasonically cleaned in isopropanol and in DI water for 15 min and dried using N_2 atmosphere. The sputtering chamber was initiated to a low 10^{-7} Torr base pressure and the pressure was maintained at 7.5×10^{-3} Torr. The distance between the target and substrates was kept constant at 75 mm. As a standard procedure, the target was pre-sputter cleaned at a power of 150 W, whereas the sputter deposition of the films was performed at 100 W. The argon gas flow rate was fixed at 10 sccm and the oxygen gas flow was varied such as 0, 0.4 and 1.0 sccm with sputter rate of 8–12 nm per minute. The sputter rate was seen to decrease with increase in oxygen flow rate. After deposition, ITO films were annealed at 300 °C for 30 min in air. ITO/Si films were subjected to the etching process using a standard chemical etchant mixture of HCl: HNO_3:H_2O (1:1:5) volume ratio. All the etching was performed at room temperature, resulting in a slow and controlled etch rate for the Si/SiO_2 films. Finally, the etched samples were thoroughly rinsed in DI water and dried under the nitrogen environment. This methodology was adapted from the previous study by Gwamuri *et al.*, 2015 [37].

The ITO films processed under different argon-oxygen ambient were chemically etched and characterized using various tools. The structural analyses of the ITO films were carried out using X-ray diffraction (XRD-Scintag-2000 PTS, Scintag Inc., Cupertino, CA, USA). Raman spectra for the ultra-thin film samples were measured at room temperature using Jobin-Yvon LabRAM HR800 Raman Spectrometer (Horiba Scientific, Edison, NJ, USA) with the excitation wavelength of 633 nm and the resolution is about ~0.1 cm^{-1}. Sheet resistance of the samples was characterized using four point probe station consisting of ITO optimized tips with 500 micron tip radii set to 60 grams pressure and an RM3000 test unit from Jandel Engineering Limited, Kings Langley, UK. The optical transmission and thickness of the films was determined using variable angle spectroscopic ellipsometry (UV-VIS V-VASE with control module VB-400, J.A. Woollam Co., Lincoln, NE, USA). Surface roughness was evaluated using a Veeco Dimension 3000 atomic force microscope (Veeco, Oyster City, NY, USA) operated in tapping mode with Budget Sensors Tap300Al-G cantilevers (Innovative Solutions Bulgaria Ltd., Sofia, Bulgaria). It should be noted that transmittance data was measured for ITO on sodalime glass (SLG) substrate and all the rest of the data was on ITO on Si/SiO_2 substrate.

2.2. Chemical Shaving: Wet Etching

In this present work, the oxygen 0, 0.4 and 1.0 sccm deposited ITO films were used for the etching process for 1, 3, 5 and 8 min, respectively. The annealed ITO/Si samples are etched at room temperature using HCl:HNO_3:H_2O (1:1:5) combination and the resistivity and thickness of the films were checked for 1, 3, 5 and 8 min etched films. For the 0 sccm ITO films, the thickness of the film was changed from 70 to 44 nm for 1 to 5 min etching time. Similarly the 0.4 sccm films thickness changed from 89 to 47 nm and 84 to 22 nm for 1.0 sccm films. The decrement of thickness was reflected in the resistivity values. The chemical reaction of the HCl and HNO_3 etching reactions are as follows [39]:

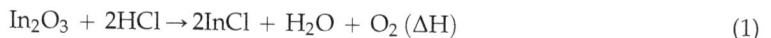

$$In_2O_3 + 2HCl \rightarrow 2InCl + H_2O + O_2 \, (\Delta H) \tag{1}$$

$$In_2O_3 + 12HNO_3 \rightarrow 2In\,(NO_3)_3 + 6NO_2 + 6H_2O \tag{2}$$

3. Results

3.1. Structural Analysis

3.1.1. XRD Analysis

XRD results for the ITO films deposited using different oxygen concentrations (0 sccm, 0.4 sccm, 1.0 sccm), annealed in air at 300 °C for 30 min and then etched for different times (1, 3, 5 and 8 min) are shown in Figure 1.

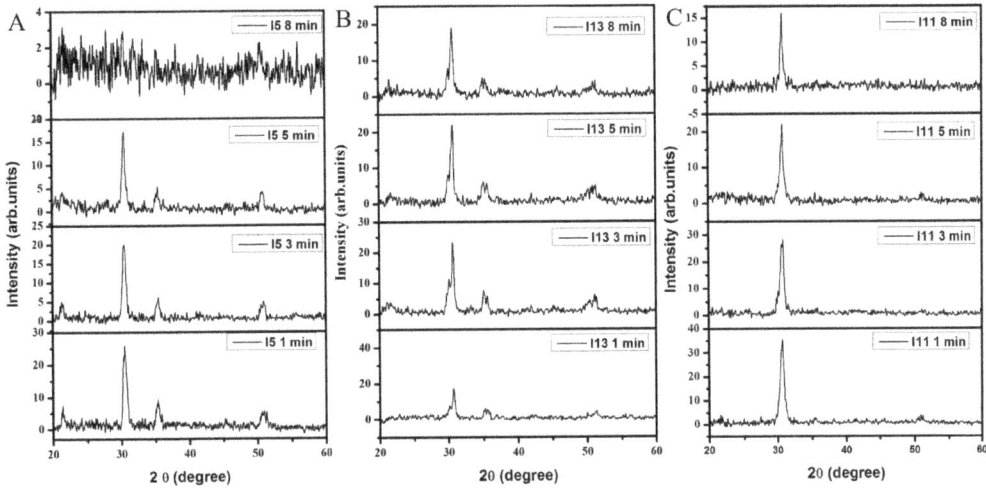

Figure 1. XRD pattern for ITO films deposited under different oxygen ambient conditions and etched for 1, 3, 5 and 8 min: (**A**) 0 sccm oxygen; (**B**) 0.4 sccm oxygen; (**C**) 1.0 sccm oxygen. Argon flow rate was maintained at 10 sccm for all materials.

In addition to that the peak shown at (222), (400) and (440) reflections are indexed to be cubic indium oxide (JCPDS No: 06-0416) [40]. All the films have a polycrystalline nature with stronger (222) reflection. No other tin phases could be identified from the cubic indium tin oxide. Normally the 30% of Sn is needed to exhibiting the SnO_2 diffraction lines in ITO. The (222) and (400) plane is ascribed for oxygen efficient and deficient nature of ITO films [41]. The effect of the oxygen flow rate on the peak intensity of the ITO films is clearly shown in the XRD spectrum. There is a general increase in the peak intensities with increased oxygen flow rate. Similarly the reflections such as (211), (400) and (440) are due to the minimum oxygen concentration in the sputter chamber. These planes are absent in the XRD pattern of ITO film processed in an oxygen-rich (1.0 sccm) environment. There is a strong evidence that for the highest oxygen ambient (1.0 sccm), (222) is the preferred growth orientation for RF sputter deposited ITO films. Varying the oxygen concentration will result in changing the preferred growth orientation of the films to other crystal lattice planes such as the (211), (400) or (440). The intensity ratios are strongly dependent on the critical level of In^{3+} and O^{2-} pairs and the pairs' density is different for different etching periods of time [42]. The presence of high oxygen concentration induce the In-O bonding networks formation and promote growth of the (222) crystal lattice planes.

During the etching, ITO films are reduced to In–Cl and In-$(NO_3)_3$ resulting in the change in crystallinity of films etched for different periods of time. The structural parameters such as d spacing, lattice constants, net lattice distortion and grain sizes are estimated and listed in Table 1 in comparison to data from the Joint Committee on Powder Diffraction Standards (JCPDS)/International Centre for Diffraction Data (ICDD) database.

The etching process also distorts the ITO structural long-range order, which has an impact on the opto-electronic properties of the films. The grain size of films did not change even after etching for 8 min., particularly for ITO films processed in an oxygen deficient environment. During the etching process the excess weakly bound oxygen atoms are removed from the ITO surfaces exposing layers with different grain sizes. The ITO structure distortion due to etching for longer periods of time (8 min) can be seen from the XRD spectra shown in Figure 1. There was however no evidence of ITO film for the results shown in Figure 1A after they were etched for 8 min. There is evidence of decreased crystallinity for the rest of the ITO films (Figure 1B,C) as the oxygen atoms are stripped from the In–O network by the HCl and HNO_3.

Table 1. Structural parameters of ITO sputtered films with 0, 0.4 and 1.0 sccm oxygen and etched at 1, 3, 5 and 8 min.

Oxygen Flow Rate (sccm)	Etching Time (min)	D Spacing (222) (Å)	Lattice Constant (222) (Å)	Net-Lattice Distortion	Grain Size (222) (nm)
Standard JCPDS for ITO 06-0416	–	2.921	10.1180	–	–
0	1	2.932	10.1552	−0.0036	16
	3	2.934	10.1629	−0.2970	16
	5	2.934	10.1629	−0.2885	17
	8	–	–	–	–
0.4	1	2.908	10.0731	−0.0075	31
	3	2.912	10.0869	−0.0157	25
	5	2.914	10.0954	−0.0153	23
	8	2.914	10.0954	−0.0169	20
1.0	1	2.917	10.1059	–	13
	3	2.913	10.0915	−0.2828	13
	5	2.911	10.0845	–	14
	8	2.908	10.0764	–	19

3.1.2. AFM Analysis

Figure 2 shows the AFM surface topology of the ITO films deposited under three different oxygen environments and etched for 1 min and 8 min.

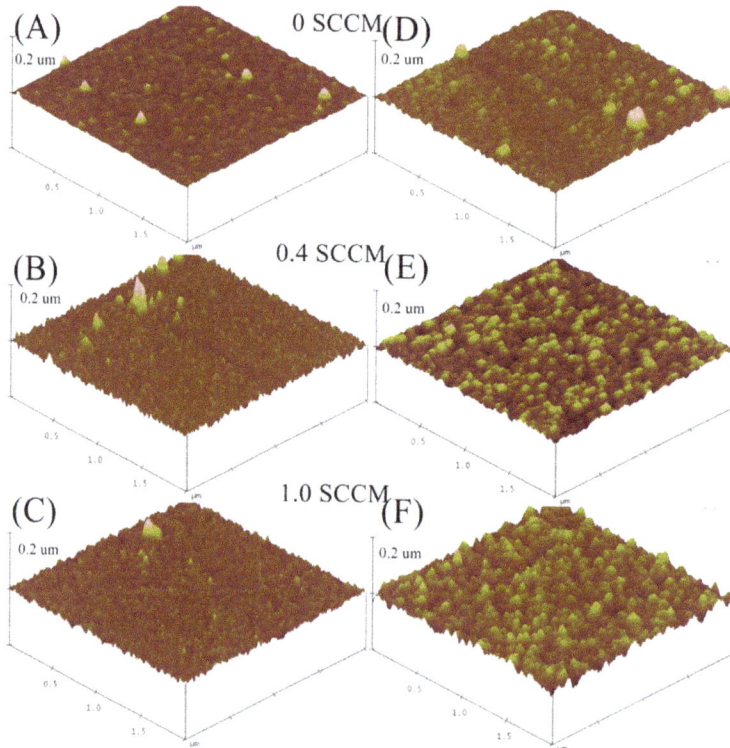

Figure 2. Surface topology image for 2 μm × 2 μm × 0.2 μm of the ITO film deposited under various oxygen environments: (**A**) 0 sccm oxygen; (**B**) 0.4 sccm oxygen; (**C**) 1.0 sccm oxygen, and etched for 8 min; (**D**) 0 sccm oxygen; (**E**) 0.4 sccm oxygen and (**F**) 1 sccm oxygen. (**A–C**) etched for 1 min and (**D–F**) films etched for 8 min. The etching was performed at room temperature.

Figure 2A,B shows the ITO films deposited in an oxygen deficient ambient and etched for 1 and 8 min, respectively. Spherical sized grains are clearly visible in all AFM images presented in Figure 2. There is variation of surface roughness of the films with both oxygen flow rate and etching time of the ITO films. The minimum value of surface roughness of 0.65 nm was measured for ITO films sputtered using 0.4 sccm oxygen flow rate and etched for 1 min, while a maximum surface roughness value of 8.9 nm was observed for films processed at 1.0 sccm oxygen flow rate and etched for 1 min. There was a slight increase in roughness with etching time observed for 0 sccm and 0.4 sccm ITO film, for etching times 1 min to 8 min. However, the 1.0 sccm films, showed the greatest variation in surface roughness even after 1 min etching process. Generally, the surface roughness of the films are observed to increase when the oxygen gas concentration is increased during processing.

3.1.3. Raman Spectroscopy

Figure 3 shows the Raman spectrum for ITO deposited at various oxygen compositions and etched at 1, 3, 5 and 8 min respectively. Raman spectroscopy is used to determine the structural conformations of the materials. Group theory predicts the Raman modes for cubic indium oxide, such as 4Ag (Raman), 4Eg (Raman), 14Tg (Raman), 5Au (inactive), and 16Tu (infrared) modes [43]. The modes observed are at 303, 621 and 675 cm^{-1} for all the films. Noticeable modes are exhibited at 302 and 621 for Eg and In–O vibrational mode [44]. The observed Raman modes in Figure 3 are in good agreement with previous reported results [40]. There are no other additional modes observable for the SnO and SnO_2 structures. In addition to that the broad band shown at 976 to 1013 cm^{-1} for all the etched films and it was not unassignable. The peak appeared at 1132, 1112, 1097 and 1120 cm^{-1} for 0, 0.4 and 1.0 sccm ITO etched films. These peaks are reported in the commercially ITO films [45]. The Raman results are correlated with XRD results. No other mixed phases were observed in the Raman spectrum indicating that etching process had no or little effect on the ITO structure.

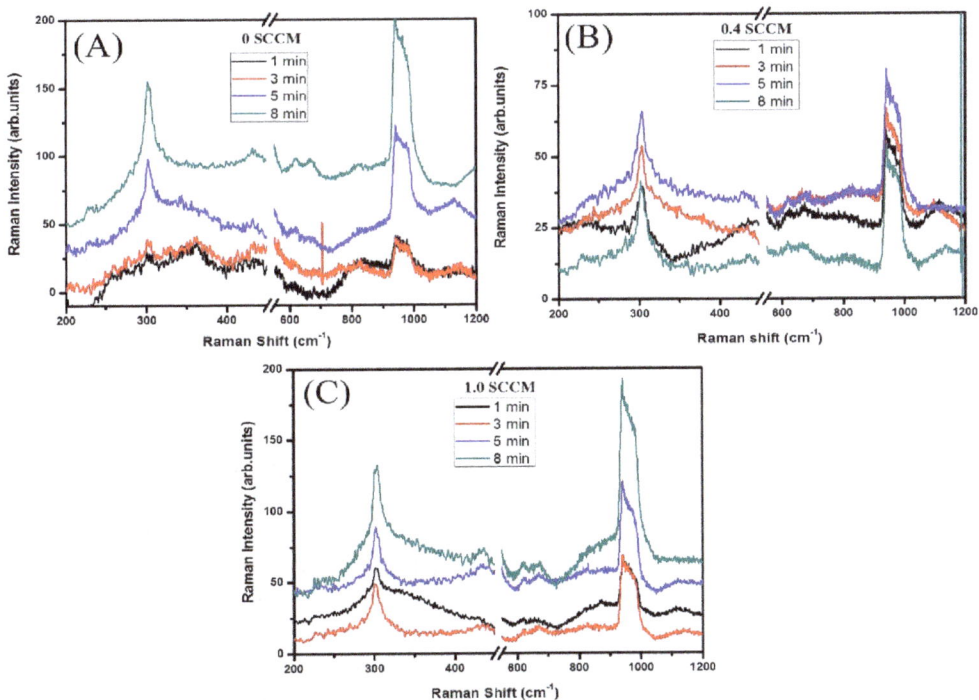

Figure 3. Raman spectra for the ITO films deposited under various oxygen concentrations and etched for 1, 3, 5 and 8 min., respectively. (**A**) 0 sccm; (**B**) 0.4 sccm; (**C**) 1.0 sccm.

3.2. Resistivity

The electrical properties of the different oxygen ambient deposited and etched ITO films were measured using a four point probe. The sheet resistance values of the ITO films are changed with respect to the oxygen ambient nature and etching time and are summarized in Table 2.

Table 2. Electrical and optical parameters of ITO films deposited under various oxygen compositions and etched for 1, 3, 5 and 8 min.

Oxygen Flow Rate (sccm)	Etching Time (min)	Sheet Resistance (Ω/square)	Thickness (nm)	Resistivity ($\Omega \cdot$cm)	Transmission (%)
0	1	83.28	70	5.83×10^{-4}	76.29
	3	103.47	59	6.11×10^{-4}	93.98
	5	209.49	44	9.22×10^{-4}	90.27
	8	–	–	–	100
0.4	1	209.02	89	1.86×10^{-3}	91.45
	3	194.23	88	1.71×10^{-3}	85.26
	5	240.08	85	2.04×10^{-3}	84.85
	8	326.9	47	1.54×10^{-3}	83.71
1.0	1	1000	84	8.4×10^{-3}	90.96
	3	2000	62	1.24×10^{-2}	89.25
	5	2400	50	1.20×10^{-2}	100
	8	7350	22	1.62×10^{-2}	100

From the obtained results, the minimum sheet resistance was observed for ITO deposited using argon ambient (0 sccm oxygen) and etched for 1 min. However, the same films exhibited the worst transmittance of about 76%. During processing in an argon rich environment, the bombardment by argon neutrals creates dangling bonds in the substrates and created the oxygen vacancies in the ITO films [46]. The argon environment (10 sccm) (*i.e.*, the oxygen deficient environment) promotes oxygen vacancies that enhance electrical resistivity while degrading the optical properties of the films. This is reflected in the XRD spectra, where the (400) and (440) lattice planes are enhanced for ITO films processed in low oxygen (0 sccm and 0.4 sccm) environments. Hence, the 1.0 sccm deposited films in which the (222) lattice plane is dominant, showed a higher electrical resistivity compared to the other films. The resistivity of the films are highly dependent on the film thickness, which is a function of the etching time. Increasing the etching time decreases the thickness of the films, and, hence, the electrical properties while improving optical properties.

3.3. Transmittance

Optical transmittances of the ITO films on glass substrates are recorded from 300 to 1000 nm at room temperature and shown in Figure 2. All the films exhibited the highest average optical transmittance in the higher wavelength range. The highest optical transmittance is attained for 0sccm oxygen ITO film etched for 8 min with and average etch rate of 5.2 nm/min (for 5 min etch) and the 1.0 sccm films etched for 5 and 8 min with average etch rates of 5.25 and 7.75 nm/min, respectively. The results are summarized in Table 2. The thickness of the film is an important parameter for determining both electrical and optical properties of the ITO films. In this work, the thickness was quantified using spectroscopic ellipsometry measurements and is shown in the Table 2. For transmittance measurements, the ITO films were deposited on SLG substrates. The SLG transmittance is measured and used as baseline data. All ITO films transmittance data involve baseline subtraction, hence 100% transmittance means that all of the ITO film has been etched off. The etch rates were much faster for the ITO on glass such that all the film was etched-off after an 8-min etch (Figure 4A), and 5 and 8 min (Figure 4C).

The results show a direct correlation between the oxygen concentration and the optical transmittances of the films and an inverse relationship with the electrical conductivity of the ITO films. These results are in agreement with observation from previous studies [47,48]. Figure 4A for 8 min etch, and Figure 4C for 5 and 8 min etch showed a transmittance cut-off wavelength around 350 nm indicating the absorption edge.

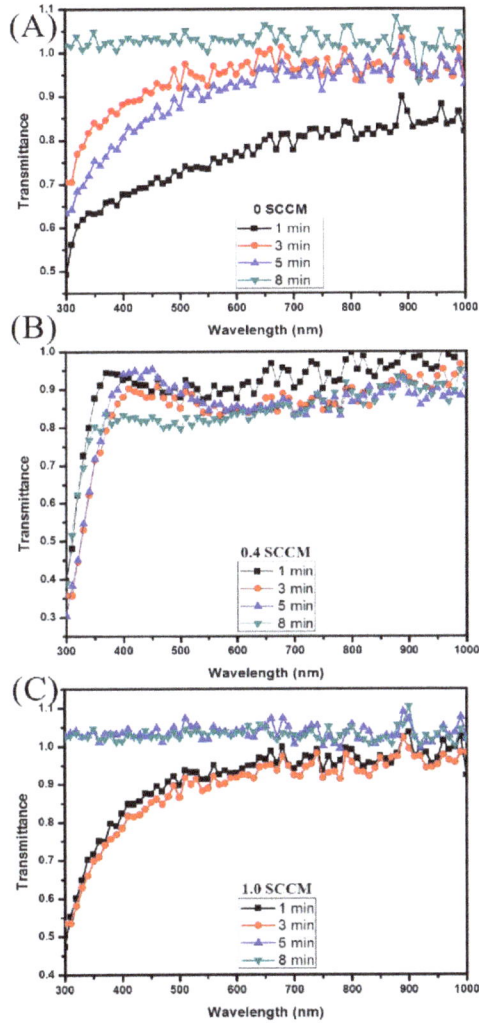

Figure 4. Optical transmission spectrum for the RF sputter deposited ITO films at different oxygen compositions and etched for min (**A**) 0 sccm; (**B**) 0.4 sccm; and (**C**) 1.0 sccm.

4. Discussion

The results presented provide further insight on the interaction of the most common fabrication variables that influence the electrical and optical properties of ITO films for PV and other opto-electronic applications. There is evidence of a strong correlation between oxygen concentration and both the resistivity and transmittances of RF sputter deposited ITO films. The processing conditions have a strong bearing on the structure of the ITO films. The (222) lattice planes are preferred in films grown in oxygen rich ambient whilst more lattice planes; (211), (222), (400), and (440) are observed in films grown under less oxygen or oxygen deficient conditions. Wan *et al.* have reported the (211), (400) and (440) planes reflections as being associated with ITO films processed under high RF power in an oxygen deficient atmosphere [35]. From the structural analysis, the exhibited (222) reflection clearly indicated the cubic indium oxide formation. The film growth rate decreased with increased O_2 concentration resulting in a much thicker ITO critical thickness (amorphous to polycrystalline transition thickness) for the 0 sccm RF sputtered films. The overall film is mixed phase crystalline and amorphous in nature. The noise is due to the ultra-thin porous and amorphous film left once the top crystalline film is etched off. This is not observed in the 0.4 and 1 sccm films because the increased oxygen composition results in reduced growth rates giving films that are more crystalline in nature with a much thinner critical thickness.

The lattice parameters and lattice distortion are seen to vary closely with oxygen concentration in the sputter chamber and the length of the etching process. The different oxygen–argon ratios sputtered ITO films have different etching behaviors, which then effected their electrical and optical properties. During the etching process, the crystal lattice of ITO films is degraded due to the exchange of bonds between indium oxide with HCl and HNO_3. This means that the indium oxide In–O and H–Cl bonds are substituted by In–Cl, In-$(NO_3)_3$ and O–H in the ITO surfaces [49]. These kinds of reactions may reduce the oxygen concentrations and distort structural long range order of the ITO films. This was reflected in the variation of electrical and optical properties of the ITO films with etching time. As no evidence of tin phases were detected, it can be concluded the reactions involving tin phases have negligible effect on the overall etch rates described in this study.

It is interesting to note that the ITO films processed at the 0.4 sccm oxygen flow rate presented the greatest resistance to acid etching in addition to exhibiting above moderate electrical and optical properties. These films show potential as candidate materials for encapsulation of PV devices or transparent conducting electrodes for varied application in acid-rich environment. However, further research into optimization of anti-acid (acid resistant) ITO films will be required before the material can be implemented in commercial PV devices.

Usually ITO is sputtered in varied combinations of reactive gas environments of argon with oxygen, hydrogen and nitrogen [40]. The oxygen ambient has been shown to be an important parameter to control electrical and optical properties. The highest oxygen concentrations enhance the transmission property and the oxygen deficient nature (oxygen vacancies) increased the electrical conductivity of the ITO thin films [47,48]. Hence, a sufficient amount of oxygen concentration can improve the opto-electronic performance of ITO thin films. High-quality ultra-thin ITO films are a needed significant step towards the realization and possible commercialization of plasmonic-based a-Si:H thin-film PV devices [15,24,36–38]. These devices have a potential to transform the thin-film based solar cells industry due to their low cost and ease of fabrication. In addition, plasmonic-enhanced PV has the potential to exhibit sophisticated light management schemes enabling unprecedented control over the trapping and propagation of light within the active region of the PV device [15], which would be expected to result in record-high device solar energy conversion efficiencies.

5. Conclusions

In this study, ultra-ITO thin films have been RF sputter deposited using different oxygen flow rates and chemical shaving is performed at room temperature for different time periods. The thicknesses of the films are altered as a result from 89 nm to 22 nm. In-between each etching process cycle, the films were characterized for both electrical and optical properties. Generally, the transmittance of the ITO films was observed to increase with decreasing film thickness, while the electrical properties were observed to degrade for the same films. This was attributed to the distortion of the In–O lattice long-range order due to the reduction reaction between the ITO and the etchants (acids). The novel method of chemical shaving further investigated here, is a simple and low-cost method with the potential to produce low loss and highly conductive ultra-thin and acid resistant ITO films for applications ranging from PV devices transparent electrodes to anti-acid materials. Using this method, ultra-thin ITO films with record low resistivity values (as low as 5.83×10^{-4} Ω·cm) were obtained and the optical transmission is generally high in the 300–1000 nm wavelength region for all films.

The etching rate strongly depends on the oxygen concentrations of RF sputtered ITO films as well as on the post process annealing. This processing has an effect on the oxygen vacancies densities even for the 0 sccm O_2 films. Surface roughness increased as the concentration of oxygen increased as expected. The etching reactions are simple redox reaction, hence the rates should increase with increases in O_2 concentration especially for non-stoichiometric films with distorted ITO matrix. The etch rate, preferred crystal lattice growth plane, d-spacing and lattice distortion were also observed to be highly dependent on the nature of growth environment for RF sputter deposited ITO films.

Acknowledgments: This work was supported by the National Science Foundation under grant award number CBET-1235750 and the Fulbright S&T award. Jeyanthinath Mayandi thanks the University Grants Commission (UGC) for providing support through Raman fellowship 2014–2015 to visit Michigan Technological University, Houghton, MI 49931, USA.

Author Contributions: Jephias Gwamuri, Jeyanthinath Mayandi and Joshua M. Pearce conceived and designed the experiments; Jephias Gwamuri, Jeyanthinath Mayandi, Murugesan Marikkannan and Patrick K. Bowen performed the experiments; all authors analyzed the data; Jeyanthinath Mayandi and Joshua M. Pearce contributed materials and tools; and all authors wrote the paper.

Conflicts of Interest: The authors declare no conflict of interest.

Abbreviations

4-PP	Four Point Probe
AFM	Atomic Force Microscopy
a-Si:H	Hydrogenated Amorphous Silicon
AZO	Aluminum doped Zinc Oxide
c-Si	Crystalline Silicon
DC	Direct Current
FTO	Fluorine doped-Tin oxide
ITO	Indium Tin Oxide
PV	Photovoltaic
RF	Radio Frequency
SCCM	Standard Cubic Centimeters per Minute (flow unit)
SLG	Sodaline Glass
SnO_2	Tin Oxide
TCOs	Transparent Conducting Oxides
VASE	Variable Angle Spectroscopic Ellipsometry
XRD	X-ray Diffraction
ZnO	Zinc Oxide

References

1. Pearce, J.M. Photovoltaics—A path to sustainable futures. *Futures* **2002**, *34*, 663–674. [CrossRef]
2. Pillai, U. Drivers of cost reduction in solar photovoltaic. *Energy Econ.* **2015**, *50*, 286–293. [CrossRef]
3. Honeyman, C.; Kimbis, T. *Solar Market. Insight Report 2014 Q2*; Solar Energy Industrial Association and GTM research: Washington, DC, USA, 2014.
4. Candelise, C.; Winskel, M.; Gross, R.J.K. The dynamics of solar PV costs and prices as a challenge for technology forecasting. *Renew. Sustain. Energy Rev.* **2013**, *26*, 96–107. [CrossRef]
5. Rubin, E.S.; Azevedo, I.M.L.; Jaramillo, P.; Yeh, S. A review of learning rates for electricity supply technologies. *Energy Policy* **2015**, *86*, 198–218. [CrossRef]
6. Branker, K.; Pathak, M.J.M.; Pearce, J.M. A Review of Solar Photovoltaic Levelized Cost of Electricity. *Renew. Sustain. Energy Rev.* **2011**, *15*, 4470–4482. [CrossRef]
7. Shah, A.; Torres, P.; Tscharner, R.; Wyrsch, N.; Keppner, H. Photovoltaic technology: The case for thin-film solar cells. *Science* **1999**, *285*, 692–698. [CrossRef] [PubMed]
8. Carlson, D.E.; Wronski, C.R. Amorphous Silicon Solar Cell. *Appl. Phys. Lett.* **1976**, *28*, 671–673. [CrossRef]

9.   Pearce, J.M. Industrial symbiosis of very large-scale photovoltaic manufacturing. *Renew. Energy* **2008**, *33*, 1101–1108. [CrossRef]

10. Wronski, C.R.; Pearce, J.M.; Koval, R.J.; Ferlauto, A.S.; Collins, R.W. Progress in Amorphous Silicon Based Solar Cell Technology. Available online: http://www.rio12.com/rio02/proceedings/pdf/067_Wronski.pdf (accessed on 13 October 2015).

11. Collins, R.W.; Ferlauto, A.S.; Ferreira, G.M.; Chen, C.; Koh, J.; Koval, R.J.; Lee, Y.; Pearce, J.M.; Wronski, C.R. Evolution of microstructure and phase in amorphous, protocrystalline, and microcrystalline silicon studied by real time spectroscopic ellipsometry. *Solar Energy Mater. Solar Cells* **2003**, *78*, 143–180. [CrossRef]

12. Deckman, H.W.; Wronski, C.R.; Witzke, H.; Yablonovitch, E. Optically enhanced amorphous silicon solar cells. *Appl. Phys. Lett.* **1983**, *42*, 968–970. [CrossRef]

13. Atwater, H.A.; Polman, A. Plasmonics for improved photovoltaic devices. *Nat. Mater.* **2010**, *9*, 205–213. [CrossRef] [PubMed]

14. Derkacs, D.; Lim, S.H.; Matheu, P.; Mar, W.; Yu, E.T. Improved performance of amorphous silicon solar cells via scattering from surface plasmon polaritons in nearby metallic nanoparticles. *Appl. Phys. Lett.* **2006**, *89*, 093103–093105. [CrossRef]

15. Gwamuri, J.; Güney, D.Ö.; Pearce, J.M. Advances in Plasmonic Light Trapping in Thin-Film Solar Photovoltaic Devices. In *Solar Cell. Nanotechnology*; Tiwari, A., Boukherroub, R., Maheshwar Sharon, M., Eds.; Wiley: Hoboken, NJ, USA, 2013; pp. 241–269.

16. Spinelli, P.; Ferry, V.E.; van de Groep, J.; van Lare, M.; Verschuuren, M.A.; Schropp, R.E.I.; Atwater, H.A.; Polman, A. Plasmonic light trapping in thin-film Si solar cells. *J. Opt.* **2012**, *14*, 024002–024012. [CrossRef]

17. Cai, W.; Salaev, V.M. *Optical Metamaterials: Fundamentals and Applications*, 1st ed.; Springer: New York, NY, USA, 2010; p. 278.

18. Maier, S.A.; Atwater, H.A. Plasmonics: Localization and guiding of electromagnetic energy in metal/dielectric structures. *J. Appl. Phys.* **2005**, *98*, 011101–011110. [CrossRef]

19. Aydin, K.; Ferry, V.E.; Briggs, R.M.; Atwater, H.A. Broadband polarization-independent resonant light absorption using ultrathin plasmonic super absorbers. *Nat. Commun.* **2011**, *2*, 517. [CrossRef] [PubMed]

20. Wu, C.; Avitzour, Y.; Shvets, G. Ultra-thin wide-angle perfect absorber for infrared frequencies. *Proc. SPIE* **2008**, *7029*. [CrossRef]

21. Ferry, V.E.; Verschuuren, M.A.; van Lare, C.; Ruud, E.I.; Atwater, H.A.; Polman, A. Optimized spatial correlations for broadband light trapping nano patterns in high efficiency ultrathin film a-Si:H solar cells. *Nano Lett.* **2011**, *11*, 4239–4245. [CrossRef] [PubMed]

22. Trevino, J.; Forestiere, C.; Di Martino, G.; Yerci, S.; Priolo, F.; Dal Negro, L. Plasmonic-photonic arrays with aperiodic spiral order for ultra-thin film solar cells. *Opt. Express* **2012**, *20*, A418–A430. [CrossRef] [PubMed]

23. Massiot, I.; Colin, C.; Pere-Laperne, N.; Roca i Cabarrocas, P.; Sauvan, C.; Lalanne, P.; Pelouard, J.-L.; Collin, S. Nanopatterned front contact for broadband absorption in ultra-thin amorphous silicon solar cells. *Appl. Phys. Lett.* **2012**, *101*, 163901–163903. [CrossRef]

24. Vora, A.; Gwamuri, J.; Pala, N.; Kulkarni, A.; Pearce, J.M.; Güney, D.Ö. Exchanging Ohmic losses in metamaterial absorbers with useful optical absorption for photovoltaics. *Sci. Rep.* **2014**, *4*, 1–13. [CrossRef] [PubMed]

25. Sato, K.; Gotoh, Y.; Wakayama, Y.; Hayashi, Y.; Adachi, K.; Nishimura, H. Highly textured SnO_2:F TCO films for a-Si solar cells. *Rep. Res. Lab. Asahi Glass Co. Ltd.* **1992**, *42*, 129–137.

26. Dixit, A.; Sudakar, C.; Naik, R.; Naik, V.M.; Lawes, G. Undoped vacuum annealed In_2O_3 thin films as a transparent conducting oxide. *Appl. Phys. Lett.* **2009**, *95*, 192105–192107. [CrossRef]

27. Lan, J.H.; Kanicki, J. ITO surface ball formation induced by atomic hydrogen in PECVD and HW-CVD tools. *Thin Solid Films* **1997**, *304*, 123–129. [CrossRef]

28. Thøgersen, A.; Rein, M.; Monakhov, E.; Mayandi, J.; Diplas, S. Elemental distribution and oxygen deficiency of magnetron sputtered indium tin oxide films. *J. Appl. Phys.* **2011**, *109*, 113532. [CrossRef]

29. Park, H.K.; Yoon, S.W.; Chung, W.W.; Min, B.K.; Do, Y.R. Fabrication and characterization of large-scale Multifunctional transparent ITO nanorod films. *J. Mater. Chem. A* **2013**, *1*, 5860–5867. [CrossRef]

30. Castaneda, S.I.; Rueda, F.; Diaz, R.; Ripalda, J.M.; Montero, I. Whiskers in Indium tin oxide films obtained by electron beam evaporation. *J. Appl. Phys.* **1998**, *83*, 1–8. [CrossRef]

31. Yao, J.L.; Hao, S.; Wilkinson, J.S. Indium Tin Oxide Films by Sequential Evaporation. *Thin Solid Films* **1990**, *189*, 221–233. [CrossRef]

32. Kobayashi, H.; Kogetsu, Y.; Ishida, T.; Nakato, Y. Increase in photovoltage of "indium tin oxide/Silicon oxide/mat-textured n–silicon" junction solar cells by silicon peroxidation and annealing processes. *J. Appl. Phys.* **1993**, *74*, 4756–4761. [CrossRef]

33. Lee, J.; Lee, S.; Li, G.; Petruska, M.A.; Paine, D.C.; Sun, S. A Facile Solution-Phase Approach to Transparent and Conducting ITO Nanocrystal Assemblies. *J. Am. Chem. Soc.* **2012**, *134*, 13410–13414. [CrossRef] [PubMed]

34. Chen, Z.; Li, W.; Li, R.; Zhang, Y.; Xu, G.; Cheng, H. Fabrication of Highly Transparent and Conductive Indium-Tin Oxide Thin films with a high figure of merit via solution processing. *Langmuir* **2013**, *29*, 13836–13842. [CrossRef] [PubMed]

35. Wan, D.; Chen, P.; Liang, J.; Li, S.; Huang, F. (211)-Orientation Preference of Transparent Conducting In$_2$O$_3$:Sn Films and Its Formation Mechanism. *ACS. Appl. Mater. Interfaces* **2011**, *3*, 4751–4755. [CrossRef] [PubMed]

36. Vora, A.; Gwamuri, J.; Pearce, J.M.; Bergstrom, P.L.; Guney, D.O. Multi-resonant silver nano-disk patterned thin film hydrogenated amorphous silicon solar cells for Staebler-Wronski effect compensation. *J. Appl. Phys.* **2014**, *116*, 093103. [CrossRef]

37. Gwamuri, J.; Vora, A.; Khanal, R.R.; Phillips, A.B.; Heben, M.J.; Guney, D.O.; Bergstrom, P.; Kulkarni, A.; Pearce, J.M. Limitations of ultra-thin transparent conducting oxides for integration into plasmonic-enhanced thin-film solar photovoltaic devices. *Mater. Renew. Sustain. Energy* **2015**, *4*, 1–12. [CrossRef]

38. Gwamuri, J.; Vora, A.; Mayandi, J.; Guney, D.O.; Bergstrom, P.; Pearce, J.M. A New Method of Preparing Highly Conductive Ultra-Thin Indium Tin Oxide for Plasmonic-Enhanced Thin Film Solar Photovoltaic Devices. **2016**, to be published.

39. Hang, C.J.; Su, Y.K.; Wu, S.L. The effect of solvent on the etching of ITO electrode. *Mater. Chem. Phys.* **2004**, *84*, 146–150. [CrossRef]

40. Marikkannan, M.; Subramanian, M.; Mayandi, J.; Tanemura, M.; Vishnukanthan, V.; Pearce, J.M. Effect of ambient combinations of argon, oxygen, and hydrogen on the properties of DC magnetron sputtered indium tin oxide films. *AIP Adv.* **2015**, *5*, 017128–017138. [CrossRef]

41. Luo, S.; Kohiki, S.; Okada, K.; Shoji, F.; Shishido, T. Hydrogen effects on crystallinity, photoluminescence, and magnetization of indium tin oxide thin films sputter-deposited on glass substrate without heat treatment. *Phys. Status Solidi A* **2010**, *207*, 386–390. [CrossRef]

42. Kato, K.; Omoto, H.; Tomioka, T.; Takamatsu, A. Changes in electrical and structural properties of indium oxide thin films through post-deposition annealing. *Thin Solid Films* **2011**, *520*, 110–116. [CrossRef]

43. Liu, D.; Lei, W.W.; Zou, B. High pressure X-ray diffraction and Raman spectra study of indium oxide. *J. Appl. Phys.* **2008**, *104*, 083506–083511. [CrossRef]

44. Berengue, O.M.; Rodrigues, A.D.; Dalmaschio, C.J.; Lanfredi, A.J.C.; Leite, E.R.; Chiquito, A.J. Structural characterization of indium oxide nanostructures: A Raman analysis. *J. Phys. D Appl. Phys.* **2010**, *43*, 045401–045404. [CrossRef]

45. Chandrasekhar, R.; Choy, K.L. Innovative and cost-effective synthesis of indium tin oxide films. *Thin Solid Films* **2001**, *398–399*, 59–64. [CrossRef]

46. Luo, S.N.; Kono, A.; Nouchi, N.; Shoji, F. Effective creation of oxygen vacancies as an electron carrier source in tin-doped indium oxide films by plasma sputtering. *J. Appl. Phys.* **2006**, *100*, 113701–113709. [CrossRef]

47. Okada, K.; Kohiki, S.; Luo, S.; Sekiba, D.; Ishii, S.; Mitome, M.; Kohno, A.; Tajiri, T.; Shoji, F. Correlation between resistivity and oxygen vacancy of hydrogen-doped indium tin oxide thin films. *Thin Solid Films* **2011**, *519*, 3557–3561. [CrossRef]

48. Ashida, T.; Miyamuru, A.; Oka, N.; Sato, Y.; Yagi, T.; Taketoshi, N.; Baba, T.; Shigesato, Y. Thermal transport properties of polycrystalline tin-doped indium oxide films. *J. Appl. Phys.* **2009**, *105*, 073709–073712. [CrossRef]

49. Van den Meerakker, J.E.A.M.; Baarslag, P.C.; Walrave, W.; Vink, T.J.; Daams, J.L.C. On the homogeneity of sputter-deposited ITO films Part II. Etching behavior. *Thin Solid Films* **1995**, *266*, 152–156. [CrossRef]

Permissions

List of Contributors

Mladena Lukovic and Guang Ye
Section of Materials and Environment, Faculty of Civil Engineering and Geosciences, Delft University of Technology, Delft 2628 CD, The Netherlands

Xian-Ming Qi, Shi-Yun Liu, Fang-Bing Chu, Shuai Pang, Yan-Ru Liang, Feng Peng and Run-Cang Sun
Beijing Key Laboratory of Lignocellulosic Chemistry, Beijing Forestry University, Beijing 100083, China

Ying Guan
School of Forestry and Landscape Architecture, Anhui Agricultural University, Hefei 230036, China

Takaharu Okada
Graduate School of Pure and Applied Sciences, University of Tsukuba, 1-1-1, Tennodai, Tsukuba, Ibaraki 305-8577, Japan
Biomaterials Unit, Nano-Life Field, International Center for Materials Nanoarchitectonics (WPI-MANA), National Institute for Materials Science (NIMS), 1-1 Namiki, Tsukuba, Ibaraki 305-0044, Japan
Japan Society for the Promotion of Science (JSPS), 8 Ichibancho, Chiyoda-ku, Tokyo 102-0083, Japan

Eri Niiyama
Graduate School of Pure and Applied Sciences, University of Tsukuba, 1-1-1, Tennodai, Tsukuba, Ibaraki 305-8577, Japan
Biomaterials Unit, Nano-Life Field, International Center for Materials Nanoarchitectonics (WPI-MANA), National Institute for Materials Science (NIMS), 1-1 Namiki, Tsukuba, Ibaraki 305-0044, Japa

Koichiro Uto and Takao Aoyagi
Biomaterials Unit, Nano-Life Field, International Center for Materials Nanoarchitectonics (WPI-MANA), National Institute for Materials Science (NIMS), 1-1 Namiki, Tsukuba, Ibaraki 305-0044, Japan

Mitsuhiro Ebara
Biomaterials Unit, Nano-Life Field, International Center for Materials Nanoarchitectonics (WPI-MANA), National Institute for Materials Science (NIMS), 1-1 Namiki, Tsukuba, Ibaraki 305-0044, Japan
Graduate School of Tokyo University of Science, 6-3-1 Niijuku, Katsushika-ku, Tokyo 125-8585, Japan

Kun Mo, Michael Pellin, Yinbin Miao and Abdellatif M. Yacout
Nuclear Engineering Division, Argonne National Laboratory, Lemont, IL 60439, USA

Di Yun
Nuclear Engineering Division, Argonne National Laboratory, Lemont, IL 60439, USA
Department of Nuclear Engineering, Xi'an Jiaotong University, Xi'an 710049, Shaanxi, China

Xiang Liu
Department of Nuclear, Plasma, and Radiological Engineering, University of Illinois at Urbana-Champaign, Urbana, IL 61801, USA

James F. Stubbins
Department of Nuclear, Plasma, and Radiological Engineering, University of Illinois at Urbana-Champaign, Urbana, IL 61801, USA; xliu128@illinois.edu (X.L
International Institute for Carbon-Neutral Energy Research (I2CNER), Kyushu University, Fukuoka 819-0395, Japan

Jonathan Almer and Jun-Sang Park
Advanced Photon Source, Argonne National Laboratory, Lemont, IL 60439, USA

Shaofei Zhu
Physics Division, Argonne National Laboratory, Lemont, IL 60439, USA

Andrei Veksha
Department of Chemical and Petroleum Engineering, University of Calgary, 2500 University Drive NW, Calgary, AB T2N 1N4, Canada
Residues and Resource Reclamation Centre, Nanyang Environment and Water Research Institute, Nanyang Technological University, 1 Cleantech Loop, Clean Tech One, Singapore 637141, Singapore

Tazul I. Bhuiyan and Josephine M. Hill
Department of Chemical and Petroleum Engineering, University of Calgary, 2500 University Drive NW, Calgary, AB T2N 1N4, Canada

Jinyang Xu and Mohamed El Mansori
MSMP — EA 7350 Laboratoire, Arts et Métiers Paris Tech, Rue Saint Dominique B.P. 508, 51006 Châlons-en-Champagne, France

M. Judith Cruz
Department of Chemical Engineering and Mineral Processing and Center for Advanced Study of Lithium and Industrial Minerals (CELiMIN), University of Antofagasta, Av. Universidad de Antofagasta 02800, Campus Coloso, Antofagasta 127300, Chile

Svetlana Ushak and Mario Grágeda
Department of Chemical Engineering and Mineral Processing and Center for Advanced Study of Lithium and Industrial Minerals (CELiMIN), University of Antofagasta, Av. Universidad de Antofagasta 02800, Campus Coloso, Antofagasta 127300, Chile
Solar Energy Research Center (SERC-Chile), Av Tupper 2007, Piso 4, Santiago 8370451, Chile

Luisa F. Cabeza
GREA Innovació Concurrent, Edifici CREA, Universitat de Lleida, Pere de Cabrera s/n, Lleida 25001, Spain

James Rouse and Christopher Hyde
Department of Mechanical, Materials and Manufacturing Engineering, University of Nottingham, Nottingham, Nottinghamshire NG7 2RD, UK

Alfredo Márquez-Herrera
Departamento de Ingeniería Agrícola, DICIVA, Campus Irapuato-Salamanca, Universidad de Guanajuato, Ex Hacienda el Copal, Carr. Irapuato-Silao km 9, Irapuato Gto 36500, Mexico

Victor Manuel Ovando-Medina and Blanca Estela Castillo-Reyes
Ingeniería Química, COARA, Universidad Autónoma de San Luis Potosí, Carr. a Cedral Km 5+600, San José de las Trojes, Matehuala, San Luis Potosí 78700, Mexico

Martin Zapata-Torres
Centro de Investigación en Ciencia Aplicada y Tecnología Avanzada, Unidad Legaría IPN, Calzada Legaría 694, Col. Irrigación, México D.F. 11500, Mexico

Miguel Meléndez-Lira
Departamento de Física, CINVESTAV-IPN, Apartado Postal 14-740, México D.F. 07000, Mexico

Jaquelina González-Castañeda
Departamento de Ingeniería Ambiental, DICIVA, Campus Irapuato-Salamanca, Universidad de Guanajuato, Ex Hacienda el Copal, Carr. Irapuato-Silao km 9, Irapuato Gto 36500, Mexico

Jiri Kudr, Lukas Richtera, Lukas Nejdl, Kledi Xhaxhiu, Branislav Rutkay-Nedecky, David Hynek, Pavel Kopel and Vojtech Adam
Department of Chemistry and Biochemistry, Mendel University in Brno, Zemedelska 1, Brno CZ-613 00, Czech Republic; george.kudr@centrum.cz (J.K
Central European Institute of Technology, Brno University of Technology, Technicka 3058/10, Brno CZ-616 00, Czech Republic

Petr Vitek
Global Change Research Institute, The Czech Academy of Sciences, v.v.i., Bˇelidla 4a, Brno CZ-603 00, Czech Republic

Rene Kizek
Department of Biomedical and Environmental Analysis, Wroclaw Medical University, Borowska 211, Wrocław PL-50 556, Poland

George S. Pappas, Stefania Ferrari, Rohit Bhagat and Chaoying Wan
Warwick Manufacturing Group, University of Warwick, Coventry CV4 7AL, UK

Xiaobin Huang
School of Aeronautics and Astronautics, Shanghai Jiao Tong University, Shanghai 200240, China

David M. Haddleton
Department of Chemistry, University of Warwick, Coventry CV4 7AL, UK

Bandar Alzahrani and Gracious Ngaile
Department of Mechanical and Aerospace Engineering, North Carolina State University, 911 Oval Drive-3160 EB3, Raleigh, NC 27695-7910, USA

Bang Yeon Lee
School of Architecture, Chonnam National University, 77 Yongbong-ro, Buk-gu, Gwangju 61186, Korea

Su-Tae Kang
Department of Civil Engineering, Daegu University, 201 Daegudae-ro, Jillyang, Gyeongsan, Gyeongbuk 38453, Korea

Hae-Bum Yun
Department of Civil, Environmental, and Construction Engineering, University of Central Florida, Orlando, FL 32816, USA

Yun Yong Kim
Department of Civil Engineering, Chungnam National University, 99 Daehak-ro, Yuseong-gu, Daejeon 34134, Korea

Michael Lorenz, Peter Schwinkendorf, Michael Bonholzer and Marius Grundmann
Institut für Experimentelle Physik II, Universität Leipzig, Leipzig D-04103, Germany; schwinkendorf@physik.uni-leipzig.de (P.S.)

GeraldWagner and Oliver Oeckler
Institut für Mineralogie, Kristallographie und Materialwissenschaft, Universität Leipzig, Leipzig D-04103, Germany

Vera Lazenka, André Vantomme and Kristiaan Temst
Instituut voor Kern- en Stralingsfysica, KU Leuven, Leuven B-3001, Belgium

Margriet J. Van Bael
Laboratorium voor Vaste-Stoffysica en Magnetisme, KU Leuven, Leuven B-3001, Belgium

Saleh A. Alkahtani and Mahmoud M. Tash
Industrial Engineering Program, Mechanical Engineering Department, College of Engineering, Prince Sattam bin AbdulAziz University, Al Kharj 11942, Saudi Arabia

Agnes M. Samuel, Fawzy H. Samuel and Emad M. Elgallad
Département des Sciences Appliquées, Université du Québec à Chicoutimi, Chicoutimi, QC G7H 2B1, Canada (A.M.S.)

Longbiao Li
College of Civil Aviation, Nanjing University of Aeronautics and Astronautics, No. 29 Yudao St., Nanjing 210016, China

Jephias Gwamuri and Patrick K. Bowen
Department of Materials Science & Engineering, Michigan Technological University, 1400 Townsend, Houghton, MI 49931, USA

Joshua M. Pearce
Department of Materials Science & Engineering, Michigan Technological University, 1400 Townsend, Houghton, MI
Department of Electrical & Computer Engineering, Michigan Technological University, 1400 Townsend, Houghton, MI 49931, USA

Murugesan Marikkannan and Jeyanthinath Mayandi
Department of Materials Science, School of Chemistry, Madurai Kamaraj University, Ta mil Nadu, Madurai 625 019, India; marikannan.mku@gmail.com (M.M.)

Index

www.ingramcontent.com/pod-product-compliance
Lightning Source LLC
Chambersburg PA
CBHW061939190326
41458CB00009B/2775